FUNCTIONAL PROTEOMICS
&
NANOTECHNOLOGY-BASED MICROARRAYS

Pan Stanford Series on Nanobiotechnology

Series Editor: Claudio Nicolini

Scientific Committee:
Christian Riekel *(ESRF, Grenoble, France)*
Wolfgang Knoll *(Max Planck Institute, Mainz, Germany)*
Josh LaBaer *(Harvard University, USA)*
Michael Kirpichnikov *(Moscow University, Russia)*

Published

Vol 1 *Nanobiotechnology and Nanobiosciences*
       Claudio Nicolini

Vol 2 *Functional Proteomics and Nanotechnology-Based Microarrays*
       Claudio Nicolini & Joshua LaBaer

Forthcoming

Vol 3 *Synchrotron Radiation and Structural Proteomics*
       Eugenia Pechkova & Christian Riekel
       (Late 2010)

Vol 4 *Nanobiotechnological Devices and Processes for Energy Compatible with the Environment*
       Claudio Nicolini & Carlo V. Bruschi
       (Spring 2011)

Vol 5 *Nanobioelectronics: Principles and Applications*
       Claudio Nicolini & Norbert Hampp
       (Winter 2011)

Claudio Nicolini
University of Genoa, Italy

Joshua LaBaer
Harvard Institute of Proteomics, USA

FUNCTIONAL PROTEOMICS & NANOTECHNOLOGY-BASED MICROARRAYS

Pan Stanford Series on **NANOBIOTECHNOLOGY**

Volume **2**

Published by

Pan Stanford Publishing Pte. Ltd.
Penthouse Level, Suntec Tower 3
8 Temasek Boulevard
Singapore 038988

Email: editorial@panstanford.com
Web: www.panstanford.com

British Library Cataloguing-in-Publication Data
A catalogue record for this book is available from the British Library.

Pan Stanford Series on Nanobiotechnology — Vol. 2
FUNCTIONAL PROTEOMICS AND NANOTECHNOLOGY-BASED MICROARRAYS

Copyright © 2010 by Pan Stanford Publishing Pte. Ltd.

All rights reserved. This book, or parts thereof, may not be reproduced in any form or by any means, electronic or mechanical, including photocopying, recording or any information storage and retrieval system now known or to be invented, without written permission from the Publisher.

For photocopying of material in this volume, please pay a copying fee through the Copyright Clearance Center, Inc., 222 Rosewood Drive, Danvers, MA 01923, USA. In this case permission to photocopy is not required from the publisher.

ISBN-13 978-981-4267-76-2
ISBN-10 981-4267-76-7

Printed in Singapore by Mainland Press Pte Ltd.

To my sons David and Peter

PREFACE

The research reported in this Volume is mostly dedicated to recent developments in "Functional Proteomics and Nanotechnology-based Microarrays" and constitutes the second Volume of the Pan Stanford Nanobiotechnology Series edited by Claudio Nicolini. It represents, with the exception of only two out of twelve chapters, the outcome of the international cooperation between the Harvard University Institute of Proteomics and the University of Genova CIRSDNNOB-Nanoworld Institute, at the conclusion of a contract granted within the framework of an agreement signed in Boston by the Italian Ministry of University – Research and the Harvard University Medical School – Harvard Armenise Foundation.

All scientific objectives (I–X) appear in the original FIRB proposal **RBIN04RXHS** and have been addressed with success (see the list below of the SCI publications) over the three years period of the project carried out between 2006 and now, namely:

I Gene Microarrays
II NAPPA via Fluorescence and DNASER
III Mass Spectrometry
IV Molecular Modelling and Bioinformatics
V Human Lymphocytes by Microarrays, MS & Bioinformatics
VI Protein Nanocrystallography
VII CHO-K1 and Osteosarcoma
VIII Label Free NAPPA
IX Atomic Force Microscopy
X Anodic Porous Alumina

With the major focus on functional proteomics and cell proliferation several correlated scientific and technological objectives have been so far carried out and summarized also in the Volume 3 of the same series entitled "*Synchrotron radiation and Structural Proteomics*" edited by Christian Riekel and Evgeniya Pechkova.

This entire volume on ***"Functional Proteomics and Nanotechnology-based Microarrays"*** is co-edited by ourselves as second volume of the Pan Stanford Nanobiotechnology Series, uniquely dedicated to a comprehensive conclusive report on the final outcome directly pertinent to the major emphasis of our international cooperation between the Harvard University Institute of Proteomics (USA) and the University of Genova CIRSDNNOB-Nanoworld Institute (Italy). The research activities have been sponsored by the research grant number **RBIN04RXHS**, within the framework of an agreement signed time ago in Boston by the Italian Ministry of University – Research and the Harvard University Medical School – Harvard Armenise Foundation.

This conclusion has being already jointly forwarded to both the Ministry of Research and University and to the Harvard University Medical School – Harvard Armenise Foundation, as a result of the original approval of the grant proposal entitled "Functional Proteomics and Cell Cycle Progression" with the project code RBIN04RXHS (Coordinator: NICOLINI Claudio).

We are glad that the coupling of Harvard HIP protein microarray technology using in house surface chemistry and medical expertise with NWI mass spectrometry expertise and NWI APA, DNASER, Mass Spectrometry, Nanocristallography and Nanogravimetry technologies has synergistically expanded our capabilities to understand the cell function and the cell cycle, as a result also of the unique complementary informations derived from the innovative structural Proteomics and Nanogenomics implemented at Genova NWI and at the European Synchrotron Radiation in Grenoble (see Volume 3 of this Series).

Josh LaBaer
Harvard University Professor at Institute of Proteomics

Claudio Nicolini
Genova University Professor at NanoWorld Institute

Boston, 13 May 2009

1. Grasso, V., Lambertini, V., Ghisellini, P., Valerio, F., Stura, E., Perlo, P. and Nicolini, C. (2006). *Nanostructuring of a porous alumina matrix for a biomolecular microarray*, **Nanotechnology 17**, 795–798.
2. Nicolini, C. and Pechkova, E. (2006). *Structure and growth of ultrasmall protein microcrystals by synchrotron radiation: I microGISAXS and microdiffraction of P450scc*, **Journal of Cellular Biochemistry 97**, 544–552.
3. Pechkova, E. and Nicolini, C. (2006). *Structure and growth of ultrasmall protein microcrystals by synchrotron radiation: II microGISAX and microscopy of lysozyme*, **Journal of Cellular Biochemistry 97**, 553–560.
4. Sivozhelezov, V., Giacomelli, L., Tripathi, S. and Nicolini, C. (2006). *Gene expression in the cell cycle of human T lymphocytes: I. Predicted gene and protein networks*, **Journal of Cellular Biochemistry 97**, 1137–1150.
5. Nicolini, C., Spera, R., Stura, E., Fiordoro, S. and Giacomelli, L. (2006). *Gene expression in the cell cycle of human T lymphocytes: II. Experimental determination by DNASER technology*, **Journal of Cellular Biochemistry 97**, 1151–1159.
6. Sivozhelezov, V. and Nicolini, C. (2006). *Theoretical framework for octopus rhodopsin crystallization*, **Journal of Theoretical Biology 240**, 260–269.
7. Sivozhelezov, V., Pechkova, E. and Nicolini, C. (2006). *Mapping electrostatic potential of a protein on its hydrophobic surface: implications for crystallization of cytochrome P450scc*, **Journal of Theoretical Biology 241**, 73–80.
8. Nicolini, C. and Pechkova, E. (2006). *Nanostructured biofilms and biocrystals (a review)*, **Journal of Nanoscience and Nanotechnology 6**, 2209–2236.
9. Nicolini, C. (2006). *Nanogenomics for medicine*, **Nanomedicine 1**, 147–151.
10. Siodmiak, J., Gadomski, A., Pechkova, E. and Nicolini, C. (2006). *Computer model of a lysozyme crystal growth with/without nanotemplate - a comparison*, **International Journal of Moder Physics C 17**, 1359–1366.
11. Giacomelli, L. and Nicolini, C. (2006). *Gene expression of human T lymphocytes cell cycle: experimental and bioinformatic analysis*, **Journal of Cellular Biochemistry 99**, 1326–1333.
12. Pechkova, E., Innocenzi, P., Malfatti, L., Kidchob, T., Gaspa, L. and Nicolini, C. (2007). *Thermal Stability of Lysozyme Langmuir-Schaefer Films by FTIR Spectroscopy*, **Langmuir 23**, 1147-1151.
13. Spera, R. and Nicolini, C. (2007). *cAMP induced alterations of Chinese hamster ovary cells monitored by mass spectrometry*, **Journal of Cellular Biochemistry 102**, 473-482.
14. Pechkova, E., Sivozhelezov, V. and Nicolini, C. (2007). *Protein thermal stability: the role of protein structure and aqueous environment*, **Archives of Biochemistry and Biophysics 466**, 40–48.
15. Pechkova, E., Sartore, M., Giacomelli, L. and Nicolini, C. (2007). *Atomic force microscopy of protein films and crystals*, **Review of Scientific Instruments 78**, 093704_1-093704-7.
16. Stura, E., Bruzzese D., Valerio, F., Grasso, V., Perlo, P. and Nicolini, C. (2007). *Anodic porous alumina as mechanical stability enhancer for LDL-cholesterol sensitive electrodes*, **Biosensors & Bioelectronics 23**, 655–660.

17. Covani, U., Giacomelli, L., Krajewski, A., Ravaglioli, A., Spotorno, L., Loria, P., Das, S. and Nicolini, C. (2007). *Biomaterials for orthopedics: a roughness analysis by atomic force microscopy*, **Journal of Biomedical Materials Research Part A 82A**, 723–730.
18. Paternolli, C., Ghisellini, P. and Nicolini, C. (2007). *Nanostructuring of heme-proteins for biodevice applications*, **IET Nanobiotechnology 1**, 22–26.
19. Adami, M., Sartore, M. and Nicolini, C. (2007). *A potentiometric stripping analyzer for multianalyte screening*, **Electroanalysis 19**, 1288–1294.
20. Sivozhelezov, V. and Nicolini, C. (2007). *Prospects for octopus rhodopsin utilization in optical ad quantum computation*, **Physics of Particles and Nuclei Letters 4**, 189–196.
21. Spera, R. and Nicolini, C., *Nappa microarrays and mass spectrometry: new trends and challenges*, **Essentials in Nanoscience Booklet Series**, (Taylor & Francis Group, LLC) 15 January 2008.
22. Vasile, F., Pechkova, E. and Nicolini, C. (2008). *Atomic structure of the beta-subunit of the translation initiation factor Aif2 from Archaebacteria sulfolobus solfataricus: high resolution NMR in solution*, **Proteins: Structure, Function and Bioinformatics 70**, 1112–1115.
23. Sivozhelezov, V., Braud, C., Giacomelli, L., Pechkova, E., Giral, M., Soulillou, J.P., Brouard, S. and Nicolini, C. (2008). *Immunosuppressive drug-free operational immune tollerance in human kidney transplant recipient: II Non-statistical gene microarray analysis*, **Journal of Cellular Biochemistry 103**, 1693-1706.
24. Braud, C., Baeten, D., Giral, M., Pallier, A., Ashton-Chess, J., Braudeau, C., Chevalier, C., Lebars, A., Lèger, J., Moreau, A., Pechkova, E., Nicolini, C., Soulillou, J.P. and Brouard, S. (2008). *Immunosuppressive drug-free operational immune tollerance in human kidney transplant recipient: I. Blood gene expression statistical analysis*, **Journal of Cellular Biochemistry 103**, 1681–1692.
25. Pechkova, E., Vasile, F., Spera, R. and Nicolini, C. (2008). *Crystallization of alpha and beta subunits of IF2 translation initiation factor from Archabacteria Sulfolobus Solfataricus*, **Journal of Crystal Growth 310**, 3767–3770.
26. Paternolli, C., Neebe, M., Stura, E., Barbieri, F., Ghisellini, P., Hampp, N. and Nicolini, C. (2009). *Photo reversibility and photo stability in films of octopus rhodopsin isolated from octopus photoreceptor membranes*, **Journal of Biomedical Materials Research Part A 88A**, 947–951.
27. Covani, U., Marconcini, S., Giacomelli, L., Sivozhelezov, V., Barone, A. and Nicolini, C. (2008). *Bioinformatic prediction of leader genes in human periodontitis*, **Journal of Periodontology 79**, 1974–1983.
28. Nicolini, C., Bruzzese, D., Sivozhelezov, V. and Pechkova, E. (2008). *Langmuir-Blodgett based lipase nanofilms of unique structure-function relationship*, **Biosystems 94**, 228–232.
29. Pechkova, E., Tripathi, S., Spera, R. and Nicolini, C. (2008). *Groel crystal growth and characterization*, **Biosystems 94**, 223–227.
30. Nicolini, C. and Nozza, F. (2008). *Objective assessment of scientific performances world wide*, **Scientometrics 76**, 527–541.

31. Nicolini, C. (2008). *Proceedings of the IASTED International Conference on Nanotechnology and Applications*, (NANA 2008), pp. 182, ISBN 978-0-88986-759-7, Acta Press, Canada.
32. Nicolini, C. (2009). **Nanobiotechnology and Nanobiosciences**, Pan Stanford Series on Nanobiotechnology, Singapore, London, New York, Volume 1, pp. 1–347.
33. Nicolini, C. (2009). **Biofisica e Propedeutica Biofisica**, Aracne Editrice, pp. 1–328.
34. Pechkova, E., Tripathi, S. and Nicolini, C. (2009). *μGISAXS of LB protein films: effect of temperature on long range order*, **Journal of Synchrotron Radiation 16**, 330–335.
35. Tosato, V., Nicolini, C. and Bruschi, C.V. (2009). *DNA bridging of homologous chromosomes in yeast leads to near-reciprocal translocation and loss of heterozygosity by deletion*, **Chromosoma 118**, 179–191.
36. Sivozhelezov. V., Bruzzese. D., Pastorino. L., Pechkova. E. and Nicolini. C. (2009). *Increase of catalytic activity of lipase towards olive oil by Langmuir-film immobilization of lipase*, **Enzyme and Microbial Technology 44**, 72–76.
37. Nicolini, C. (2009). *Nanogenomics in medicine*, **Wires Nanomedicine and Nanobiotechnology** (Wiley Interdisciplinary Reviews).
38. Pechkova, E., Tripathi, S., Ravelli, R.G., McSweeney, S. and Nicolini C. (2009). *Radiation stability of proteinase K grown by LB nanotemplate method*, **Journal of Structural Biology**, in press.
39. Riekel, C. and Pechkova, E. (2009). **Synchrotron Radiation and Structural Proteomics**. Pan Stanford Series on Nanobiotechnology., Singapore, London, New York.
40. Nicolini, C. and LaBaer, J. (2009). *Nucleic Acid Programmable Protein Array For Functional Proteomics In Medicine*, in **Functional Proteomics and Nanotechnology-based Microarrays**, Pan Stanford Series on Nanobiotechnology., Singapore, London, New York, pp. 1–29.
41. Sivozhelezov, V., Spera, R., Giacomelli, L., Hainsworth, E., LaBaer, J., Bragazzi, N. L. and Nicolini, C. (2009). *Bioinformatics and fluorescence DNASER for NAPPA studies on cell transformation and cell cycle*, in **Functional Proteomics and Nanotechnology-based Microarrays**, Pan Stanford Series on Nanobiotechnology, Singapore, London, New York, pp. 31–59.
42. Spera, R., Badino, F., Hainsworth, E., Fuentes, M., Srivastava, S., LaBaer, J. and Nicolini, C. (2009). *Label Free Detection of NAPPA Via Mass Spectrometry*, in **Functional Proteomics and Nanotechnology-based Microarrays**, Pan Stanford Series on Nanobiotechnology, Singapore, London, New York, pp. 61–78.
43. Adami, M., Eggenhoffner, R., Sartore, M., Hainsworth, E., Labaer, J. and Nicolini, C. (2009). *Label Free NAPPA via Nanogravimetry*, in **Functional Proteomics and Nanotechnology-based Microarrays**, Pan Stanford Series on Nanobiotechnology, Singapore, London, New York, pp. 79–93.
44. Stura, E., Larosa, C., Bezerra, T., Hainsworth, E., Ramachandran, N., LaBaer, J. and Nicolini, C. (2009). *Label-Free NAPPA: Anodic Porous Alumina*, in **Functional Proteomics and Nanotechnology-based Microarrays**, Pan Stanford Series on Nanobiotechnology, Singapore, London, New York, pp. 95–108.

45. Sartore, M., Eggenhoffner, R., Bezerra, T., Stura, E., Hainsworth, E., Labaer, J. and Nicolini, C. (2009). *Label-free detection of NAPPA via Atomic Force Microscopy*, in **Functional Proteomics and Nanotechnology-based Microarrays**, Pan Stanford Series on Nanobiotechnology, Singapore, London, New York, pp. 109–120.
46. Pechkova E., Chong, R., Tripathi, S. and Nicolini, C. (2009). *Cell free expression and APA for NAPPA and protein nanocrystallography*, in **Functional Proteomics and Nanotechnology-based Microarrays**, Pan Stanford Series on Nanobiotechnology, Singapore, London, New York, pp. 121–147.
47. Spera, R. and Nicolini, C. (2009). *Overall proteome alterations during reverse transformation of growing CHO-K1 cells*, in **Functional Proteomics and Nanotechnology-based Microarrays**, Pan Stanford Series on Nanobiotechnology, Singapore, London, New York, pp. 183–228.
48. Bezerra, T., Egghenoffner, R., Sartore, M., Hainsworth E., LaBaer, J. and Nicolini C. (2009). *Atomic force microscopy analysis of Nucleic Acid Programmable Protein Array*, **Nanotechnology,** submitted.
49. Eggenhoffner, R., Sallam, S., Pechkova, E., Adami, M., Hainsworth E., LaBaer, J. and Nicolini, C. (2009). *Real time protein expression analysis by innovative multiple-nanogravimeter.* **Proteins,** submitted.
50. Larosa C., Egghenoffner R., Stura E., Ramachandran N., LaBaer J., Nicolini C., (2009). *Functionalize Anodic Porous Alumina and gene expression during Nucleic Acid Programmable Protein Array*, **Proteome Research** submitted.
51. Tripathi, S., Pechkova, E. and Nicolini, C. (2009). *Radiation stability of LB thaumatin crystals grown by LB nanotemplate method.* **Journal of Synchrotron Radiation,** submitted.
52. Spera, R., Sivozhelezov, V., Hainsworth, E., and Nicolini, C. (2009). *Protein detection in NAPPA via MALDI-TOF and Bioinformatics.* **Journal of Mass Spectrometry** submitted.
53. Ramachandran, N., Hainsworth, E., Demirkan, G. and LaBaer, J. (2006). *On-chip protein synthesis for making microarrays*, **Methods in Molecular Biology 328**, 1–14.
54. Braun, P. and LaBaer, J. (2006). *High Throughput Purification of Proteins from Escherichia Coli*, **Cell Biology: A Laboratory Handbook**, 3rd Edition, Volume 4. Elsevier Academic Press: Burlington, MA, pp. 73–77.
55. Butcher, R.A., Bhullar, B.S., Perlstein, E.O., Marsischky, G., LaBaer, J. and Schreiber, S.L. (2006). *Microarray-based method for monitoring yeast overexpression strains reveals small-molecule targets in TOR pathway.* **Nature Chemical Biology 2**, 103–109.
56. Witt, A.E., Hines, L.M., Collins, N.L., Hu, Y., Gunawardane, R.N., Moreira, D., Raphael, J., Jepson, D., Koundinya, M., Rolfs, A., Taron, B., Isakoff, S.J., Brugge, J.S. and LaBaer, J. (2006). *Functional proteomics approach to investigate the biological activities of cDNAs implicated in breast cancer.* **Journal of Proteome Research 5**, 599–610.
57. Park, J., Labaer, J. (2006). *Recombinational cloning.* **Current Protocol in Molecular Biology**, Chapter 3:Unit 3.20.

58. Cooper, A.A., Gitler, A.D., Cashikar, A., Haynes, C.M., Hill, K.J., Bhullar, B., Liu, K., Xu, K., Strathearn, K.E., Liu, F., Cao, S., Caldwell, K.A., Caldwell, G.A., Marsischky, G., Kolodner, R.D., Labaer, J., Rochet, J.C., Bonini, N.M. and Lindquist, S. (2006). *Alpha-synuclein blocks ER-Golgi traffic and Rab1 rescues neuron loss in Parkinson's models*. **Science 313**, 324–328.
59. Haab, B.B., Paulovich, A.G., Anderson, N.L., Clark, A.M., Downing, G.J., Hermjakob, H., Labaer, J. and Uhlen, M. (2006). *A reagent resource to identify proteins and peptides of interest for the cancer community: a workshop report*. **Molecular & Cellular Proteomics 5**, 1996–2007.
60. Chatterjee, D.K. and LaBaer, J. (2006). *Protein Technologies*. **Current Opinion in Biotechnology** 17, 334–336.
61. Zuo, D., Mohr, S.E., Hu, Y., Taycher, E., Rolfs, A., Kramer, J., Williamson, J. and LaBaer, J. (2007). *PlasmID: a centralized repository for plasmid clone information and distribution*. **Nucleic Acids Research 35**, D680–D684.
62. Hu, Y., Rolfs, A., Bhullar, B., Murthy, T.V., Zhu, C., Berger, M.F., Camargo, A.A., Kelley, F., McCarron, S., Jepson, D., Richardson, A., Raphael, J., Moreira, D., Taycher, E., Zuo, D., Mohr, S., Kane, M.F., Williamson, J., Simpson, A., Bulyk, M.L., Harlow, E., Marsischky, G., Kolodner, R.D. and LaBaer, J. (2007). *Approaching a complete repository of sequence-verified protein-encoding clones for Saccharomyces cerevisiae*. **Genome Research 17**, 536–543.
63. Taycher, E., Rolfs, A., Hu, Y., Zuo, D., Mohr, S.E., Williamson, J. and LaBaer, J. (2007). *A novel approach to sequence validating protein expression clones with automated decision making*. **BMC Bioinformatics 8**, 198.
64. Murthy, T., Rolfs, A., Hu, Y., Shi, Z., Raphael, J., Moreira, D., Kelley, F., McCarron, S., Jepson, D., Taycher, E., Zuo, D., Mohr, S.E., Fernandez, M., Brizuela, L. and LaBaer, J. (2007). *A full-genomic sequence-verified protein-coding gene collection for Francisella tularensis*. **PLoS ONE 2**, e577
65. Langlais, C., Guillaume, B., Wermke, N., Scheuermann, T., Ebert, L., LaBaer, J. and Korn, B. (2007). *A systematic approach for testing expression of human full-length proteins in cell-free expression systems*. **BMC Biotechnology 7**, 64.
66. Xu, L., Zhu, J., Hu, X., Zhu, H., Kim, H.T., LaBaer, J., Goldberg, A. and Yuan, J. (2007). *c-IAP1 cooperates with Myc by acting as a ubiquitin ligase for Mad1*. **Molecular Cell 28**, 914–922.
67. Slagowski, N.L., Kramer, R.W., Morrison, M.F., LaBaer, J. and Lesser, C.F. (2008). *A functional genomic yeast screen to identify pathogenic bacterial proteins*. **PLoS Pathog. 4**, e9.
68. Rolfs, A., Hu, Y., Ebert, L., Hoffmann, D., Zuo, D., Ramachandran, N., Raphael, J., Kelley, F., McCarron, S., Jepson, D.A., Shen, B., Baqui, M.M., Pearlberg, J., Taycher, E., DeLoughery, C., Hoerlein, A., Korn, B. and LaBaer, J. (2008). *A biomedically enriched collection of 7000 human ORF clones*. **PLoS ONE. 3**, e1528.
69. Rolfs, A., Montor, W.R., Yoon, S.S., Hu, Y., Bhullar, B., Kelley, F., McCarron, S., Jepson, D.A., Shen, B., Taycher, E., Mohr, S.E., Zuo, D., Williamson, J., Mekalanos, J. and Labaer, J. (2008). *Production and sequence validation of a complete full length ORF collection for the pathogenic bacterium Vibrio cholerae*. **Proceedings of the National Academy of Sciences USA 105**, 4364–4369.

70. Anderson, K.S., Ramachandran, N., Wong, J., Raphael, J.V., Hainsworth, E., Demirkan, G., Cramer, D., Aronzon, D., Hodi, F.S., Harris, L., Logvinenko, T. and LaBaer, J. (2008). *Application of protein microarrays for multiplexed detection of antibodies to tumor antigens in breast cancer.* **Journal of Proteome Research 7**, 1490-1499.
71. Hughes, S.R., Dowd, P.F., Hector, R.E., Panavas, T., Sterner, D.E., Qureshi, N., Bischoff, K.M., Bang, S.S., Mertens, J.A., Johnson, E.T., Li, X.L., Jackson, J.S., Caughey, R.J., Riedmuller, S.B., Bartolett, S., Liu, S., Rich, J.O., Farrelly, P.J., Butt, T.R., Labaer, J. and Cotta, M.A. (2008). *Lycotoxin-1 insecticidal peptide optimized by amino acid scanning mutagenesis and expressed as a coproduct in an ethanologenic Saccharomyces cerevisiae strain.* **Journal of Peptide Science 14**, 1039–1050.
72. Ramachandran, N., Raphael, J.V., Hainsworth, E., Demirkan, G., Fuentes, M.G., Rolfs, A., Hu, Y. and Labaer, J. (2008). *Next-generation high-density self-assembling functional protein arrays.* **Nature Methods 5**, 535–538.
73. Ramachandran, N., Srivastava, S. and Labaer, J. (2008). *Applications of protein microarrays for biomarker discovery.* **Proteomics Clinical Applications 2**, 1444–1459.
74. Ramachandran, N., Anderson, K.S., Raphael, J.V., Hainsworth, E., Sibani, S., Montor, W.R., Pacek, M., Wong, J., Elianne, M., Sanda, M.G., Hu, Y., Lovinenko, T. and Labaer, J. (2008). *Tracking humoral responses using self assembling protein microarrays.* **Proteomics Clinical Applications 2**, 1518–1527.
75. Berman, H.M., Westbrook, J.D., Gabanyi, M.J., Tao, Y., Shah, R., Kouranov, A., Schwede, T., Arnold, K., Kiefer, F., Bordoli, L., Kopp, J., Podvinec, M., Adams, P.D., Carter, L.G., Minor, W., Nair, R. and LaBaer, J. (2009). *The Protein Structure Initiative Structural Genomics Knowledgebase.* **Nucleic Acid Research 37**, D365–D368.
76. Hughes, S.R., Sterner, D.E., Bischoff, K.M., Hector, R.E., Dowd, P.F., Qureshi, N., Bang, S.S., Grynaviski, N., Chakrabarty, T., Johnson, E.T., Dien, B.S., Mertens, J.A., Caughey, R.J., Liu, S., Butt, T.R., LaBaer, J., Cotta, M.A. and Rich, J.O. (2009). *Engineered Saccharomyces cerevisiae strain for improved xylose utilization with a three-plasmid SUMO yeast expression system.* **Plasmid 61**, 22–38.
77. Srivastava, S. and LaBaer, J. (2008). *Nanotubes light up protein arrays.* **Nature Biotechnology 26**, 1244–1246.
78. Murphy, R.F. and Labaer, J. (2009). *Locations Everyone: Lights, Camera, Action!* **Journal of Proteome Research 8**, 1.
79. Zhu, C., Byers, K.J., McCord, R.P., Shi, Z., Berger, M.F., Newburger, D.E., Saulrieta, K., Smith, Z., Shah, M.V., Radhakrishnan, M., Philippakis, A.A., Hu, Y., De Masi, F., Pacek, M., Rolfs, A., Murthy, T., Labaer, J. and Bulyk, M.L. (2009). *High-resolution DNA binding specificity analysis of yeast transcription factors.* **Genome Research 19**, pp. 556–566.

CONTENTS

Preface vii

1. **NANOTECHNOLOGY APPLICATIONS OF NUCLEIC ACID PROGRAMMABLE PROTEIN ARRAYS** 1
   C. Nicolini and J. LaBaer
   1.1 Introduction 1
   1.2 Materials and Methods 5
       1.2.1 The FLEXGene repository 6
       1.2.2 Nucleic Acid-Programmable Protein Array (NAPPA) 7
       1.2.3 Protein–protein interactions 8
   1.3 Results 11
       1.3.1 Atomic force microscopy (AFM) 13
       1.3.2 Nanogravimetry 14
       1.3.3 Mass spectrometry (MS) 15
       1.3.4 Anodic porous alumina (APA) 16
       1.3.5 Fluorescence via DNASER and bioinformatics 18
       1.3.6 Transcription translation kit 18
   1.4 Discussion 19
   1.5 Conclusions 23
   Acknowledgments 24
   References 25

2. **BIOINFORMATICS AND FLUORESCENCE DNASER FOR NAPPA STUDIES ON CELL TRANSFORMATION AND CELL CYCLE** 31
   V. Sivozhelezov, R. Spera, L. Giacomelli, E. Hainsworth,
   Joshua LaBaer, N. L. Bragazzi and C. Nicolini
   2.1 Introduction 31
   2.2 Materials and Methods 33
       2.2.1 The experimental layout 34
       2.2.2 Procedure of NAPPA expression and DNASER analysis 36
   2.3 Results 37

2.3.1 NAPPA DNASER imaging	37
2.3.2 NAPPA-targeted prediction of gene interactions relevant to progressing between phases of cell cycle of human T lymphocytes	40
2.3.3 NAPPA-targeted prediction of protein interactions relevant to differences between lymphoma and normal T cells	43
2.3.4 Planned experimentation	53
2.4 Conclusions	56
Acknowledgments	57
References	57

3. LABEL FREE DETECTION OF NAPPA VIA MASS SPECTROMETRY 61

R. Spera, F. Badino, E. Hainsworth, M. Fuentes, S. Srivastava, J. LaBaer and C. Nicolini

3.1 Introduction	61
3.2 Materials and Methods	63
3.2.1 The experimental layout	63
3.2.2 Fabrication of NAPPA	64
3.2.3 NAPPA expression	65
3.2.4 Autoflex analysis	66
3.3 Results	68
3.3.1 Human kinase NAPPA	68
3.3.2 NAPPA for MS	70
3.3.2.1 Matching algorithm	71
3.4 Conclusions	76
Acknowledgments	77
References	77

4. LABEL FREE NAPPA VIA NANOGRAVIMETRY 79

M. Adami, S. Sallam, R. Eggenhöffner, M. Sartore, E. Hainsworth, J. LaBaer and C. Nicolini

4.1 Introduction	79
4.2 Materials and Methods	81
4.2.1 QCM-frequency technique	81
4.2.2 QCM-D dissipation (quality) factor technique	83
4.3 Results	88
4.4 Conclusions and Suggestions	91
Acknowledgments	92
References	92

5. **LABEL-FREE NAPPA: ANODIC POROUS ALUMINA** 95
   E. Stura, C. Larosa, T. Bezerra Correia Terencio, E. Hainsworth,
   N. Ramachandran, J. LaBaer and C. Nicolini
   5.1 Introduction 95
   5.2 Materials and Methods 96
       5.2.1 Experimental details 97
       5.2.2 Mechanical tests for anodic porous alumina 99
             5.2.2.1 Grip test 99
             5.2.2.2 Ball-crush test 100
             5.2.2.3 APA over glass slides and reversibility test 101
   5.3 Results 103
       5.3.1 Electric impedance spectroscopy set up for NAPPA
             analysis 103
       5.3.2 Sponge effect and application to existing spotting
             systems 104
       5.3.3 AFM analysis of APA samples 104
   5.4 Conclusions 106
   Acknowledgments 107
   References 107

6. **LABEL FREE DETECTION OF NAPPA VIA ATOMIC FORCE MICROSCOPY** 109
   M. Sartore, R. Eggenhöffner, T. Bezerra, E. Stura, E. Hainsworth,
   J. LaBaer and C. Nicolini
   6.1 Introduction 109
   6.2 Materials and Methods 110
       6.2.1 Technical details for NAPPA imaging by AFM 111
   6.3 Results 113
       6.3.1 NAPPA investigations with the functionalization of
             cantilever and the measurement of interaction forces 116
   6.4 Conclusions and Advancements 118
   Acknowledgments 119
   References 119

7. **CELL FREE EXPRESSION AND APA FOR NAPPA AND PROTEIN NANOCRYSTALLOGRAPHY** 121
   E. Pechkova, S. Chong, S. Tripathi and C. Nicolini
   7.1 Introduction 121
   7.2 APA and NAPPA Microarray 122
   7.3 Background of Cell-Free Protein Synthesis 123
   7.4 APA-Cell Free and Protein Nanocrystallography 124

7.5 Materials and Methods 126
    7.5.1 The PURExpress system and its advantages 126
    7.5.2 APA template preparation 129
        7.5.2.1 APA process 130
        7.5.2.2 APA process 131
    7.5.3 In vitro applications of the PURExpress system 133
        7.5.3.1 High throughput functional genomics and Proteomics 133
        7.5.3.2 Protein engineering 133
        7.5.3.3 Study of protein expression, translation and folding 134
        7.5.3.4 Incorporation of unnatural amino acids 134
    7.5.4 Crystal growth on APA template and in nanovolume 135
        7.5.4.1 APA protein crystals X-ray diffraction 136
        7.5.4.2 Nanovolume protein crystals X-ray diffraction 137
        7.5.4.3 Data processing and reduction 137
        7.5.4.4 Nanovolume-grown protein crystal structure: LB versus classical 139
7.6 Conclusions and Future Prospectives 140
Acknowledgments 144
References 144

8. STRUCTURAL AND FUNCTIONAL STUDIES ON THE *HELICOBACTER PYLORI* PROTEOME: THE STATE OF THE ART 149
G. Zanotti and L. Cendron
8.1 Introduction 149
8.2 The cag Pathogenicity Islands 151
8.3 Enzymes 154
    8.3.1   Enzymes involved in fatty acids metabolism 157
    8.3.2   Enzymes of the shikimate pathway 158
    8.3.3   Enzymes with antioxidant properties 160
    8.3.4   Enzymes involved in sugar metabolism 162
    8.3.5   Enzymes involved in cell wall biosynthesis 162
    8.3.6   Vitamins precursors 163
    8.3.7   Enzymes involved in NAD biosynthesis 164
    8.3.8   Enzymes involved in peptides biosynthesis or degradation 164
    8.3.9   Urea metabolism 166
    8.3.10 Miscellanea 167

8.4	Proteins Interacting with DNA and Regulators of Gene Expression	169
8.5	Toxins and Other Proteins Directly Involved in Pathogenicity	172
8.6	Unknown Function Proteins	173
8.7	Conclusions	174
	References	175

9. OVERALL PROTEOME ALTERATIONS DURING REVERSE TRANSFORMATION OF GROWING CHO-K1 CELLS ... 183
R. Spera and C. Nicolini

9.1	Introduction	183
9.2	General Effects of cAMP on CHO-K1 Cells	185
	9.2.1 Involvement of vimentin	188
	9.2.2 cAMP-induced gene expression	188
	9.2.3 Puck model	190
9.3	Overall CHO Proteome Investigation	191
	9.3.1 LC-ESI MS analysis of nucleus fraction	194
	9.3.2 HPLC analysis	201
	9.3.3 Gel electrophoresis	206
	9.3.4 MALDI-TOF MS	213
9.4	Conclusions	220
	References	225

10. ORGAN TRANSPLANTS AND GENE MICROARRAYS ... 229
R. Danger, J. P. Soulillou1, S. Brouard and C. Nicolini

10.1	Introduction	229
10.2	A Glass of Genes?	230
	10.2.1 Microarray principle	230
	10.2.2 Sample source	232
	10.2.3 Overview of microarray analysis in organ transplantation	233
10.3	Organ Transplantation Under Gene Magnifying Glass	237
	10.3.1 Pre- or peri-transplantation related factors	237
	10.3.2 Microarray analysis in acute allograft rejection	239
	10.3.3 Microarray analysis in chronic allograft rejection	242
	10.3.4 Immunosuppressive drugs in transplantation	243
10.4	Conclusion	246
	References	249
	Abbreviations	256

11. SIGNALING NETWORKS. SIMULATIONS OF BIOCHEMICAL INTERACTIONS. APPLICATIONS TO MOLECULAR ONCOLOGY 257
L. Tortolina, N. Castagnino, R. Montagna, R Pesenti, A. Balbi1 and S. Parodi
11.1 Introduction 258
11.2 Molecular Interaction Maps (MIMs) 264
11.3 Implementation of a Dynamic Modeling of Cell Signaling Pathways and Networks: What It Is About? What Do We Expect From Dynamic Modeling? 270
    11.3.1 Model construction is error prone 272
    11.3.2 Stochastic simulation approach 274
    11.3.3 Sufficient/adequate MIM size 276
    11.3.4 Forward and backward kinetics 278
    11.3.5 Problem of validation of a given dynamic simulation 279
    11.3.6 Multidimensional sensitivity analysis 280
11.4 Conclusions 282
References 284

12. LABEL-FREE DETECTION OF NAPPA: SURFACE PLASMON RESONANCE 287
M. Fuentes, S. Svrivastava, N. Ramachandran, E. Hainsworth and J. LaBaer
12.1 Nucleic Acids Programmable Protein Arrays 287
12.2 Surface Plasmon Resonance 290
    12.2.1 What is a Surface Plasmon and what is Surface Plasmon resonance? 291
    12.2.2 The principle of measurement-towards a sensorgram 292
12.3 Materials and Methods 296
    12.3.1 NAPPA array production 296
    12.3.2 NAPPA expression 296
    12.3.3 Detection of in situ expressed protein by SPR 297
12.4 Results 298
    12.4.1 Detection of GST purified protein by SPR 298
    12.4.2 Detection of in situ expressed protein by SPR 300
12.5 Conclusions and Future Directions 302
References 302

Index 305

CHAPTER 1

NANOTECHNOLOGY APPLICATIONS OF NUCLEIC ACID PROGRAMMABLE PROTEIN ARRAYS

Claudio Nicolini[1] and Joshua LaBaer[2]

[1]*Nanoworld Institute and CIRSDNNOB, Genova University*
Corso Europa 30, 16132 Genoa, Italy
[2]*Harvard University, Institute of Proteomics, Cambridge USA*

NAPPA slides printed using the four genes JUN, P53, CDK2 and CDK4, were analyzed by means of fluorescence labeling in conjunction with DNASER and Bioinformatics and by means of several innovative Label Free approaches based on Atomic Force Microscopy (AFM), Nanogravimetry, Mass Spectrometry (MS) and Anodic Porous Alumina (APA), strictly using all technologies being developed in house. The experimental activities part of this Volume and summarized in this review, are proved capable to underline and assess the feasibility of Nucleic Acid Programmable Protein Arrays (NAPPA) and of its Label Free approaches, as well the traditional fluorescence technique utilizing the DNASER, in contributing to the progress of functional proteomics towards the understanding of the molecular events controlling cell cycle and cell transformation.

1.1 Introduction

The post-genomic era has left us with numerous novel proteins of unknown function. For a systematic functional analysis of these proteins, highly parallelised and cost–effective methods are required, namely microarrays proven capable to be miniaturized and to carry out high throughout (HT) analysis of genes (Nicolini *et al.*, 2002; Nicolini 2006; Sivozhelezov *et al.*, 2006; Nicolini, 2009). For proteins, however, the peculiar physico-chemical requirements of the different proteins for activity have severely hampered the use of protein microarrays (LaBaer and Ramachandran, 2005). Protein microarrays focus on protein-protein interactions, key elements in the control of cell function (Figure 1.1).

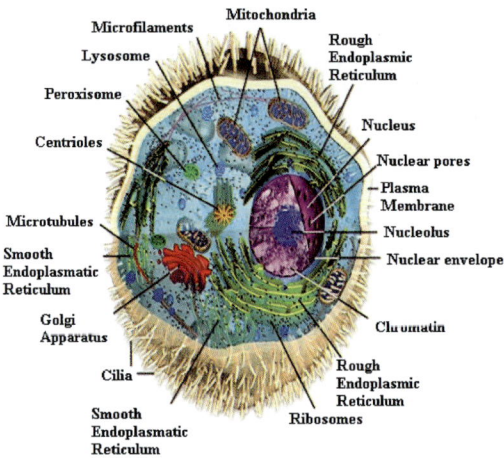

Figure 1.1. Protein–protein interactions.

The concept of protein microarrays has stirred a great deal of excitement in the proteomics community (Mitchell 2002). Once the technology is fully realized, it promises to enable the study of a broad variety of protein features at an unprecedented pace and scale. At present, all protein microarrays fall into two general classes: *antibody* arrays (Haab *et al.*, 2001) and *test protein* arrays (MacBeath and Schreiber, 2000; Zhu *et al.*, 2001). The concept behind an antibody array (Borrebaeck, 2000) is analogous to DNA microarrays. The test protein array includes NAPPA, main object of this Volume. In contrast to the antibody arrays, the test proteins themselves are indeed arrayed. There are relatively few studies published regarding the development or exploitation of protein microarrays, although this has become an active area of investigation. MacBeath and Schreiber (2000) have demonstrated the feasibility of printing proteins on a high-density array. A density of approximately 20,000 spots per slide was achieved by spotting nanoliter volumes of sample. A limitation of this work was that it was performed on a relatively small set of proteins, reflecting the current challenge of producing thousands of proteins to spot on a microarray. Subsequently, Zhu *et al.* (2001) developed a highly dense array containing most of the yeast proteome. Here the authors exploited homologous recombination in yeast to create a complete collection of strains where each gene was replaced with a copy that had a His-GST double tag. The proteins were purified, then immobilized by spotting on aldehyde-treated slides or using His binding to nickel coated glass slides. Functionality of the proteins was demonstrated by screening the yeast proteome for calmodulin and phosphoinositide interactors. Although this effort was a dramatic demonstration of the power of protein microarrays, it is not clear how universal the methods will be. The ability to replace genes by gap repair and to express and purify the proteins from their natural cell types does not easily extrapolate to mammalian proteins. Additional reports have used this approach to screen for protein function (Newman and Keating, 2003; Espejo *et al.*, 2002) which is also reviewed in (Jona and Snyder, 2003).

In the development of protein microarrays there are several key issues that need to be addressed:

1. *Binding chemistry;* where contrary to DNA microarrays where poly–lysine slides were preferred as they are able to bind negatively charged DNA, for with proteins there are a number of factors to consider such as:
   (a) ability to bind all spotted proteins;
   (b) ability to bind large amounts of protein;
   (c) orientation-specific or random orientation;
   (d) surface chemistry formulated to contain hydrophobic or hydrophilic residues, to support the binding of non-denatured or denatured protein.

2. *Availability of a large collection of proteins,* being a major challenge because most proteins are expressed from cloned cDNAs and cDNA repositories in expression-ready formats have not been available at large scale. In addition, the high throughput (HT) expression and purification of thousands of proteins needed to make a microarray requires access to the robotics and equipment needed for implementing HT purification.

3. *Protein integrity,* where protein folding and modification is best preserved if mammalian proteins are expressed in mammalian cell.

4. *Protein stability*, whereby proteins are notoriously fragile raising concerns about the storage conditions and shelf life of purified protein arrays.

Here, we present progress on the use of DNA micro-arrays for *in situ* generation of protein micro arrays. Our method is based on the Nucleic Acid Programmed Protein Array (NAPPA) — technology for a large scale assay of protein–protein interactions (Ramachandran *et al.*, 2004, 2008a). In the NAPPA method (Figure 1.2), plasmids directing the expression of GST-fusion proteins are spotted along with an antibody against GST.

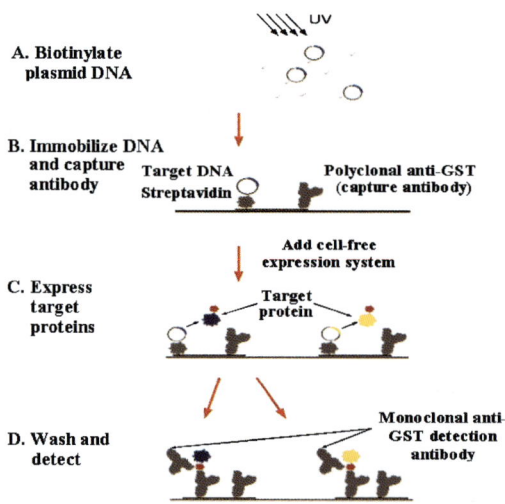

Figure 1.2. NAPPA approach.

Presently the fluorescence labeled fusion proteins produced by use of a coupled transcription/translation reaction are captured and immobilized via their GST fluorescent tag (LaBaer and Ramachandran, 2005). Since arrays are produced just before their use, problems of protein stability can be avoided with fluorescent NAPPA, but with Label Free detection will be possible to search for far more effective and unperturbed means to carry out the analysis.

Two main fields of application of NAPPA technology are of large interest among several emerging in the last few years (Butcher *et al.*, 2006; Ramachandran *et al.*, 2008b,c; Anderson 2008; Berman *et al.*, 2009; Rolfs *et al.*, 2008; Zuo *et al.*, 2007; Cooper *et al.*, 2006; Witt *et al.*, 2006): first, these chips are suited for the detection of protein interactions, where an array of potential binding partners are probed with a soluble protein for binding; secondly, since the proteins on the array should in principle be in a native state, these chips should allow the parallel detection of protein activity. With Label Free detection this could be exploited even more effectively in drug discovery research, where the interaction of a family of drug target proteins with a given candidate therapeutic substance can be examined simultaneously. The protein-protein interactions leading the given physiological and pathological process can be monitored as well in human biopsies. The most important potential applications of NAPPA technologies include:
- the detection of protein–protein binding;
- the detection of protein binding to small drug-like molecules;
- the detection of protein activity;
- testing of serum samples or antibodies for their reactivity against human proteins.

Ultimately, we believe that the ability to synthesize active protein in a micro-format in combination with systematic collections of DNA constructs for protein expression and with Label Free approaches, ranging from Atomic Force Microscopy (Sartore *et al.*, 2000; Pechkova *et al.*, 2007, Sartore *et al.*, 2009) to Nanogravimtery (Adami *et al.*, 2009) and Mass Spectrometry (Spera and Nicolini, 2007, 2008; Spera *et al.*, 2009), will open up tremendous possibilities in post-genomic research.

1.2 Materials and Methods

The basics of the NAPPA arrays and the FLEXGene Repository and the underlying technology of recombinational cloning are summarized along with the protocols for the HT assembly of cDNA clones and subsequently.

1.2.1 *The FLEXGene repository*

The FLEXGene Repository is a collection of at least one representative coding sequence (CDS) for every human gene or organism studied (LaBaer, CSHL, 2005). The FLEXGene Repository consists of a collection of individually bar-coded plasmids, each containing a unique full–length CDS. Fundamental aspect of the FLEXGene clones is that they are assembled using recombinational cloning. In brief, recombinational cloning permits the directional and in-frame shuttling of inserts from a "master–clone" into any functional vector, without the use of restriction enzymes. In its final form, the FLEXGene repository will be a collection of such master clones containing sequence–verified full–length CDSs. Each CDS in the repository is captured in two different forms: one construct has the native STOP-codon for the expression of native protein. The second construct has the STOP-codon removed for the expression of C-terminal fusion proteins. Detailed information of each assembled construct is available through a web–accessible searchable database that tracks the cloning history and relevant links to other biological databases (http://www.hip.harvard.edu). The FLEXGene Repository will be a public resource accessible to all scientists — academic, governmental or commercial. Pursuant to this principle, the FLEXGene Repository is being constructed in collaboration with scientists worldwide, both in academia and in industry-based laboratories.

Presently, two recombinational cloning systems are commercially available: the Gateway™ system (Invitrogen) and the Creator™ system (Clontech). The Gateway system is based on the site-specific integration of bacteriophage lambda into the genome of *Escherichia coli*. As soon as the FLEXGene clones are in possession of the researcher, they need to be transferred into the expression vector that is required in his or her experiment. Both Invitrogen and Clontech offer a wide diversity of expression vectors that cover most expression systems and, which are compatible with recombinational cloning.

1.2.2 *Nucleic Acid-Programmable Protein Array (NAPPA)*

The *Nucleic Acid-Programmable Protein Array* (NAPPA) is a novel protein microarray technique. In conventional protein microarrays, purified proteins are spotted onto glass slides. NAPPA eliminates the need for purified proteins by spotting plasmid DNA instead of proteins. All of the genes on the array are then simultaneously transcribed/translated in a cell–free system, and the resulting proteins are immobilized *in situ*. This minimizes direct manipulation of the proteins, and enables the expression and interactions of proteins in a mammalian milieu. Current approaches to constructing protein arrays generally involve the laborious process of expressing proteins in living cells followed by purifying, stabilizing, and spotting the samples. Our approach, called *Nucleic Acid-Programmable Protein Array* (NAPPA), begins with spotting plasmids encoding target proteins. All genes are then simultaneously transcribed/translated in a cell-free system and the resulting proteins are immobilized *in situ*, minimizing direct manipulation of the proteins and enabling the expression and interactions of proteins in a mammalian milieu.

The design of NAPPA has been geared to overcome the limitations of current technology mentioned above:

1. *Availability of a comprehensive protein library.* With access to clones in a recombinational system, such as the Gateway system or Clontech Creator system, coding regions can be readily moved into the necessary expression vectors that append the appropriate tags to the proteins.

2. *Protein Integrity.* Expression of proteins in NAPPA can be carried out by a cell–free mammalian system, such as reticulocyte lysate systems. Thus, the conditions are excellent for mammalian protein expression and folding. Natural post-translational modifications will occur in these extracts and/or can be supplemented with activated lysates (Walter and Blobel 1983; Starr and Hanover 1990).

3. *Protein Stability.* On a NAPPA microarray, proteins are produced just in time: once the array is activated for analysis, all reactions occur in solution and in real time so stability is not an issue.

In the NAPPA version the *chemistry* consists of several steps:
- Plasmid DNA is mixed with psoralen-biotin, and cross–linked using UV light.
- The DNA is mixed with streptavidin, linker, and the anti-GST capture antibody, and this mix is arrayed on the aminosilane-coated glass slides.
- After blocking, cell–free expression mix is applied to the slide, and during a temperature–programmed incubation the proteins are produced and bind to the capture antibody.
- The slide is washed, and the proteins are detected by detecting the GST tag (using a monoclonal anti-GST antibody, an HRP-labeled anti-mouse antibody, and Cy3-tyramide (TSA) HRP substrate).

In theory, NAPPA can provide the same *applications* as other test protein arrays including detection of interactions with proteins, nucleic acids, lipids, small molecules, antibodies and enzymes, as discussed above. A valuable aspect of NAPPA lies in its *flexibility* to allow a broad range of target immobilization and query detection schemes, using various biochemical methods for target binding and query detection.

1.2.3 *Protein–protein interactions*

A powerful application of protein microarrays is the ability to detect and map protein-protein interactions. It is becoming increasingly clear that all proteins function as part of complexes. These multimeric protein complexes are thought to regulate nearly all cellular processes. Recent reports have pursued HT approaches to map protein interactions within the yeast proteome (Ho *et al.*, 2002; Gavin *et al.*, 2002). Immunoprecipitation followed by mass spectrometric analysis of tandem affinity purified complexes showed that 78% of 589 bait proteins were associated with partners (Gavin *et al.*, 2002). Understanding protein-protein interactions provides important clues to the function of proteins.

The identification of interactions with known proteins may suggest the functional role played by a novel protein. To date, two major techniques have been used for the HT analysis of protein–protein interactions: yeast two-hybrid (Ito *et al.*, 2001; Uetz *et al.*, 2000;

Table 1.1. Advantages and Disadvantages of HT approaches

HT Approaches	Advantages	Disadvantages
Yeast Two Hybrid	• Easy to execute • No need for expensive equipment • Ability to screen large libraries	• High false positive/negative rates • Occurs only in nucleus • Limited to pair-wise interaction (high order complexes missed) • Sensitive to toxic genes • Proper folding of mammalian proteins is not assured
Immunoprecipitation	• Proteins in native state • Interactions are natural • Large order complexes can be observed	• Sticky proteins appear regularly • Unclear whether interaction is direct or indirect • Expensive (need for massive equipment both analytical and computational) • Applications in organisms other than yeast may be challenging

Walhout *et al.*, 2000) and immunoprecipitation followed by mass spectrometry (Ho *et al.*, 2002; Gavin *et al.*, 2002).

The relative advantages and disadvantages are shown in Table 1.1. The use of well–executed protein arrays such as NAPPA complement both of the above approaches and avoid some of their pitfalls, warranting:

- Interaction not limited to context of nucleus.
- Expression of mammalian proteins in mammalian milieu.
- No need of gene introduction into mammalian cells.
- Gene toxicity and auto activation of reporter genes not an issue.
- Not required large mass spectrometers.
- Able to examine multimeric complexes.
- Able to perform post translational modifications.
- Able to produce protein and obtain data in real time.
- Plasmid templates sequence-verified providing good control.
- Able to change to different attachment/detection scheme if needed.

Hardware and software that were developed are the ones typically used for protein microarrays (Figure 1.3). Arrayers are machines that pick up sample from 96 or 384 well plates, and deposit tiny uniform drops in a uniform pattern on many identical glass slides. Most arrayers come with humidity control.

Figure 1.3. Pins most commons and least expensive.

Most surface chemistries are aimed at modifying standard glass microscope slides (1 × 3 inches) as they can readily used with standard DNA microarrayers and scanners. Several chemistries are available to immobilize proteins, all offering several advantages and disadvantages for 2D surfaces and 3D surfaces.

Most scanners use lasers for the light source, and filters and photomultiplier tubes (PMTs) for detection; the image is built up from many tiny detection points. An alternate approach is to use white light and a filter for illumination, and a CCD camera for detection of the whole array simultaneously, as in the case of the DNASER (Nicolini *et al.*, 2002). Scanners developed for glass-slide DNA microarrays (see later DNASER in Figure 1.11) have at least two colors, for the two most common fluorophores: Cy3 (green) and Cy5 (red). Only a single color is needed for the protein interactions that we are doing, since we use one tag for all query proteins. Multiplexed applications, doing multiple queries simultaneously, will use two or more fluorophores, and therefore need multicolor scanners.

Detection can be through the slide, or from same side as the illumination, which allows use of opaque slides. Some arrayers can scan wet slides.

The NAPPA protocol (LaBaer, CSHL Proteomics 2005) go through various steps:
- aminosilane slide coating
- slide coating
- DNA preparation
- DNA quantitation
- sample preparation
- expression of the NAPPA slides
- detection of the NAPPA slides
- detection of the DNA on NAPPA slides

1.3 Results

The protein and corresponding genes chosen for this study and their characteristics are:
- P53_Human: 43,653 kD with Gene of 1182 bp http://plasmid.med.harvard.edu/PLASMID/GetCloneDetail.do?cloneid=613&species=Homo%20sapiens.
- CDKN1A_Human (called also P21): 18,119 kDa with Gene of 495 bp http://plasmid.med.harvard.edu/PLASMID/GetCloneDetail.do?cloneid=1447&species=Homo%20sapiens.
- CdK2_Human: 33,930 kDa with Gene of 897 bp http://plasmid.med.harvard.edu/PLASMID/GetCloneDetail.do?cloneid=1475&species=Homo%20sapiens.
- JUN_Human: 35,676 kDa with Gene of 996 bp http://plasmid.med.harvard.edu/PLASMID/GetCloneDetail.do?cloneid=666&species=Homo%20sapiens.

Only the four Genes JUN, P53, CDK2 and CDK4 were utilized in all experimental studies and placed in the proper fashion according to the technique being used (Figure 1.4 and Figure 1.5) to monitor label–free protein–protein interactions in a high-throughput format using self–

assembling protein NAPPA microarrays and AFM, nanogravimetry QCM, MS and APA in the control of cell proliferation and neoplastic transformation. Also fluorescence with our DNASER (Nicolini *et al.*, 2002) for prompt comparison with well established analysis (LaBaer and Ramachadran, 2005).

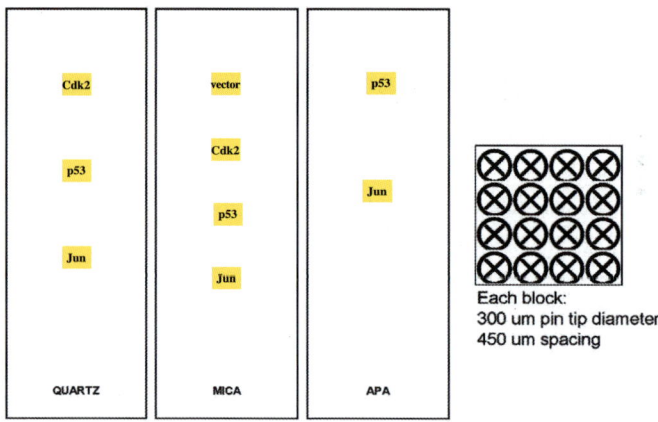

Figure 1.4. Slides for quartz Mica and APA.

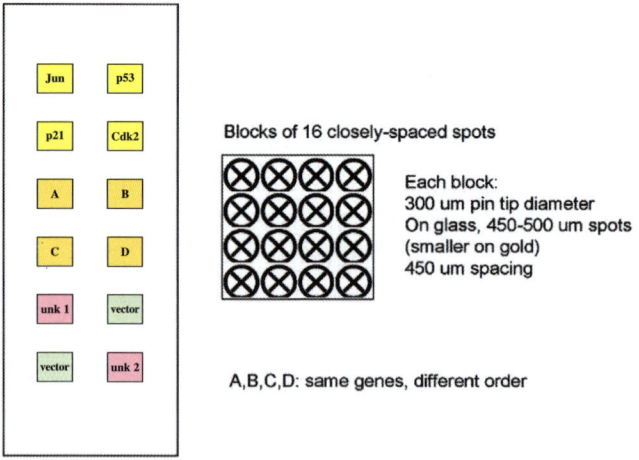

Figure 1.5. MALDI arrays.

Details are provided in the flowchart below (Figure 1.6), beginning with the QCM-Q technology needing printing on the metal pad of quartz crystals, and with the AFM needing pure Mica or gold–coated Mica substrates and optical markers. Here reference is made to AFM and QCM, but remain valid for MS and APA.

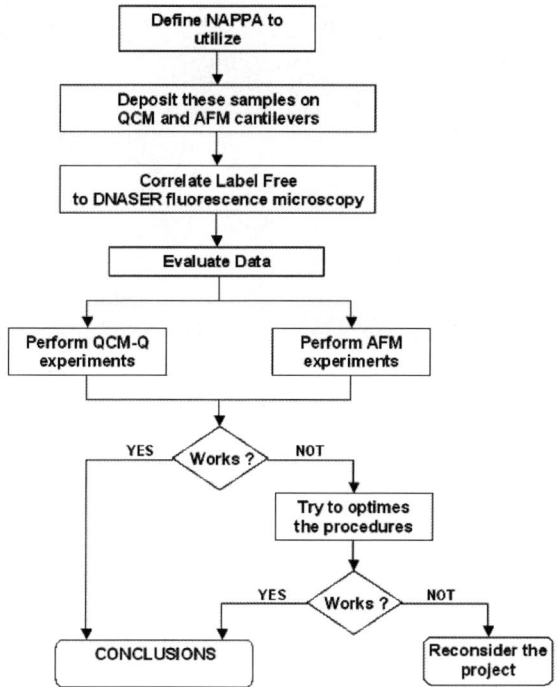

Figure 1.6. Flowchart A.

1.3.1 *Atomic force microscopy (AFM)*

It appears that our experiments show that to couple NAPPA to AFM (Figure 1.7) the optimal choice was to work on a plain mica surface and to fully undertake an independent atomic characterization on the proteins being expressed in each gene-containing spot. Subsequently the three genes JUN (1), P53 (2) and CDK2 respectively were spotted on Golden

Mica Slide to utilize the optimal gold chemistry, similarly yielding excellent results (Sartore *et al.*, 2009).

The mica have been positioned on the glass slide in such a way to make marginal the thickness increase of the sample and to allow their subsequent removal after NAPPA printing spots and/or their expression.

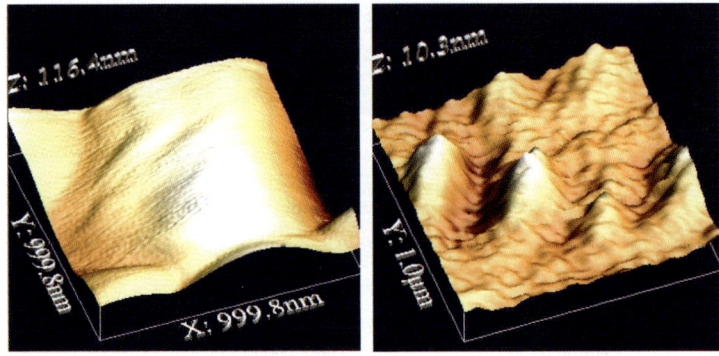

Figure 1.7. AFM of P53 before and after NAPPA expression (Sartore *et al.*, 2009).

Obviously if one is looking for an alternative high-throughput devices to the actual optical reading used in NAPPA, the answer is not certainly AFM, because many different optical techniques (*i.e.*, scanning laser or one-shot imaging) have demonstrated to be more flexible, reliable and much less costly than AFM. Building multi-tip measuring heads is both difficult and extremely costly (*i.e.*, to get a matrix of tips).

1.3.2 *Nanogravimetry*

The progress is not only in the sample holder, but in the actual identification of the novel highly sensitive parameter (quality factor D) in addition to the frequency and its hardware/software implementation (Figure 1.8).

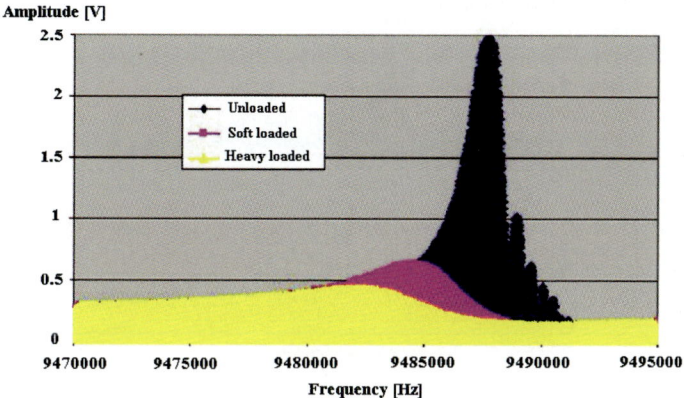

Figure 1.8. Quartz dumping. (Nicolini 2009).

The prototype is already built and is in the optimizing and validation phase with glycerol and various protein solution. The quartz have been positioned on the glass slide in such a way to make marginal the thickness increase of the sample and to allow their subsequent removal after NAPPA printing spots and/or their expression. For what concerns specific applications this label–free technique could be utilized to investigate the kinetics of protein–protein interaction and to optimize NAPPA printing as function of: pH, temperature, reagent composition and concentration. Such technique could also be used to optimize the conditions for cell–free expression of a given protein.

1.3.3 *Mass spectrometry (MS)*

This method (Spera *et al.*, 2009; Spera and Nicolini, 2008) has made the most progress in terms of coupling a label–free method to NAPPA, clearly demonstrating production of NAPPA protein (Figure 1.9) and few MS spectra of total protein.

In this case, the end game is to demonstrate that we can identify proteins, in particular proteins that bind to the target proteins on the array. Towards this end we intend to show that we can identify the printed and expressed proteins.

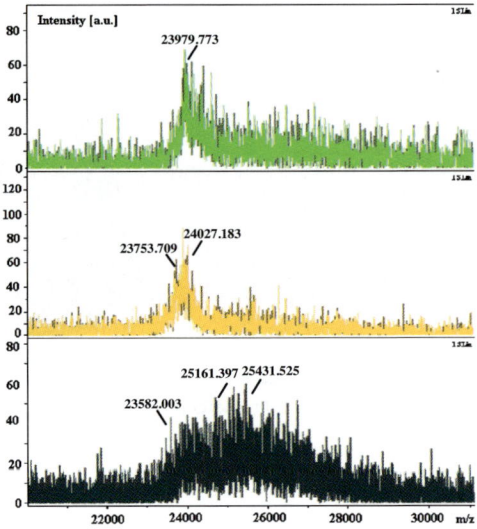

Figure 1.9. MS spectrum at high molecular weights (20–35 kDa) for 3 different spots/genes (Reprinted with the permission from Spera and Nicolini, Nappa microarray and mass spectrometry: new trends and challenges, Essential in Nanoscience Booklet Series, © 2008, underwritten by Taylor & Francis Group, LLC).

Thus, the plan is to print some gold slides with the 4 different genes (each one with 16 × 300 micron spots) — first in a known configuration, then in a configuration unknown to NWI but known to HIP. Towards this end the samples have gold faces, namely microscopy slides coated with gold: it's necessary to spot only on the tagged face for MS (Spera *et al.*, 2009). After this MS analysis was conducted successfully by searching peptides on a database. Key to success was to do tryptic digests and get the peptides to fly and to be identified.

1.3.4 *Anodic porous alumina (APA)*

The mechanical implications of grip and ball crush tests are clear, but the fluid exchanges on the arrays (mixing reagents and washing them away) closed design of the APA can be made optimal for NAPPA (Figure 1.10), even if the open pore design to force fluid through the device and exchange it with new fluid appears a complicated task never so far attempted by anyone (namely nanomanipulation without breakage and differential corrosion).

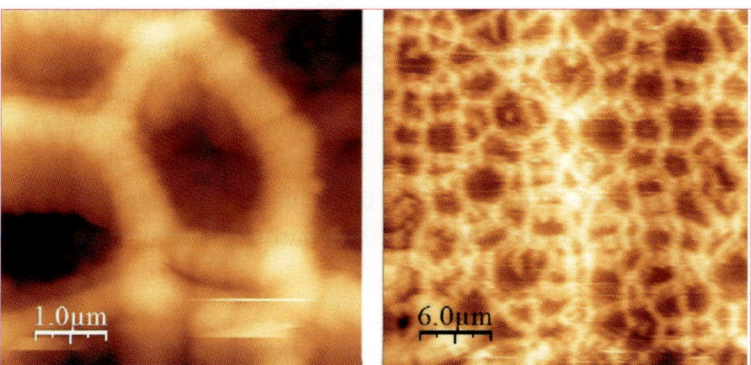

Figure 1.10. Typical APA pores for subsequent NAPPA.

At the same time we have successfully attempted simple NAPPA printing on APA (Stura *et al.*, 2009), after proving:
 (i) The ability to spot a colored fluid on the APA surface in discrete spots.
 (ii) The ability to rapidly exchange that fluid with a different fluid.
 (iii) The ability to repeat these manipulations as needed.
 (iv) The production of the APA slides in a format compatible also with a fluorescence reader and not only with electrochemistry (Stura *et al.*, 2009).
 (v) The prove tat the APA slide format (either alone or over aluminum) has acceptable structural integrity for "routine" manipulation and especially during fluorescence readings, mainly relying for the standalone configuration also on alternative printing which exploit capillary forces.

Once we have these basics worked out, it would be extremely worth trying with NAPPA whatever is the result considering what being already achieved for the high potential of this technology to be used in a high through output setting.

1.3.5 *Fluorescence via DNASER and bioinformatics*

We have reproduced fluorescence analysis using our in house DNASER technology (Figure 1.11), but to optimally image at highest resolution only a part of the fluorescence-labeled NAPPA slide for proper data analysis. Maximum area which can be imaged in a single shot with excellent resolution is 14 × 14 spots, for a total of 196 spots which appear adequate for the control of Label Free data. We definitely need such NAPPA slides with rows of the four existing genes to send to Genova (on glass), in addition with a control glass similar to the one above for mass spectrometry.

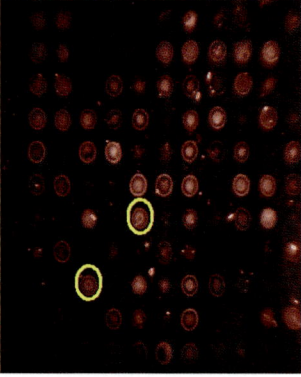

Figure 1.11. NAPPA diagnostics.

1.3.6 *Transcription translation kit*

NEBL Pure System is identified as the optimal system to monitor the proteins/interactions by a Label Free such as AFM, QCM, MS, and mostly with APA (this last pointed out for high–throughoutput), and is seen as very useful since it can bridge these NAPPA activities to LB nanocrystallography where this cooperation NWI-NEBL was initiated over one year ago to be now integrated in a coherent project. It would be advantageous to make the proteins using the Pure System as the system has defined components, and are less likely to interfere with the measurements or the interference. The eventual disadvantage of Pure System compared to RRL used by Harvard is protein translation/folding of larger proteins, a problem being solved sooner rather than later. Chong and NEBL have indeed already tried with success eukaryotic proteins 62 kDa and >100 kDa (Pechkova *et al.*, 2009b). The defined components of the Pure System which we should test in Label Free NAPPA with your consensus are: ribosomes for transcription, recombinant T7 RNA polymerase for translation, aminoacyl-tRNA synthetases, GTP and ATP for generating GDP and AMP, KGluc (100 mM), MgOAc (13 mM) and spermidine (2 mM) used to stabilize nucleic acids.

1.4 Discussion

High density functional protein arrays allow the functional testing of thousands of protein simultaneously. A key remaining challenge for producing protein microarrays has been uniting high content (many different proteins) with high density and functionality. Most approaches rely on expressing and purifying proteins to print on the array surfaces and have succeeded at displaying many different proteins at high density.

Considerable challenges, however, accompany the use of purified protein for printing microarrays. Variable yields of proteins result in dynamic ranges in the amount of protein displayed that cover several orders properties. Array batch-to-batch variation may affect experimental reproducibility and the folding and function of some proteins may also be lost during purification, printing and storage.

Nucleic Acid Programmable Protein Array (NAPPA) allows for functional proteins to be synthesized *in situ* directly from printed cDNAs just in time for assay. The proteins are translated using a T7-coupled rabbit reticulocyte lysate *in vitro* transcription–traslation (IVTT) system and presently we are attempting to use the Pure System described above and in Pechkova *et al.* (2009b). The expressed proteins are captured locally with an antibody to a C-terminal glutathione S-transferase (GST) tag on each protein (anti-GST). This approach eliminates the need for high-throughput protein isolation and ensure that all proteins are produced fresh (that is, coincident with or minutes before use) for each experiment. Many experiments have confirmed that NAPPA produces functional protein.

Here we have described label-free NAPPA method, for detection fresh-protein, *in situ*; printed substrate for NAPPA is purified DNA, which is simpler to prepare, quantify, print and store than protein. In performing optimisation experiments, we observed that high-quality supercoiled DNA provided the best substrate for cell–free protein expression, and that commercial chemistries had insufficient yield and purity for this purpose (Murphy and LaBaer, 2009).

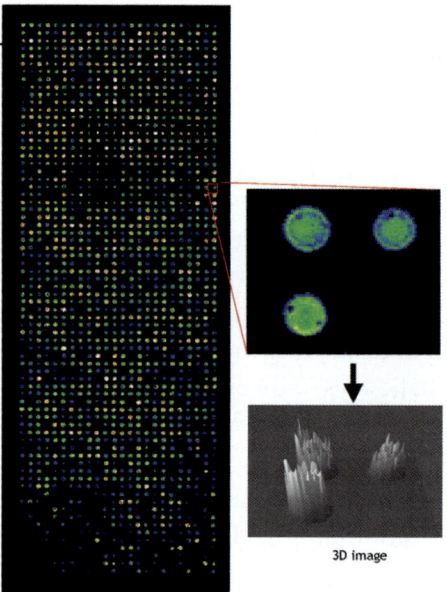

Figure 1.12. Human kinase array on gold surface, using amine terminating SAMs on gold surface (LaBaer and Ramachadran, 2005).

Our present NAPPA approach (Figure 1.12) routinely produced 9 fmol of protein per feature with about 90% success for a broad variety of proteins of different sizes, including membrane proteins and proteins > 100 kDa. To our knowledge, this is the first non-protein printing method to produce over a thousand unique proteins on a microarray surface. Notably, the range of protein signals for nearly all proteins was narrow:

92% of displayed protein yields were within twofold of the mean. This limited variation may be due to the saturation of the capture sites on the array by the expressed target. The method was highly reproducible from array to array and simple (CV = 6%) which compares favourably with that reported for DNA microarray chemistries in which coefficients of variation range from 20 to 40%. This is particularly important considering that NAPPA entails not only printing cDNA but also transcription, translation and protein capture. The ability to array proteins at high density will be well suited for testing protein-protein interactions, screening for enzyme substrates and measuring selectivity of small-molecule drug binding. Particular attention are presently directed with NAPPA to the over 600 human kinases (Figure 1.12) and to their mutual interaction during mammalian cell cycle progression (Sivozhelezov *et al.*, 2009). These function-based microarrays are indeed used to examine protein interactions with other proteins, nucleic acids, lipids, small molecules and other biomolecules (LaBaer and Ramachandran, 2005; Ramachandran *et al.*, 2004). In addition, function-based microarrays can be used to examine enzyme activity and substrate specificity associated to cell transformation, to cell cycle and to cell differentiation. These microarrays are produced by printing the proteins of interest on the array using methods designed to maintain the integrity and activity of the protein, allowing hundreds to thousands of target proteins to be simultaneously screened for function using a wide range of gene collections (Figure 1.13). The list of potential applications of such microarrays is large, including cancer application (Paweletz *et al.*, 2001). A microarray of a particular class of enzymes such as kinases could be screened with a candidate inhibitor to examine binding selectivity. A candidate drug could be used to probe a broad range of enzymes to look for unintended binding targets that might suggest possible toxicities. Proteins expressed by pathogenic organisms can be screened with serum from convalescent patients to identify immunodominant antigens, leading to good vaccine candidates.

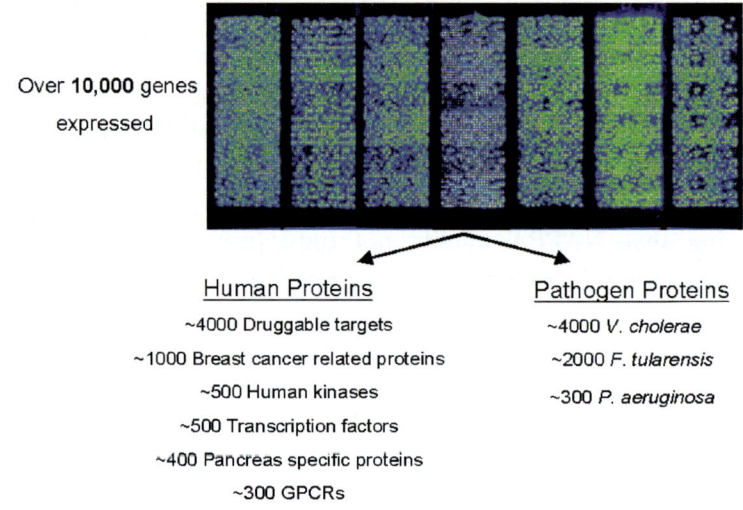

Figure 1.13. Different gene collections. (Nicolini 2009).

Protein interaction networks, including the assembly of multiprotein complexes, can shed light on biochemical pathways and networks in the control of cell transformation and cell differentiation. Eventually, it may even be possible to use with proper inert surface these high–density microarrays as a MALDI source for mass spectrometry, allowing to probe complex samples for binding partners to many proteins simultaneously. However, as with the abundance-based microarrays, there still remain challenges in building and using function based protein microarrays. First, the notorious liability of proteins raises concerns about their stability and integrity on the microarray surface. Second, it is time consuming and costly to produce proteins of good purity and yield, and many proteins cannot be purified at all. Finally, the methods used to attach proteins to the array surface may affect the behaviour of the proteins. Despite these challenges, there has been some success in building and using function–based protein microarrays in medicine (LaBaer and Ramachandran, 2005; Ramachandran *et al.*, 2004). In this chapter we have described different approaches, protein spotting microarrays and self–assembling microarrays, focusing on the recent most promising advances utilizing Label Free technologies (Figure 1.14).

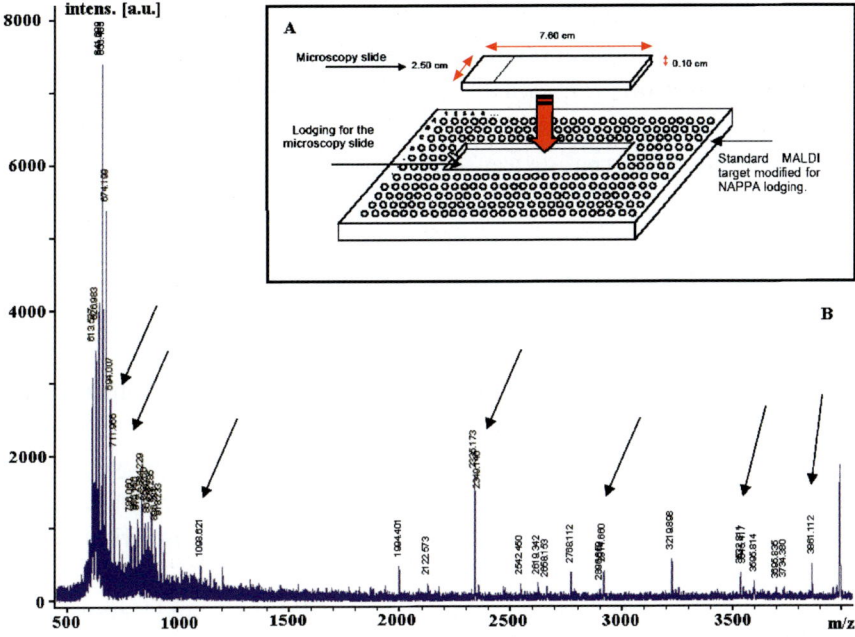

Figure 1.14. (A) MALDI-TOF target modification project. (B) MALDI TOF MS Spectrum of Human Kinase NAPPA array acquired after protein synthesis (spot gene NA 7-A 12), low masses region. The arrows indicate the peaks identified also in the Harvard Institute of Proteomics spectrum (namely $m/z = 711 \pm 12$, $m/z = 865 \pm 18$, $m/z = 1090 \pm 20$, $m/z = 2900 \pm 40$, $m/z = 3540 \pm 40$, $m/z = 3860 \pm 60$) (Reprinted with the permission from Spera and Nicolini, Nappa microarray and mass spectrometry: new trends and challenges, Essential in Nanoscience Booklet Series, © 2008, underwritten by Taylor & Francis Group, LLC).

1.5 Conclusions

In the context of present activity on functional proteomics and cell cycle within the last three years of our existing international Harvard HIP (USA) — Genova NWI (Italy) cooperation we have achieved several significant positive results which open new avenues worldwide, namely:
- proved the feasibility to a different degree and for different applications of four alternative Label-Free approaches on NAPPA, namely nanogravimetric QCM, mass spectrometry (MS), anodic porous alumina (APA) and atomic force microscopy (AFM), even

with respect to Surface Plasmon based Sensing methods or to any of the other Label Free approached developed;
- optimized with DNASER and Bioinformatics the existing fluorescence Labeling technologies;
- introduced an alternative transcriptional/translational system;
- identified an innovative optimal approach to cell cycle, cell transformation and reverse transformation;
- defined the field of possible applications of each above four Label Free Technologies.

In this context it has been extremely important to establish also a baseline, by addressing the functional proteomics of cell cycle in the cell systems used in Genova (CHO-K1 and Human Lymphocytes) with traditional fluorescent NAPPA as suggested by the Bioinformatics being provided and initially with the four genes NAPPA being printed as a result of this cooperation.

Correlation with the numerous relevant and correlated nanotechnologies implemented (Stura *et al.*, 2007; Pechkova *et al.*, 2007, Pechkova and Nicolini 2006; Nicolini *et al.*, 2008), nanogenomic applications to Medicine (Nicolini, 2006; Sivozhelezov *et al.*, 2006, Giacomelli and Nicolini, 2006, Braud *et al.*, 2008, Sivozhelezov *et al.*, 2008) and structural proteomics (Pechkova and Nicolini, 2006; Pechkova *et al.*, 2007, 2008a,b, 2009b) being carried out within this very same grant and cooperation.

Acknowledgments

This project was supported by grants to Fondazione EL.B.A. by MIUR for "Funzionamento" and by a FIRB International Grant on Proteomics and Cell Cycle (RBIN04RXHS) from MIUR (Ministero dell'Istruzione, Università e Ricerca) to CIRSDNNOB-Nanoworld Institute of the University of Genova.

References

Adami, M., Sallam S., Eggenhöffner, R., Sartore, S., Hainsworth, G., LaBaer, J. and Nicolini,C., (2009). Label free NAPPA via nanogravimetry, in *Functional Proteomics and Nanotechnology-based Microarrays*, Nicolini C. and LaBaer J. Eds, Pan Stanford Series on Nanobiotechnology Volume 2, London, New York and Singapore, pp. 79–93.

Anderson, K. S., Ramachandran, N., Wong, J., Raphael, J. V., Hainsworth, E., Demirkan, G., Cramer, D., Aronzon, D., Hodi, F. S., Harris, L., Logvinenko, T. and LaBaer, J. (2008). Application of protein microarrays for multiplexed detection of antibodies to tumor antigens in breast cancer. *Journal of Proteome Research*, 7, pp. 1490–1499.

Berman, H. M., Westbrook, J. D., Gabanyi, M. J., Tao, Y., Shah, R., Kouranov, A., Schwede, T., Arnold, K., Kiefer, F., Bordoli, L., Kopp, J., Podvinec, M., Adams, P. D., Carter, L. G., Minor, W., Nair, R. and LaBaer, J. (2009). The Protein Structure Initiative Structural Genomics Knowledgebase. *Nucleic Acid Research*, 37, pp. D365–D368.

Borrebaeck, C. (2000). Antibodies in diagnostics - from immunoassays to protein chips. *Immunology Today*, 21, pp. 379–382.

Braud, C., Baeten, D., Giral, M., Pallier, A., Ashton-Chess, J., Braudeau, C., Chevalier, C., Lebars, A., Léger, J., Moreau, A., Pechkova, E., Nicolini, C., Soulillou, J. P. and Brouard, S. (2008). Immunosuppressive drug–free operational immune tolerance in human kidney transplants recipients: I. Blood gene expression statistical analysis. *Journal of Cellular Biochemistry*, 103, pp. 1681–16902.

Butcher, R. A., Bhullar, B. S., Perlstein, E. O., Marsischky, G., LaBaer, J. and Schreiber, S. L. (2006). Microarray-based method for monitoring yeast overexpression strains reveals small-molecule targets in TOR pathway. *Nature Chemical Biology,* 2, 103–109.

Cooper, A. A., Gitler, A. D., Cashikar, A., Haynes, C. M., Hill, K. J., Bhullar, B., Liu, K., Xu, K., Strathearn, K. E., Liu, F., Cao, S., Caldwell, K. A., Caldwell, G. A., Marsischky, G., Kolodner, R.D., LaBaer, J., Rochet, J. C., Bonini, N. M. and Lindquist, S. (2006). Alpha-synuclein blocks ER-Golgi traffic and Rab1 rescues neuron loss in Parkinson's models. *Science*, 313, pp. 324–328.

Espejo, A., Cote, J., Bednarek, A., Richard, S. and Bedford, M. T. (2002). A protein-domain microarray identifies novel protein-protein interactions. *Biochemical Journal*, 367, pp. 697–702.

Gavin, A., Bosche, M., Krause, R., Grandi, P., Marzioch, M., Bauer, A., Schultz, J., Rick, J., Michon, A., Cruciat, C., Remor, M., Höfert, C., Schelder, M., Brajenovic, M., Ruffner, H., Merino, A., Klein, K., Hudak, M., Dickson, D., Rudi, T., Gnau, V., Bauch, A., Bastuck, S., Huhse, B., Leutwein, C., Heurtier M. A., Copley, R. R., Edelmann, A., Querfurth, E., Rybin, V., Drewes, G., Raida, M., Bouwmeester, T., Bork, P. Seraphin, B., Kuster, B., Neubauer, G. and Superti-Furga1, G. (2002). Functional organization of the yeast proteome by systematic analysis of protein complexes. *Nature*, 415, pp. 141–147.

Giacomelli, L. and Nicolini, C. (2006). Gene expression of human T lymphocytes cell cycle: Experimental and bioinformatic analysis. *Journal of Cellular Biochemistry*, 99, pp. 1326–1333.

Haab, B., Dunham, M. and Brown, P. (2001). Protein microarrays for highly parallel detection and quantitation of specific proteins and antibodies in complex solutions. *Genome Biology*, 2, pp. 4.1–4.13.

Ho, Y., Gruhler, A., Heilbut, A., Bader, G., Moore, L., Adams, S., Millar, A., Taylor, P., Bennett, K., Boutilier, K., Yang, L., Wolting, C., Donaldson, I., Schandorff, S., Shewnarane, J., Vo, M., Taggart, J., Goudreault, M., Muskat, B., Alfarano, C., Dewar, D., Lin, Z., Michalickova, K., Willems, A.R., Sassi, H., Nielsen, P.A., Rasmussen, K.J., Andersen, J.R., Johansen, L.E., Hansen, L.H., Jespersen, H., Podtelejnikov, A., Nielsen, E., Crawford, J., Poulsen, V., Sørensen, B.D., Matthiesen, J., Hendrickson, R.C., Gleeson, F., Pawson, T., Moran, M.F., Durocher, D., Mann, M., Hogue, C.W.V., Figeys, D. and Tyers, M. (2002). Systematic identification of protein complexes in Saccharomyces cerevisiae by mass spectrometry. *Nature*, 415, pp. 180–183.

Ito, T., Chiba, T., Ozawa, R., Yoshida, M., Hattori, M., and Sakaki, Y. (2001). A comprehensive two-hybrid analysis to explore the yeast protein interactome. *Proceedings of the National Academy of Sciences USA*, 98, pp. 4569–4574.

Jona, G. and Snyder, M. (2003). Recent developments in analytical and functional protein microarrays. *Current Opinion in Molecular Therapeutics*, 5, pp. 271–277.

LaBaer, J. and Ramachandran, N. (2005). Protein microarrays as tools for functional proteomics. *Current Opinion in Chemical Biology*, 9, pp. 14–19.

MacBeath, G. and Schreiber, S. (2000). Printing proteins as microarrays for high-throughput function determination. *Science*, 289, pp. 1760–1763.

Mitchell, P. (2002). A perspective on protein microarrays. *Nature Biotechnology*, 20, pp. 225–229.

Murphy, R. F. and LaBaer, J. (2009). Locations Everyone: Lights, Camera, Action!. *Journal of Proteome Research*, 8, pp. 1.

Newman, J. R. and Keating, A. E. (2003). Comprehensive identification of human bZIP interactions with coiled-coil arrays. *Science*, 300, pp. 2097–2101.

Nicolini, C. (2006). Nanogenomics for medicine. *Nanomedicine*, 1, pp. 147–151.

Nicolini, C. (2009). *Nanobiotechnology and Nanobiosciences*, Pan Stanford Series on Nanobiotechnology, Volume 1, Ed. C. Nicolini, London, New York and Singapore, pp. 1–376.

Nicolini, C., Bruzzese, D., Sivozhelezov, V. and Pechkova, E. (2008). Langmuir-Blodgett based lipase nanofilms of unique structure-function relationship. *Biosystems*, 94, pp. 228–232.

Nicolini, C., Malvezzi, A. M., Tomaselli, A., Sposito, D., Tropiano, G. and Borgogno, E. (2002). DNASER I: Layout and Data Analysis, *IEEE Transactions on Nanobioscience*, 1, pp. 67–72.

Paweletz, C. P., Charboneau, L., Bichsel, V. E., Simone, N. L., Chen, T., Gillespie, J. W., Emmert-Buck, M. R., Roth, M. J., Petricoin, I. E. and Liotta, L. A. (2001). Reverse phase protein microarrays which capture disease progression show activation of pro-survival pathways at the cancer invasion front. *Oncogene* 20, pp. 1981–1989.

Pechkova, E. and Nicolini, C. (2006). Structure and growth of ultrasmall protein microcrystals by synchrotron radiation: II. µGISAXS and microscopy of lysozyme. *Journal of Cellular Biochemistry*, 97, pp. 553–560.

Pechkova, E., Chong, S., Tripathi, S. and Nicolini, C. (2009a). Cell free expression and APA for NAPPA and protein nanocrystallography. In *Functional Proteomics and Nanotechnology-based Microarrays*, Nicolini C. and LaBaer J. Eds, Pan Stanford Series on Nanobiotechnology Volume 2, London, New York and Singapore, pp. 121–147.

Pechkova, E., Sartore, M., Giacomelli, L. and Nicolini, C. (2007). Atomic force microscopy of protein films and crystals. *Review of Scientific Instruments*, 78, pp. 093704-1–093704-7.

Pechkova, E., Tripathi, S. and Nicolini, C. (2009b). µGISAXS of LB protein films: effect of temperature on long range order. *Journal of Synchrotron Radiation*, 16, pp. 330–335.

Pechkova, E., Tripathi, S., Spera, R. and Nicolini, C. (2008b). Groel crystal growth and characterization. *Biosystems*, 94, pp. 223–227.

Pechkova, E., Vasile, F., Spera, R. and Nicolini, C. (2008a). Crystallization of alpha and beta subunits of IF2 translation initiation factor from Archabacteria Sulfolobus Solfataricus. *Journal of Crystal Growth*, 310, pp. 3767–3770.

Ramachandran, N., Anderson, K. S., Raphael, J. V., Hainsworth, E., Sibani, S., Montor, W. R., Pacek, M., Wong, J., Elianne, M., Sanda, M. G., Hu, Y., Lovinenko, T. and LaBaer, J. (2008c). Tracking humoral responses using self assembling protein microarrays. *Proteomics Clinical Application*, 2, pp. 1518–1527.

Ramachandran, N., Hainsworth, E., Bhullar, B., Eisenstein, S., Rosen, B., Lau, A. Y., Walter, J. C. and LaBaer, J. (2004). Self-assembling protein microarrays. *Science*, 305, pp. 86–90.

Ramachandran, N., Raphael, J.V., Hainsworth, E., Demirkan, G., Fuentes, M.G., Rolfs, A., Hu, Y.H., LaBaer, J. (2008a). Next-generation high-density self-assembling functional protein arrays. *Nature Methods*, 5, pp. 535–538.

Ramachandran, N., Srivastava, S. and LaBaer, J. (2008b). Applications of protein microarrays for biomarker discovery. *Proteomics Clinical Application*, 2, 1444–1459.

Rolfs, A., Montor, W. R., Yoon, S. S., Hu, Y., Bhullar, B., Kelley, F., McCarron, S., Jepson, D. A., Shen, B., Taycher, E., Mohr, S. E., Zuo, D. M., Williamson, J., Mekalanos, J. and LaBaer, J. (2008). Production and sequence validation of a complete full length ORF collection for the pathogenic bacterium Vibrio cholerae. *Proceedings of the National Academy of Sciences USA*, 105, pp. 4364–4369.

Sartore, M., Bezerra Correia Terencio, T., Eggenhöffner, R., Stura, E., Hainsworth, G., LaBaer, J. and Nicolini, C. (2009). Label free detection of NAPPA: atomic force microscopy. In *Functional Proteomics and Nanotechnology-based Microarrays*, Nicolini C. and LaBaer J. Eds, Pan Stanford Series on Nanobiotechnology Volume 2, London, New York and Singapore, pp. 109–120.

Sartore, M., Pace, R., Faraci, P., Nardelli, D., Adami, M., Ram, M. K. and Nicolini, C. (2000). Controlled-atmosphere chamber for atomic force microscopy investigations. *Review of Scientific Instruments*, 71, pp. 1–5.

Sivozhelezov, V., Braud, C., Giacomelli, L., Pechkova, E., Giral, M., Soulillou, J. P., Brouard, S. and Nicolini C. (2008). Immunosuppressive drug-free operational immune tolerance in human kidney transplants recipients: II. Nonstatistical gene microarray analysis. *Journal of Cellular Biochemistry*, 103, pp. 1693–1706.

Sivozhelezov, V., Giacomelli, L., Tripathi, S. and Nicolini, C. (2006). Gene expression in the cell cycle of human T lymphocytes: I. Predicted gene and protein networks. *Journal of Cellular Biochemistry*, 97, pp. 1137–1150.

Sivozhelezov, V., Spera, R., Giacomelli, L., Hainsworth, G., LaBaer, J., Bragazzi N. and Nicolini, C. (2009). Bioinformatics and fluorescence DNASER for NAPPA studies on cell transformation and cell cycle. In *Functional Proteomics and Nanotechnology-based Microarrays*, Nicolini C. and LaBaer J. Eds, Pan Stanford Series on Nanobiotechnology Volume 2, London, New York and Singapore, pp. 31–59.

Spera, R. and Nicolini, C. (2007). cAMP induced alterations of chinese hamster ovary cells monitored by mass spectrometry. *Journal of Cellular Biochemistry*, 102, pp. 473–482.

Spera, R. and Nicolini, C. (2008). Nappa microarrays and mass spectrometry: new trends and challenges, *Essential in Nanoscience Booklet Series* (Taylor & Francis Group/CRC Press).

Spera, R., Badino, F., Hainsworth, G., Fuentes, M., Srivastava, S., LaBaer, J. and Nicolini, C. (2009). Label free detection of NAPPA via mass spectrometry. In *Functional Proteomics and Nanotechnology-based Microarrays*, Nicolini C. and LaBaer J. Eds, Pan Stanford Series on Nanobiotechnology Volume 2, London, New York and Singapore, pp.61–78.

Starr, C. and Hanover, J. (1990). Glycosylation of nuclear pore protein p62. Reticulocyte lysate catalyzes O-linked N-acetylglucosamine addition in vitro. *Journal of Biological Chemistry*, 265, pp. 6868–6873.

Stura, E., Bruzzese, D., Grasso, V., Valerio, V., Perlo, P. and Nicolini, C. (2007), Anodic porous alumina as mechanical stability enhancer for Ldl–cholesterol sensitive electrodes. *Biosensors & Bioelectronics*, 23, pp. 655–660.

Stura, E., Larosa, C., Bezerra Correia Terencio, T., Hainsworth, E., Ramachandran, N., LaBaer, J. and Nicolini, C. (2009). Label-free NAPPA: anodic porous alumina. In *Functional Proteomics and Nanotechnology-based Microarrays*, Nicolini C. and LaBaer J. Eds, Pan Stanford Series on Nanobiotechnology Volume 2, London, New York and Singapore, pp. 95–108.

Uetz, P., Giot, L., Cagney, G., Mansfield, T., Judson, R., Knight, J., Lockshon, D., Narayan, V., Srinivasan, M., Pochart, P., Qureshi-Emili, A., Li, Y., Godwin, B., Conover, D., Kalbfleisch, T., Vijayadamodar, G., Yang, M. J., Johnston, M., Fields, S. and Rothberg, J. M. (2000). A comprehensive analysis of protein-protein interactions in Saccharomyces cerevisiae. *Nature*, 403, pp. 623–627.

Walhout, A., Sordella, R., Lu, X., Hartley, J., Temple, G., Brasch, M., Thierry-Mieg, N. and Vidal, M. (2000). Protein interaction mapping in C. elegans using proteins involved in vulval development. *Science*, 287, pp. 116–122.

Walter, P. and Blobel, G. (1983). Preparation of microsomal membranes for cotranslational protein translocation. *Methods Enzymology*, 96, pp. 84–93.

Witt, A. E., Hines, L. M., Collins, N. L., Hu, Y., Gunawardane, R. N., Moreira, D., Raphael, J., Jepson, D., Koundinya, M., Rolfs, A., Taron, B., Isakoff, S. J., Brugge, J. S. and LaBaer J. (2006). Functional proteomics approach to investigate the biological activities of cDNAs implicated in breast cancer. *Journal of Proteome Research*, 5, pp. 599–610.

Zhu, H., Bilgin, M., Bangham, R., Hall, D., Casamayor, A., Bertone, P., Lan, N., Jansen, R., Bidlingmaier, S., Houfek, T., Mitchell, T., Miller, P., Dean, R. A., Gerstein, M. and Snyder, M. (2001). Global analysis of protein activities using proteome chips. *Science*, 293, pp. 2101–2105.

Zuo, D. M., Mohr, S. E., Hu, Y. H., Taycher, E., Rolfs, A., Kramer, J., Williamson, J. and LaBaer, J. (2007). PlasmID: a centralized repository for plasmid clone information and distribution. *Nucleic Acids Research*, 35, pp. D680–D684.

CHAPTER 2

BIOINFORMATICS AND FLUORESCENCE DNASER FOR NAPPA STUDIES ON CELL TRANSFORMATION AND CELL CYCLE

Victor Sivozhelezov[1], Rosanna Spera[1], Luca Giacomelli[1], Eugenie Hainsworth[2], Joshua LaBaer[2], Nicola L. Bragazzi[1] and Claudio Nicolini[1]

[1]*Nanoworld Institute, CIRSDNNOB-University of Genova and Fondazione EL.B.A., Italy*
[2]*Institute of Proteomics, Harvard University, Cambridge MA USA*

We present the results obtained analyzing for the first time a fluorescent-labeled self-assembling protein microarray of new conception, called nucleic acid programmable protein array (NAPPA), by DNASER and by bioinformatics. The DNASER is a novel bioinstrumentation for real-time acquisition and elaboration of images from fluorescent DNA and protein microarrays that permits to acquire images faster than the traditional systems. The fluorescent microarrays images are processed to recognize the DNA and/or protein spots, to analyze their superficial distribution on the glass slide and to evaluate their geometric and intensity properties. NAPPA is finally here utilized in conjunction with a bioinformatic analysis in order to predict protein interactions relevant to progressing between phases of cell cycle of human T lymphocytes and to compare between lymphoma versus normal T cells. Work is in progress about synchronized mitotic HeLa cells compared to logarithmically growing HeLa cells, and about cAMP reverse transformed versus transformed CHO-K1 cells.

2.1 Introduction

We employed a novel bioinstrumentation for real-time acquisition and elaboration of images from fluorescent DNA and protein microarrays realized in our laboratories, the DNASER, acronym of DNA analySER, (equipped with the ORCA II Hamamatsu CCD camera) (Nicolini et al., 2002) to analyze Nucleic Acid Programmable Protein Array (NAPPA; U.S. Patent No. 6,800,453) (Ramachandran et al., 2004, 2008 Ramachandran et al., 2005).

A white light beam illuminates the target sample allowing the images grabbing on a high sensibility and wide–band charge-coupled device camera (ORCA II, Hamamatsu). This high–performance device permits to acquire images faster and of higher quality than the traditional systems. The DNA microarrays images are processed to recognize the DNA chip spots, to analyze their superficial distribution on the glass slide and to evaluate their geometric and intensity properties. Differently form conventional techniques, the spots analysis is fully automated and the DNASER does not require any additional information about the DNA microarray geometry (Nicolini et al., 2002).

The nucleic acid programmable protein array (NAPPA) is a self-assembling protein microarray. Building upon the successful use of in vitro translated protein in standard scale applications, Ramachandran et al. substituted the use of purified proteins with the use of cDNAs encoding the target proteins at each feature of the microarray. The proteins are transcribed and translated by a cell-free system and immobilized *in situ* by means of epitope tags fused to the proteins (Ramachandran et al., 2004, 2005; Spera and Nicolini, 2008).

Through testing a variety of cDNA printing schemes, Ramachandran et al. found that an optimum balance was required between binding DNA efficiently and maintaining a DNA conformation that supported efficient transcription and translation. The addition of a C-terminal glutathione S-transferase (GST) tag to each protein enabled its capture to the array through an antibody to GST printed simultaneously with the expression plasmid (Ramachandran et al., 2004).

To activate and use the array, a cell-free coupled transcription and translation system was added as a single continuous layer covering the arrayed cDNAs on the microscope slide. Expression of target protein was confirmed with an antibody to GST (different from the capture GST antibody), and the signals were measured with a standard glass slide DNA-microarray scanner (Ramachandran et al., 2004). It is the first time that the DNASER technology is employed on NAPPA array, previously we tested its performance on commercial DNA microarray (Nicolini *et al.*, 2006). Spots intensities (14 × 14 spots) were collected, in order to give a preliminary idea of intensity distribution among different spots.

For the first time the NAPPA approach is used herein for particular cellular systems, where tasks for NAPPA analysis are formulated starting from genomics. Namely we first study gene expression using DNA microarray techniques and/or gene interactions, and then pick proteins coded by the emerging genes either as query proteins for the existing NAPPA microarrays, or to design novel NAPPA microarrays.

To analyze gene interactions, we introduce a novel bioinformatic approach (Sivozhelezov *et al.*, 2006, 2008; Nicolini *et al.*, 2006, Nicolini, 2006) which as we will further see will provide a link from genomics to NAPPA technology, for the cellular systems of particular interest. Earlier we performed such an analysis for cell cycle studies of T lymphocytes (Sivozhelezov *et al.*, 2006) and validated it via DNA microarray (Nicolini *et al.*, 2006), and later for kidney graft tolerance versus chronic rejection (Sivozhelezov *et al.*, 2008). Similar analysis is herein performed for T cell lymphoma in comparison with normal T cells. Other cellular systems under study include the cAMP-induced reverse transformation of CHO K1, for which cell geometry was earlier related to changes in chromatin structure pertinent to change from transformed to reverse transformed cells (Nicolini and Beltrame, 1982), and HeLa cells with respect to their cell cycle, following earlier studies of Nicolini *et al.* (1975), where HeLa cells were subjected to synchronization, and chromatin was obtained at several time points from synchronized cell populations at various times during the cell cycle, showing marked differences in the shape of chromatin DNA. Those differences were detected from DNA's CD spectra, used as indicators of DNA molecule helicity, and also the ability of DNA to bind an intercalator molecule, which contacts DNA in its grooves so its binding is strongly dependent on the DNA shape.

2.2 Materials and Methods

To analyse NAPPA array we utilized DNASER in two different settings. First, we set DNASER to acquire a maximum area, with good resolution, of 1 × 1 cm. In such setting it is not possible to acquire a whole image of the NAPPA slide in a single shot. We, then, acquired an image of the center of the slide. Maximum area which can be imaged in a single shot with good resolution (1 × 1 cm) includes 14 × 14 spots, for a total of 196 spots. It should be noticed that resolution is usually better in central columns and rows and in central area of the slide than in peripheral area.

The NAPPA we analyzed was a Human Kinase NAPPA spotted with 24 × 72 200 microns spots of kinase genes from HIP FLEXGene human kinase collection (pDNR-Dual, complete set) — total number of clones in the collection: 645 (http://plasmid.med.harvard.edu/PLASMID/GetCollection.do?collectionName=HIP%20FLEXGene%20human%20kinase%20collection%20(pDNR-Dual,%20complete%20set)).

Analysis of the DNA microarrays were performed as described in (Sivozhelezov *et al.*, 2008), and bioinformatic approach for identification of interaction-based leader genes is described in Sivozhelezov *et al.* (2006).

2.2.1 *The experimental layout*

The DNASER generates wide area images in a single shot using samples with very low fluorescence intensity. The sample is illuminated with a beam produced by a white light source (a 150-W Xenon lamp). The beam is accurately filtered and focused on the sample and then on the CCD camera sensitive area. Either of the two interchangeable interference filters are placed between two achromatic doublets so as to work in parallel illumination conditions. These two filters are different because are tuned respectively on the excitation and emission frequency of the DNA microarray fluorescent spots. This geometry is used both for illumination of the sample and collection of the fluorescence signal. The spectrally selected light is then focused onto the sample at 45 degrees of incidence. A supervisor PC controls DNA microarray images acquisition and elaboration. Figure 2.1 shows the main hardware blocks of the bioinstrumentation. The illumination system, the optical filter system and the motorized holder blocks constitute the opto-mechanical components.

Three motorized linear translation stages, which are controlled by the computer, are used to fine position the sample on the focal plane (z axis) of the CCD camera and to select the appropriate portion of area to be imaged (x, y axis). An ad hoc software has been developed in order to position the sample with micrometer precision.

Figure 2.1. The DNASER (Nicolini *et al.* 2002).

The ORCA II Dual Scan Cooled 12–14 bit B/W CCD Camera combines high-resolution, low noise and high sensitivity, high quantum efficiency, a wide spectral response (300–700 nm) and a high dynamic range (full well capacity of 16,000 electrons and of 40,000 with binning). The ORCA II incorporates a CCD with an effective 1280 × 1024 (1.3 M pixel) pixel array and progressive scan interline design to eliminate the need for a mechanical shutter. On the supervisor PC runs the software routines (C language) that implement the image processing algorithms for the spot analysis.

Microarrays are built laying spots equidistantly in a grid. While the orientation error of the grid is negligible, its spatial position can significantly vary.

Furthermore, depending on the arrayer used, the distance between spots is different.

In order to develop a robust system, the image processing algorithm developed to perform the spot analysis estimates these data instead of supposing them given.

The input data to the spot analysis are the raw DNA microarray images. The output data of this processing are all the spot features:
- Brightness features (spot foreground and background intensity).
- Geometric features (spot foreground and background area, circularity) (Nicolini *et al.*, 2002).

The obtained images were analysed by GenePix Software. The image was acquired above an intensity threshold, considered significant. The identification of the spots using GenePix is usually realized by comparing the image with a grid, which contains every information needed (*e.g.*, coordinates and gene names) to recognize the spots.

2.2.2 *Procedure of NAPPA expression and DNASER analysis*

To express and detect proteins we follow the protocol (Ramachandran *et al.*, 2008):
- Blocking: 1 hr on rocking shaker at room temperature with SuperBlock. We used 30 mL in a pipette box for 3 slides.
- Rinse with milli-Q water. Dry with nitrogen.
- Apply HybriWell (Grace Biolabs) gasket to the slide.
- Pre-heat the incubator to be used for IVT at 30 °C.
- Prepare IVT: 130 µL of IVT lysate mix for slide. Each tube after component addition will contain 400 µL of lysate mix and the lysate tubes cannot be re-frozen, *e.g.*, 1 tube = 3 slides = 400 µl → 16 µl TNT buffer, 8 µl T7 polymerase, 4 µl of –Met, 4 µl of –Leu or –Cys, 8 µl of RNaseOUT, 160 µl of DEPC water, 200 µl of reticulocyte lysate.
- Add IVT mix. Pipette the mix in slowly. Gently massage the HybriWell to get the IVT mix to spread out and cover all of the area of the array. Apply the small round port seals to both ports.
- Place the slides on a bioassay dish with divider on top of the leveling shelf inside the incubator. Incubate for 1.5 hr at 30 °C for protein expression, followed by 30 min at 15 °C for the query protein to bind to the immobilized protein.

- Remove the HybriWell and immerse each slide in 5% (wt/vol) defatted dry milk solution in phosphate buffer saline (PBS) immediately; wash 3 times, 3 minutes each, in a pipette box. Dry with nitrogen.
- Block with milk on rocking shaker at room temperature for 1 hr.
- Prepare antibody solutions in SuperBlock;
- Prepare the TSA stock solution;
- Apply primary AB (mouse anti-GST) by adding 150 μl to the slide, then apply a coverslip. Incubate for 1 hr at RT; wash with milk on a rocking shaker (3 times, 5 min each). Drain.
- Apply secondary AB (anti-mouse HRP) by adding 150 μl to the slide, place a coverslip. Incubate for 1 hr at RT; wash with PBS (pH 7.4) 3 times, 5 min each. Quickly rinse with Milli-Q water. Drain.
- Apply 150 μl TSA mix and place coverslip. Incubate for 10 minutes. Rinse in Milli-Q water; dry with filtered compressed air.
- Scan with DNASER.

2.3 Results

2.3.1 *NAPPA DNASER imaging*

We first acquired an image of NAPPA array of human kinases genes, after their expression with subsequent kinases protein synthesis and their detection with the anti-GST antibody, of an area located in the center of the slide (Figure 2.2). In Figure 2.3 is reported the scheme of the genes spotted in the area acquired in Figure 2.2.

By MS we analyzed the genes NA7-A12, NA7-E12 and NA8-D12 (that are spotted in the last line of the array, where is not possible to perform image acquisition with DNAser) in this area are present some homologues of these genes, namely NA3 (spots 11–31, 12–31, 7–32, 9–32, 8–34, 9–35).

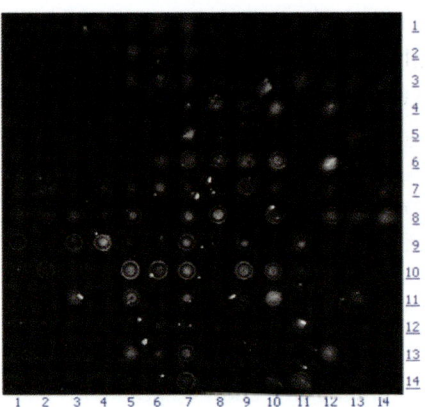

Figure 2.2. DNASER image of central area (reported are the Kinase NAPPA coordinates).

rows

columns	3	4	5	6	7	8	9	10	11	12	13	14	15	16
26	KIAA20 02 2-H9	PRKAC A 6-B3	No DNA	MLKL 6-D3	PTK9 1-F9	FLAG1	CDK5 1-H9	MASK 5-B3	No DNA	STK38 5-D3	IPMK 2-E9	Mouse IgG1	PFKFB 1 2-G9	LOC91 461 6-A3
27	PDIK1L 6-B9	AK7 6-D9	STK25 6-F9	ULK3 6-H9	PIPSK2 C 5-F3	VRK2 5-H3	HIPK4 5-B9	PAK6 5-D9	BUB1B 5-F9	SRPK2 5-H9	RAGE 6-E3	STYK1 6-G3	GRK6 6-A9	MELK 6-C9
28	PTK6 2-F9	TEX14 2-B9	AK2 2-F3	SGK 2-B3	No DNA	CDK5 1-H9	PTK9 1-F9	NME4 1-B9	GNE 1-F3	PRKCA 1-B3	No DNA	PFKFB 1 2-G9	IPMK 2-E9	SYK 2-A9
29	ACVR2 A 2-D9	Mouse IgG1	PRKAC A 6-B3	MLKL 6-D3	CDC2L 1 1-D3	STK40 1-H3	GUK1 1-D9	FLAG1	MASK 5-B3	STK38 5-D3	MAPK1 0 2-C3	MAPK1 2-G3	EPHA1 0 2-C9	Mouse IgG1
30	PFKFB 3 6-H3	STK25 6-F9	PDIK1L 6-B9	PRKAA 2 6-F3	SRPK2 5-H9	PAK6 5-D9	VRK2 5-H3	BUB1B 5-F9	HIPK4 5-B9	PIPSK2 C 5-F3	APEG1 6-G9	MELK 6-C9	STYK1 6-G3	TNK1 6-E9
31	FGR 4-F3	PDXK 4-H3	RPS6K B1 4-B9	CLK4 4-D9	KIT 3-B3	BRD2 3-D3	RIPK5 3-F3	AURKC 3-H3	NA 3-B9	NA 3-D9	GALK2 4-A3	NME7 4-C3	PFKL 4-E3	ADK 4-G3
32	MKNK2 4-H9	TSSK6 8-B3	Non Spots	PBK 8-D3	NA 3-F9	FLAG1	NA 3-H9	PMVK 7-B3	Non Spots	MAP3K 7 7-D3	MAPKA PK2 4-E9	Mouse IgG1	PRKAA 1 4-G9	STK3 8-A3
33	HSPB8 8-B9	FLJ109 86 8-D9	RPS6K B2 8-F9	DTYMK 8-H9	CSNK1 A1 7-F3	CDK9 7-H3	CDK6 7-B9	MAPK1 0 7-D9	CDK5 7-F9	CSNK1 D 7-H9	CSNK1 A1 8-E3	UCK1 8-G3	EIF2AK 1 8-A9	COASY 8-C9
34	MAP3K 8 4-F9	RPS6K B1 4-B9	FGR 4-F3	PCTK1 4-B3	Non Spots	NA 3-H9	NA 3-F9	NA 3-B9	RIPK5 3-F3	KIT 3-B3	Non Spots	PRKAA 1 4-G9	MAPKA PK2 4-E9	STK24 4-A9
35	CLK4 4-D9	Mouse IgG1	TSSK6 8-B3	PBK 8-D3	BRD2 3-D3	AURKC 3-H3	NA 3-D9	FLAG1	PMVK 7-B3	MAP3K 7 7-D3	NME7 4-C3	ADK 4-G3	YES1 4-C9	Mouse IgG1
36	AK5 8-H3	RPS6K B2 8-F9	HSPB8 8-B9	PIP5K1 C 8-F3	CSNK1 D 7-H9	MAPK1 0 7-D9	CDK9 7-H3	CDK5 7-F9	CDK6 7-B9	CSNK1 A1 7-F3	AKT1 8-G9	COASY 8-C9	UCK1 8-G3	PI4KII 8-E9
37	ALS2C R7 2-F4	BRAF 2-H4	AK3L2 2-B10	MAPK1 0 2-D10	ABL1 1-B4	TBK1 1-D4	PHKA2 1-F4	CAMK2 D 1-H4	CSNK2 A2 1-B10	CDK9 1-D10	ERBB2 2-A4	UHMK1 2-C4	DAPK2 2-E4	CHKA 2-G4
38	CALM3 2-H10	MAP2K 1 6-B4	No DNA	SRC 6-D4	LCK 1-F10	FLAG1	CSNK1 G2 1-H10	ACVR1 5-B4	No DNA	PIM1 5-D4	BUB1B 2-E10	Mouse IgG1	CALM2 2-G10	PRKAC G 6-A4
39	ACVR2 B 6-B10	GRK1 6-D10	IRAK2 6-F10	NADK 6-H10	GK 5-F4	AMHR2 5-H4	CMPK 5-B10	STK32 A 5-D10	CLK1 5-F10	MARK3 5-H10	TIE1 6-E4	FGFR4 6-G4	FLJ250 06 6-A10	CDK3 6-C10

Figure 2.3. Scheme of the spots of the Figure 2.2 (there are reported the NAPPA coordinates).

Spots intensities were analyzed using GenePix software. Background noise was calculated and subtracted from each spot intensity (Table 2.1).

Among measured values, VC is maybe the most significant, since it can be an index of capability to identify different expression.

Table 2.1. Intensities of identified spots.

Column	Mean intensity (AU)	SD	VC
1	4710,57	2165,67	45,97
2	5049,67	2251,92	44,60
3	8049,25	3995,24	49,63
4	7052,54	4549,69	64,51
5	8692,42	6144,82	70,69
6	6590,54	4356,29	66,10
7	7317,69	3673,27	50,20
8	9815,00	8399,18	85,57
9	11098,92	8433,78	75,99
10	10177,31	6767,52	66,50
11	10482,30	6550,38	62,49
12	10243,56	5122,66	50,01
13	8730,89	3509,14	40,19
14	6346,29	2982,59	47,00

Spots were then divided into groups of similar intensities. This process was performed by grouping spots according to intensity range (2000–2999; 3000–3999 and so on) and then by means of *K-means* clustering. Results are shown in Figure 2.4.

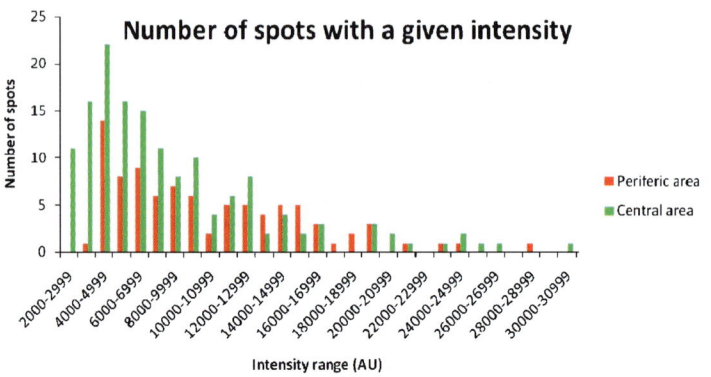

Figure 2.4. Number of spots with a given intensity (see only central area ones).

2.3.2 *NAPPA-targeted prediction of gene interactions relevant to progressing between phases of cell cycle of human T lymphocytes*

To properly construct NAPPA or determine query proteins for existing NAPPA, we determined list of genes involved in G2 and M phases of T lymphocytes (Table 2.2), and then found which genes were *common* for the two lists. Both sets of Table 2.2 were verified by the BioGene database. Out of 35 and 38 genes we found relevant for G2 and M phases, respectively, 27 and 34 were assigned by BioGene, "First degree" associations with the corresponding MeSH terms, namely "G2 phase" and "mitosis" (no general MeSH term is available for "M phase". Only 2 and 3 genes of our set, respectively, were termed irrelevant "New to Biological term"). We use NCBI Gene database in combination with the Pubmed database instead of Protein database since the latter contains too many duplicate records preventing the use of links between NCBI databases.

Between 35 G2 phase and 38 M phase genes, 11 were in common, out of which 4, namely CDC2, NFKB1, CCNB1, and PPP2CA were experimentally confirmed to participate in cell cycle progression from G2 to M phase, mainly via the regulatory roles of the proteins encoded by those genes.

Table 2.2. Sets of proteins involved in G2 and M phase and their interaction scores within each set. In bold are proteins common in the two sets. Among them, underlined are the seven proteins experimentally proven to participate in progression of cell cycle from G2 to M phases.

Protein involved in G2 phase	Interaction score among "G2 phase" proteins	Class among G2 phase proteins	Protein involved in M phase	Interaction score among "M phase" proteins	Class among M phase proteins
LCK	8.117	A	TP53	20.67	A
KHDRBS1	7.78	A	CDC2	17.22	A
FYN	7.431	A	BRCA1	12.30	A
PLCG1	6.899	A	CDKN1A	11.20	A
CDC2	6.694	A	CASP3	10.65	A
ABL1	6.605	A	ATM	9.47	B
TP53	6.585	A	CHEK2	8.71	B
PTPN6	5.593	B	PARP1	8.45	B
GRB2	5.562	B	*CCNB1*	7.95	B
LYN	5.455	B	NFKBIA	7.79	B
ZAP70	4.953	B	*NFKB1*	7.47	B
ITK	4.498	B	KPNA2	7.13	B
CDKN1A	3.144		GADD45A	6.97	B
ATM	2.582		CHEK1	6.90	B
RAF1	2.346		LMNB1	6.54	B
JAK3	2.227		LMNA	6.36	B
KPNA2	2.025		CCNB2	6.20	B
CCNB1	1.963		WEE1	6.07	B
CCNB2	1.778		SLC25A4	6.06	B
THY1	1.674		RPA2	5.98	B
XPO1	1.344		HSPB1	5.76	B
TERT	1.344		CASP9	5.59	B
TERF1	1.344		BAX	5.42	B
PPP2CA	0.845		*PPP2CA*	5.32	B
PRF1	0.672		COPS6	5.22	B
COPS6	0.672		ATR	4.83	B
GZMB	0.672		TNF	4.82	B
NFKB1	0.516		TNFSF10	4.60	B

Interaction scores (connectivities) for the 11 proteins participating in both G2 and M phase are shown in the Figure 2.5, where scores were calculated within each of the G2 and M phase protein sets.

To evaluate interactions between the proteins comprising the "G2 phase" set (35 proteins) and "M phase" set (38 proteins), we calculated interaction scores according to our LeaderGene technique (Sivozhelezov et al., 2006, 2008; Nicolini et al., 2006), using only the protein-protein interaction subset of STRING database (Tables 2.2).

Table 2.2. Continuous from previous page.

Protein involved in G2 phase	Interaction score among "G2 phase" proteins	Class among G2 phase proteins	Protein involved in M phase	Interaction score among "M phase" proteins	Class among M phase proteins
IL2	0		CCNB3	4.50	B
CCNB3	0		CCL4	3.30	B
PPM1D	0		PPP2R5A	3.15	B
HGF	0		CCL5	2.91	B
FADD	0		HSPA1A	2.70	B
RHOH	0		LMNB2	2.41	B
IL18	0		PPIA	2.37	B
CDC20	0			2.26	
HIST1H1B	0			2.23	
RAB38	0			2.05	

Expansion of this list via interactions predicted by STRING was not performed to avoid using the same database for forming the protein list and scoring (source of circular reasoning).

For the 16 proteins in common for the two sets, we plotted their interaction scores between the two sets and found that those scores are almost proportionally related to one another.

That is, for a protein involved in both G2 phase and M phase, the weighted number of its partner proteins in M phase is proportional to that in the G2 phase.

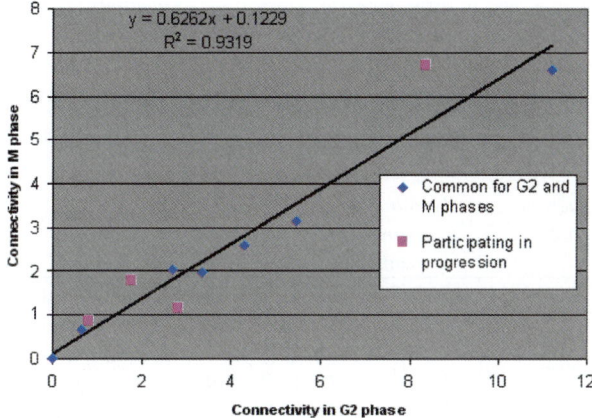

Figure 2.5. Interaction scores (connectivities) for the 11 proteins participating in both G2 and M phase, with 4 being correlated to G2/M transition. Scores were calculated within each of the G2 and M phase protein sets.

2.3.3 *NAPPA-targeted prediction of protein interactions relevant to differences between lymphoma and normal T cells*

Genes involved in interactions differing between lymphoma and normal T cells are shown in Table 2.3. From interaction-based bioinformatics alone (Sivozhelezov *et al.*, 2006), it is not possible to see which of those genes are pro-lymphoma or pro-normal, because only the fact of interaction is recorded between two genes/proteins, without specifying the direction of action of the given interaction.

Note that, even though manually curated interaction maps (Parodi *et al.*, 2004) in principle allow to define such direction, it takes too long to complete them.

Patterns of gene expression (Franzke *et al.*, 2006) allow to specify the pro-lymphoma or pro-normal nature of the gene (Table 2.3), unless or the data are inconclusive or *neutral*, *i.e.*, insignificant difference in expression levels between lymphoma or normal datasets, or opposite results for copies of the same gene (Table 2.4). Furthermore, frequently (as in this case) the gene is not specified in the microarray because of nomenclature problem or incompleteness of microarray (Table 2.5).

Table 2.3. Genes involved in differences in interactions (Sivozhelezov et al., 2006, 2008) between lymphoma and normal T cells. Pro-lymphoma or pro-normal nature of the gene, inferred from microarray data (Franzke et al., 2006) is marked by or L (lymphoma) or N (normal). Classification of interaction leader gene is described in Sivozhelezov et al., (2006).

Gene symbol	Interaction score	Class	Pro-lymphoma or pro-normal	Gene description
CD4	4.57	B	L	CD4 - T-cell surface glycoprotein CD4 precursor (T-cell surface antigen T4/Leu-3)
SYK	3.699	B	N	SYK – Tyrosine-protein kinase SYK (EC 2.7.10.2) (Spleen tyrosine kinase)
SH2D2A	2.832	B	L	SH2D2A - SH2 domain-containing protein 2A (T cell-specific adapter protein) (TSAd) (VEGF receptor-associated protein) (SH2 domain-containing adapter protein)
CCL3	2.514	B	L	CCL3 - CCL3) (Macrophage inflammatory protein 1-alpha) (MIP-1-alpha) (Tonsillar lymphocyte LD78 alpha protein)
CCL11	1.872		N	CCL11 - Eotaxin precursor (Small inducible cytokine A11) (CCL11) (Eosinophil chemotactic protein)
P53	1.871		L	P53 - Cellular tumor antigen p53 (Tumor suppressor p53) (Phosphoprotein p53) (Antigen NY-CO-13)
REL	1.642		N	REL - C-Rel proto-oncogene protein (C-Rel protein)
E2F2	1.542		L	E2F2 - Transcription factor E2F2 (E2F-2)
FB1	0		N	FB1 - TCF3 fusion partner (Protein FB1)
FMOD	0		N	FMOD - Fibromodulin precursor (FM) (Collagen-binding 59 kDa protein) (Keratan sulfate proteoglycan fibromodulin) (KSPG fibromodulin)
NOS2A	0		N	NOS2A - Nitric oxide synthase, inducible (EC 1.14.13.39) (NOS type II) (Inducible NO synthase) (Inducible NOS) (iNOS) (Hepatocyte NOS) (HEP- NOS)
PTK6	0		N	PTK6 - Tyrosine-protein kinase 6 (EC 2.7.10.2) (Breast tumor kinase) (Tyrosine-protein kinase BRK)

Table 2.4. *Neutral* genes involved in differences in interactions (Sivozhelezov et al., 2006, 2008) between lymphoma and normal T cells for which available expression data suggest that the genes are neither pro-lymphoma nor pro-normal.

Gene symbol	Interaction score	Class	Gene description
LCK	8.567	A	LCK - LCK (EC 2.7.10.2) (p56-LCK) (Lymphocyte cell-specific protein-tyrosine kinase) (LSK) (T cell- specific)
PRKCZ	3.082	B	PRKCZ - Protein kinase C zeta type (EC 2.7.11.13) (nPKC-zeta)
PRKCQ	2.531	B	PRKCQ - Protein kinase C theta type (EC 2.7.11.13) (nPKC-theta)
TXK	1.932		TXK - Tyrosine-protein kinase TXK (EC 2.7.10.2)
ARID3A	1.514		ARID3A - AT-rich interactive domain-containing protein 3A (ARID domain- containing protein 3A) (B-cell regulator of IgH transcription) (Bright) (E2F-binding protein 1)
ATM	0.999		ATM - Serine-protein kinase ATM (EC 2.7.11.1) (Ataxia telangiectasia mutated) (A-T, mutated)
VEGFA	0.9		VEGFA - Vascular endothelial growth factor A precursor (VEGF-A) (Vascular permeability factor) (VPF)
PAX5	0.644		PAX5 - Paired box protein Pax-5 (B-cell-specific transcription factor) (BSAP)
IL4R	0.638		IL4R - Interleukin-4 receptor alpha chain precursor (IL-4R-alpha) (CD124 antigen)
ENSP00000355331	0		ENSP00000355331 - CDNA FLJ12684 fis, clone NT2RM4002460, weakly similar to ENV POLYPROTEIN (Hypothetical protein FLJ12684) (LP9056 protein)
ENSP00000363993	0		ENSP00000363993 - Proteasome subunit beta type 9 precursor (EC 3.4.25.1) (Proteasome chain 7) (Macropain chain 7) (Multicatalytic endopeptidase complex chain 7) (RING12 protein) (Low molecular mass protein 2)
ERVK6	0		ERVK6 - HERV-K_1q22 provirus ancestral Env polyprotein (Envelope polyprotein) (HERV-K102 envelope protein) (HERV-K(III) envelope protein)
IKZF1	0		IKZF1 - DNA-binding protein Ikaros (IKAROS family zinc finger protein 1) (Lymphoid transcription factor LyF-1)

Generally, microarray studies are often neglecting the nomenclature of genes, let alone the available information about gene interactions. This underestimation makes microarray data difficult to properly evaluate. In Tables 2.3–2.5, kinases are marked in bold. Intriguingly, kinases seem abundant (5) in Table 2.4 compared to Tables 2.3 and 2.5 (one in each).

As seen from Figure 2.6, the subset of the genes shown in Table 2.3 that is ones showing marked difference in expression in lymphoma versus normal T cells, only shows one connection, namely CD4-SYK, which, as closer examination shows, was inferred only for peripheral T cells, and even then the SYK expression was, under similar conditions, about 10 fold lower than in peripheral B cells (Chan *et al.*, 1994).

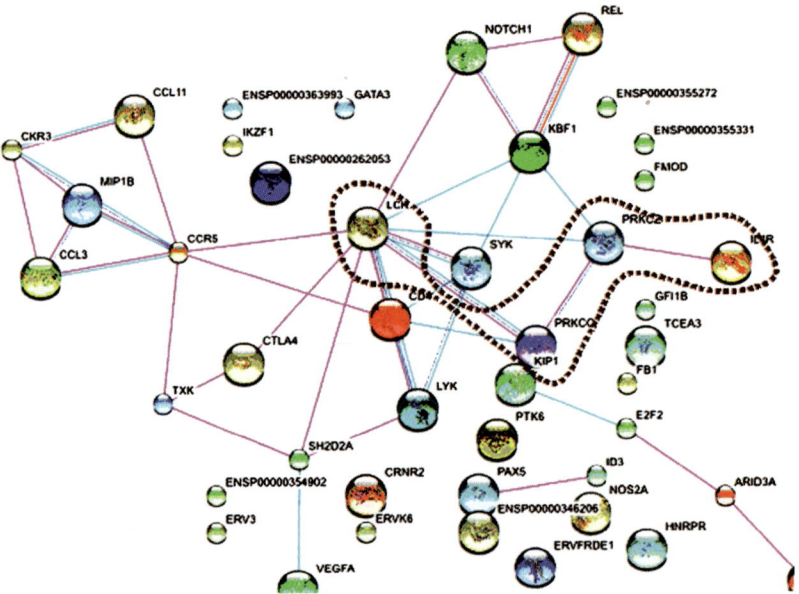

Figure 2.6. Interaction network for genes distinguishing lymphoma from normal T cells. Sub-network connecting the four leader genes which are "neutral" according to their expression pattern are shown with a dotted line.

Note also that SYK is the only kinase belonging to the genes from Table 2.3. In contrast, the genes shown in Table 2.4 ("*neutral*" with respect to lymphoma from expression data), show a distinct interaction sub-network (Figure 2.6, dotted line).

Table 2.5. *Absent* genes involved in differences in interactions (Sivozhelezov *et al.*, 2006, 2008) between lymphoma and normal T cells for which expression data are not available because descriptions of commercial microarray spots used in Franzke *et al.*, 2006) do *not* contain the HUGO-HGNC-specified gene symbols (continue on next page).

Gene symbol	Interaction score	Class	Gene description
CCR5	5.929	A	CCR5 - type 5 (C-C CKR-5) (CC-CKR-5) (CCR-5) (CCR5) (HIV-1 fusion coreceptor) (CHEMR13) (CD195 antigen)
KBF1	4.652	B	KBF1 - subunit (DNA-binding factor KBF1) (EBP- 1)
LYK	3.428	B	LYK - kinase ITK/TSK (EC 2.7.10.2) (T-cell-specific kinase) (Tyrosine-protein kinase Lyk) (Kinase EMT)
CKR3	3.412	B	CKR3 - type 3 (C-C CKR-3) (CC-CKR-3) (CCR-3) (CCR3) (CKR3) (Eosinophil eotaxin receptor) (CD193 antigen)
MIP1B	2.284		MIP1B - Small inducible cytokine A4 precursor (CCL4) (Macrophage inflammatory protein 1-beta) (MIP-1-beta) (MIP-1-beta(1-69)) (T-cell activation protein 2) (ACT-2) (PAT 744) (H400) (SIS-gamma) (Lymphocyte activation gene 1 protein) (LAG-1) (HC21)
NOTCH1	2.237		NOTCH1 - Neurogenic locus notch homolog protein 1 precursor (Notch 1) (hN1) (Translocation-associated notch protein TAN-1) [Contains
CTLA4	1.515		CTLA4 - Cytotoxic T-lymphocyte protein 4 precursor (Cytotoxic T-lymphocyte- associated antigen 4) (CTLA-4) (CD152 antigen)
KIP1	0.9		KIP1 - Cyclin-dependent kinase inhibitor 1B (Cyclin-dependent kinase inhibitor p27) (p27Kip1)
ID3	0.644		ID3 - DNA-binding protein inhibitor ID-3 (Inhibitor of DNA binding 3) (ID- like protein inhibitor HLH 1R21) (Helix-loop-helix protein HEIR-1)
CRNR2	0		CRNR2 - Protocadherin alpha 6 precursor (PCDH-alpha 6)
ENSP00000262053	0		ENSP00000262053 - Cyclic AMP-dependent transcription factor ATF-1 (Activating transcription factor 1) (TREB36 protein)

Among the four proteins belonging to the this "neutral" but strongly connected network, three are kinases.

Table 2.5. Continuous from previous page.

Gene symbol	Interaction score	Class	Gene description
ENSP00000346206	0		ENSP00000346206 - Antigen peptide transporter 1 (APT1) (Peptide transporter TAP1) (ATP- binding cassette sub-family B member 2) (Peptide transporter PSF1) (Peptide supply factor 1) (PSF-1) (Peptide transporter involved in antigen processing 1)
ENSP00000354902	0		ENSP00000354902 - HERV-K_1q23.3 provirus ancestral Env polyprotein (Envelope polyprotein) (HERV-K18 envelope protein) (HERV-K110 envelope protein) (IDDMK1,2 22 envelope protein) (HERV-K(C1a) envelope protein) (HERV- K18 superantigen) (IDDMK1,2 22 superantigen) [Contains
ENSP00000355272	0		ENSP00000355272 - HERV-K_1p13.3 provirus ancestral Env polyprotein (Envelope polyprotein) [Includes
ERV3	0		ERV3 - ERV3 envelope protein) (ERV-3 envelope protein) (HERV-R envelope protein) (ERV-R envelope protein
ERVFRDE1	0		ERVFRDE1 - HERV-FRD_6p24.1 provirus ancestral Env polyprotein precursor (Envelope polyprotein) (HERV-FRD) (Syncytin-2) [Contains
GATA3	0		GATA3 - Trans-acting T-cell-specific transcription factor GATA-3 (GATA-binding factor 3)
GFI1B	0		GFI1B - growth factor independent 1B (potential regulator of CDKN1A, translocated in CML)
HNRPR	0		HNRPR - Heterogeneous nuclear ribonucleoprotein R (hnRNP R)
TCEA3	0		TCEA3 - Transcription elongation factor A protein 3 (Transcription elongation factor S-II protein 3) (Transcription elongation factor TFIIS.h)

As for genes from Table 2.5, which are missing from the expression data, there are only two isolated connections, *i.e.*, between CRK3 and CCR5 (left part of Figure 2.6), and NOTCH with KBF1 (top of Figure 2.6), so this case can be considered as intermediate between the genes of Table 2.3 and Table 2.4, which was to be expected considering that it is indeed intermediate, since Table 2.5 contains unrecognized genes that

Table 2.6. Domain organization of "interaction leader" kinases distinguishing lymphoma and normal T cells.

Gene symbol	Expression trend lymphocyte/normal	Classification	Domains		
SYK	Pro-normal	Protein tyrosine kinase	PTKc_Syk: Protein Tyrosine Kinase (PTK) family; Spleen tyrosine kinase (Syk); catalytic (c) domain.	SH2: Src homology 2 domain	-
LCK	Neutral	Protein tyrosine kinase	PTKc_Lck_Blk: Protein Tyrosine Kinase (PTK) family; Lck and Blk kinases; catalytic (c) domain.	SH2: Src homology 2 domain	SH3: Src homology 3 domains
PRKCZ	Neutral	Serine/Threonine protein kinases	STKc_aPKC_zeta: Serine/Threonine Kinases (STKs), Atypical Protein Kinase C (aPKC) subfamily, zeta isoform, catalytic (c) domain	PB1_aPKC: PB1 domain is an essential modular domain of the atypical protein kinase C (aPKC)	C1: Protein kinase C conserved region 1 (C1)
PRKCQ	Neutral	Serine/Threonine protein kinases	STKc_nPKC_theta: Serine/Threonine Kinases (STKs), Novel Protein Kinase C (nPKC), theta isoform, catalytic (c) domain.	-	C1: Protein kinase C conserved region 1 (C1)

may show either a positive, negative, or neutral with respect to lymphoma.

In Figure 2.6, interaction network for lymphoma versus normal are shown. It is immediately visible that the map is dominated by kinases and cytokines. One of the kinases, namely ATM, forms its own sub-network together with the tumor suppressor p53, while the kinase PTK6 remains the only orphan.

Taken together, our data indicate that the more strongly genes are connected into an interaction network, the less likely they will tend to

Table 2.6. Continuous from previous page.

Gene symbol	Expression trend lymphocyte/ normal	Classification	Domains		
TXK	Neutral	Protein tyrosine kinase	PTKc_Tec_Rlk: Protein Tyrosine Kinase (PTK) family; Tyrosine kinase expressed in hepatocellular carcinoma (Tec) and Resting lymphocyte kinase (Rlk); catalytic (c) domain	SH2: Src homology 2 domain	SH3: Src homology 3 domains
ATM	Neutral	Phosphoinositide 3-kinase (either lipid or protein)	PIKKc_ATM: Ataxia telangiectasia mutated (ATM), catalytic domain	-	-
LYK	Unknown	Protein tyrosine kinase	PTKc_Itk: PTKc_Itk: Protein Tyrosine Kinase (PTK) family; Interleukin-2 (IL-2)- inducible T-cell kinase (Itk); catalytic (c) domain	SH2: Src homology 2 domain	SH3: Src homology 3 domains

show a difference in expression pattern, which is especially true for kinases controlling the neoplastic transformation of lymphocytes.

In terms of kinase classification, the highly-interacting kinases among the lymphoma/normal distinguishing ones are protein tyrosine kinases (SYK, LCK, TXK, LYK), serine/threonine protein kinases (PRKCZ, PRKCQ), and one phosphoinositide 3-kinase (ATM). Conservative domain composition of those kinases (Table 2.6) shows that neither of them contains a receptor (presumably transmembrane) domain. Notably, the pro-normal SYK kinase lacks one of the domains present in other kinases, which suggests separate NAPPA analysis of for different kinase domains, *e.g.*, by using the domains as separate query proteins.

We are performing analysis of the highly interacting kinases in terms of their involvement in regulatory networks. Figure 2.7 shows how the highly interacting kinases discriminating lymphoma versus normal T cells are involved in regulatory networks, as follows from NCI-Nature curated data from the Pathway Interaction Database from National Cancer Institute (Schaefer *et al.*, 2009).

We immediately note that our absolute "leader" gene, LCK, appears in this network in two different regulatory complexes. It tentatively follows from the map in Figure 2.7 that the diacylglycerol signaling can be important in mediating gene and protein interactions determining the differences between lymphoma and normal T cells.

Finally, we can make the conclusion with respect to the overall issue of gene transcription, translation, and post-translation that the emerging protein is actually controlling the whole process in the living cell proliferation, differentiation, and translation. In this context NAPPA technology which has been emerging in the last several years as summarized in Nicolini and LaBaer (2009) is especially relevant.

Possible NAPPA application for solution of above problem come from the following considerations.

In Tables 2.3–2.5, kinases are marked in bold. Also, the entire interaction map (Figure 2.6) is dominated by kinases and the complete NAPPA array containing likely the complete set of human kinases will be a tremendous source of information. In particular, it seems of interest

(i) Immediately use our existing NAPPA kinase microarray, with some of the interaction leaders shown above used as individual query proteins.

(ii) In the future, to manufacture a special NAPPA microarray containing "interaction leader" proteins shown above (*e.g.*, Classes A + B), and then to apply labeled proteins from lymphoma/normal T cell lysate. Thus overall binding capacity of each of those proteins will be evaluated, in comparison between lymphoma and normal T cells.

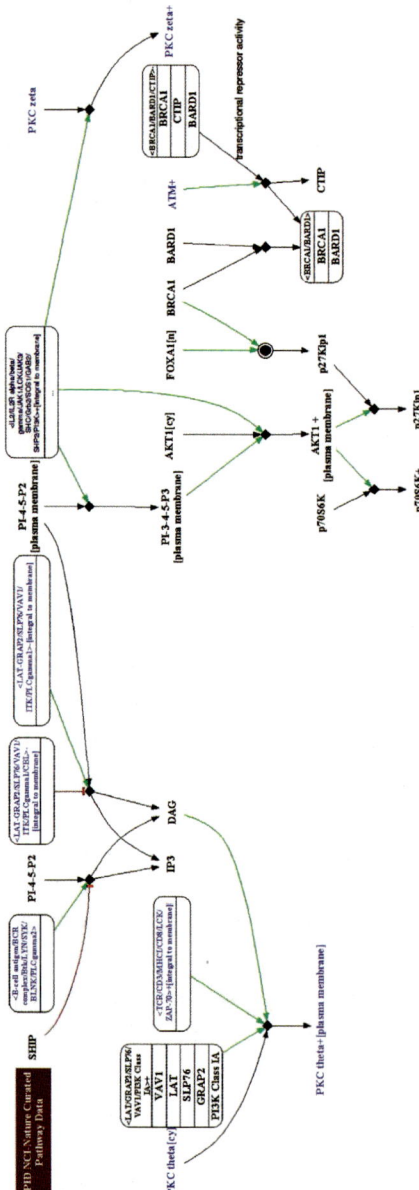

Figure 2.8. Involvement of highly interacting kinases in lymphoma versus normal T cells in regulatory networks. Complexes involving the kinases in question are shown in blue. Symbols indicate processes: diamond for general noncovalent interaction such as activation or inhibition, circle for transcription, triangle for translocation. Arrows in and out are input and output substances for processes, green arrows indicate positive regulation, red blunt-ended line negative regulation.

To evaluate the representativity of NAPPA kinase microarray, we evaluated the number of all currently know genes coding human kinases; we used NCBI Entrez and NCBI GenBank as search engine and database respectively, using boolean expressions for refining the search. We checked results of this search against dedicated databases, such as KinWeb (http://kinweb.ceinge.unina.it/) or Kinase.com. As a result, we found circa 700 unique genes. Further we checked which of them are available in the HIP PlasmID repository and in the NAPPA. Totally, PlasmID contains 506 human kinases, of which 444 are available in the NAPPA. The total number of kinase spots on the NAPPA is 982, but there are many duplicates. Only 347 kinases are recognized via HUGO HGNC gene symbols, which is a consequence of a nomenclature problem we immediately noticed with PlasmID and NAPPA. Namely, some HIP PlasmID genes Ids are aliases or are even withdrawn symbols. For example, SUDD instead of RIOK3 (withdrawn symbol), or TYK1 (alias for LTK).

Activities following similar bioinformatics approaches are in progress for mitotic HeLa cells compared to those at other stages of cell cycle (Nicolini *et al.*, 1975; Nicolini, 1974; Kendall *et al.*, 1977), and cAMP reverse transformed versus transformed CHO-K1 cells (Spera and Nicolini, 2007; Nicolini and Beltrame, 1982).

2.3.4 *Planned experimentation*

On the above premises we are presently carrying out with DNASER and Bioinformatics technologies a study for better understanding with NAPPA the genes and the proteins being critically in control of cell cycle progression of human T lymphocytes.

With Bioinformatics we are *ab initio* identifying leader genes and leader proteins in human T lymphoma and *ab initio* identifying leader proteins in human T lymphocytes, in order to subsequently both plan proper NAPPA experimentations and optimal interpretation. Alternatively studies CHO-K1 versus cAMP CHO-K1 reverse transformed cells us function of cell cycle are carried out in parallel.

Particular attention is being paid to CDK2, TP53, and JUN as possible candidate NAPPA targets and the query protein to be identified

54 V. Sivozhelezov et al.

tentatively interacting with the above three proteins within the pathways relevant to the given biological process. These calculations will be performed manually, for each interaction discovered automated and included into a software package for NAPPA data verification to yield a pattern similar to what shown below in Figure 2.9.

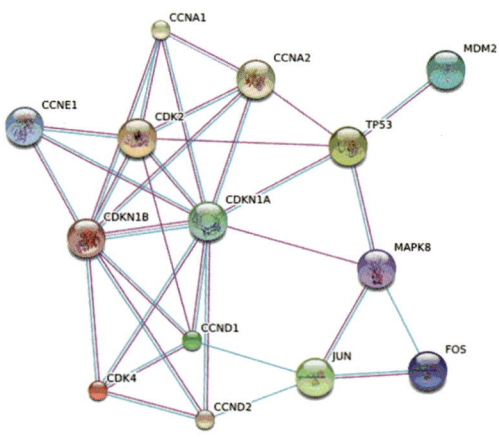

Figure 2.9. Network of interactions (connectivities) for the proteins participating in both G2 and M phase of T lymphocytes.

The typical approach for NAPPA protein interactions analysis (Ramachandran *et al.*, 2004) includes immobilization of plasmids for the given set of proteins, where N copies of the same microarray are manufactured, N being the number of query proteins. The query proteins are expressed *in situ* meaning that the same expression system (transcription and translation) is used for expression of both target (immobilized) and query (analyzed) proteins. Assuming (until now confirmed) that this dependence is not caused by a bias of the STRING database, **the 11 proteins identified (Figure 2.9) can be used as query proteins in NAPPA experiments**. Besides, it is of interest to identify the interactions missed by the STRING database. For that purpose, proteins with zero interaction scores should be used as query proteins. Should indeed NAPPA study reveal previously unknown interactions within the given 35 and 38-protein datasets, then a statistical analysis could be used to determine if they are specific to each of the phases or to the G2/M transitions.

An obvious drawback of the approach is that the proteins used as queries will not reproduce, primarily in terms of protein concentrations, differences in the in vivo expression profiles typical for either G2 versus M phases T lymphocyte cell cycle or lymphoma versus normal T lymphocytes simply because concentration of cDNA of query protein used for its for expression is arbitrary.

An alternative NAPPA approach is pursued herein with the DNASER fluorescence that is free of the above drawback:

(i) taking T Lymphomas (*) and T lymphocytes cells (+) kinetically matched, then purifying the proteins being green fluorescence labeled (*) and do the same purification but with the red fluorescence labeled (+) proteins; then add them to the human kinase NAPPA, in order to determine their interaction by monitoring which is dominant, namely if RED is the tumor, if GREEN is the normal, if YELLOW they are equally interacting.

(ii) carrying out microarray genes experiments using 10,000-spots slide and analysis of data, in healthy subjects and lymphoma patients, isolating T Lymphomas and T lymphocytes cells. Blood samples of healthy subjects and lymphoma patients will be obtained and T lymphocytes will be isolated by Ficoll centrifugation.

Alternatively the same study is being performed with transformed CHO-K1 (green) versus cAMP reverse transformed CHO-K1 (red) cells at different stages of cell cycle, in correlation with Mass Spectrometry data already available extensively (Spera and Nicolini, 2007, 2008).

The conventional method of protein fluorescent labeling used in NAPPA in which the proteins are labeled indirectly, via labeling an antibody that is specific to each protein, is unviable. The problem is that actually there are known only a very limited number of specific antibodies, not for all the existing protein (especially human proteins). However, newer method of fluorescent labeling is available using the antibiotic puromycin, which is specific to C-terminus of the protein. The fluorescent label such as Cy3 or Cy5 is attached to puromycin, which then attaches to C-terminus of each protein. The expected result will be, identification of kinases showing the interaction bias towards lymphoma or normal proteins (without specifying however which proteins from lysate are responsible for the bias). Such kinases should exist, there is already one proven example, ALK (Benharroch et al., 1998; Pulford et al., 2004).

The difficulty here is, *in situ* NAPPA protein expression will not work, because then the immobilized proteins will also be labeled. Therefore the correct technique would be to purify proteins from the cell lysate, to label them with Cy3 or Cy5-puromycin, and to use this preparation as the NAPPA query, preferentially with different concentrations of protein and/or added salt.

The alternative to the above approach is first using the DNA microarrays to identify query proteins (those coded by genes showing maximal change of expression, plus bioinformatically identified interactors of "maximal expression difference" genes) and then use the standard, "one by one query" NAPPA approach. For the tasks of "T cell G2/M phases" and "T lymphoma versus normal T cells", such "maximal expression difference" genes are already available.

Use the DNA/mRNA preparations as used in Approach C, and express proteins in it using the same cell-free expression system as for the NAPPA slide (but separately), and proceed with Approach B.

2.4 Conclusions

In conclusion the combination of our DNASER technology and of our implemented Leader genes (Sivozhelezov et al., 2006, 2008; Nicolini et al., 2006) bioinformatic approach can be effectively used to quantitatively probe the expression of the key gene and key protein being involved in the control of mammalian cell cycle, including transformed and reverse transformed cells, thus allowing to identify query proteins for NAPPA and/or design of specific NAPPA microarrays. Maximum area here imaged in a single shot with good resolution is only 14 × 14 spots, for a total of 196 spots, but proves anyhow successfully capable to verify the possibility to employ on NAPPA array the new DNASER technology implemented in our laboratories. Quite better performances appeared recently possible when imaging the whole area of the slide.

The results obtained and work in progress indicate that the application of NAPPA may have a much wider range and provide deeper insights in cellular processes than could be expected in its original formulation. Namely, instead of studying details of protein interactions of specific protein subsets/cascades, opportunities are now provided to study processes on the cellular level, for which only genomic (roughly speaking abundance based) approaches were available. This, in combination of genomics and bioinformatics as specified herein, abundance-based and interaction-based techniques will be united into a single approach.

Acknowledgments

This project was supported by grants to Fondazione EL.B.A. by MIUR for "Funzionamento" and by a FIRB International Grant on Proteomics and Cell Cycle (RBIN04RXHS) from MIUR (Ministero dell'Istruzione, Università e Ricerca) to CIRSDNNOB-Nanoworld Institute of the University of Genova.

References

Aladjem, M. I., Pasa, S., Parodi, S., Weinstein, J. N., Pommier, Y. and Kohn, K. W. (2004). Molecular interaction maps--a diagrammatic graphical language for bioregulatory networks. *Sci STKE* 2004(222), pp. pe8.

Benharroch, D., Meguerian-Bedoyan, Z., Lamant, L., amin, C., Brugieres, L., Terrier-Lacombe, M. J., Haralambieva, E., Pulford, K., Pileri, s., Morris, S. W., Mason, D.Y. and Delsol, G. (1998). ALK-positive lymphoma: a single disease with a broad spectrum of morphology. *Blood*, 91, pp. 2076–2084.

Chan, A. C., van Oers, N. S., Tran, A., Turka, L., Law, C. L., Ryan, J. C., Clark, E. A., Weiss, A. (1994). Differential expression of ZAP-70 and Syk protein tyrosine kinases, and the role of this family of protein tyrosine kinases in TCR signaling. *Journal of Immunology* 15, pp. 4758–4766.

De Paul, S. *et al.* (2001). Measuring bound water in protein-resistant coatings. Poster AVS Conference.

Höök, F., Vörös, J., Rodahl, M., Kurrat, R., Böni, P., Ramsden, J. J.. Textor, M., Spencer, N. D., Tengvall, P., Gold, J. and Kasemo, B. (2002). A comparative study of protein adsorption on titanium oxide surfaces using in situ ellipsometry, optical waveguide lightmode spectroscopy, and quartz crystal microbalance/dissipation. *Colloids and Surfaces B: Biointerfaces*, 24, pp. 155–170.

Kendall, F., Swenson, R., Borun, T., Rowinski, J. and Nicolini, C. (1977). Nuclear morphometry during the cell cycle. *Science*, 196, pp. 1106–1109.

Koopmann, J. O. and Blackburn, J. (2003). High affinity capture surface for matrix-assisted laser desorption/ionisation compatible protein microarrays. *Rapid Communications in Mass Spectrometry*, 17, pp. 455–462.

Manning, G., Whyte, D. B., Martinez, R., Hunter, T. and Sudarsanam, S. (2002). The Protein Kinase Complement of the Human Genome. *Science*, 298, pp. 1912–1934.

Nicolini, C. (1975). The discrete phases of the cell cycle: autoradiographic physical and chemical evidences. *Journal of the National Cancer Institute*, 55, pp. 821–826.

Nicolini, C. (2006). Nanogenomics for medicine. *Nanomedicine*, 1, pp. 147–152.

Nicolini, C. and Beltrame, F. (1982). Coupling of chromatin structure to cell geometry during the cell cycle: transformed versus reverse-transformed CHO. *Cell Biology International Reports*, 6, pp. 63–71.

Nicolini, C. and LaBaer, J. (2009). Nucleic acid programmable protein array for functional proteomics in medicine. In *NAPPA microarrays and functional proteomics*, Nicolini C. and LaBaer J. Eds, Pan Stanford Series on Nanobiotechnology Volume 2, London, New York and Singapore.

Nicolini, C., Ajiro, K., Borun, T. W. and Baserga, R. (1975). Chromatin changes during the cell cycle of HeHa cells. *Journal of Biological Chemistry*, 250, pp. 3381–3385.

Nicolini, C., Malvezzi, M., Tomaselli, A., Sposito, D., Tropiano, G. and Borgogno, E. (2002). DNASER I: layout and data analysis. *IEEE Transaction on Nanobiosciences*, 1, pp. 67–72.

Nicolini, C., Spera, R., Stura, E., Fiordoro, S. and Giacomelli, L. (2006). Gene expression in the cell cycle of human T lymphocytes: II. Experimental determination by DNASER technology. *Journal of Cellular Biochemistry*, 97, pp. 1151–1159.

Pulford, K., Lamant, L., Espinos, E., Jiang, Q., Xue, L., Turturro, F., Delsol, G. and Morris, S. W. (2004). The emerging normal and disease-related roles of anaplastic lymphoma kinase. *Cellular and Molecular Life Science*, 61, pp. 2939–2353.

Ramachandran, N., Raphael, J. V., Hainsworth, E., Demirkan, G., Fuentes, M. G., Rolfs, A., Hu, Y. and LaBaer, J. (2008). Next-generation high-density self-assembling functional protein arrays. *Nature Methods*, 5, pp. 535–538.

Ramachandran, N., Hainsworth, E., Bhullar, B., Eisenstein, S., Rosen, B., Lau, A. Y., Walter, J. C. and LaBaer, J. (2004). Self-assembling protein microarrays. *Science*, 305, pp. 86–90.

Ramachandran, N., Larson, D. N., Stark, P. R., Hainsworth, E. and LaBaer, J. (2005). Emerging tools for real-time label-free detection of interactions on functional protein microarrays. *FEBS Journal*, 272, pp. 5412–5425.

Rodahl, M., Höök, F., Fredriksson, C., Keller, C. A., Krozer, A., Brzezinski, P., Voinova, M., and Kasemo B. (1997). Simultaneous frequency and dissipation factor QCM measurements of biomolecular adsorption and cell adhesion. *Faraday Discussions*, 107, pp. 229–246.

Schaefer, C. F., Anthony, K., Krupa, S., Buchoff, J., Day, M., Hannay, T. and Buetow, K. H. (2009). PID: the Pathway Interaction Database. *Nucleic Acids Research*, 37, pp. D674–D679.

Sivozhelezov, V., Giacomelli, L., Tripathi, S. and Nicolini, C. (2006). Gene expression in the cell cycle of human T lymphocytes: I. Predicted gene and protein networks. *Journal of Cellular Biochemistry*, 97, pp. 1137–1150.

Sivozhelezov, V., Braud, C., Giacomelli, L., Pechkova, E., Giral, M., Soulillou, J.P., Brouard. S. and Nicolini, C. (2008). Immunosuppressive drug-free operational immune tolerance in human kidney transplants recipients. Part II. Non-statistical gene microarray analysis. *Journal of Cellular Biochemistry*, 103, pp. 693–706.

Spera, R. and Nicolini, C. (2007). cAMP induced alterations of Chinese hamster ovary cells monitored by mass spectrometry. *Journal of Cellular Biochemistry*, 102, pp. 473–482.

Spera, R. and Nicolini, C. (2008). Nappa microarrays and mass spectrometry: new trends and challenges. *Essential in Nanoscience Booklet Series* (Taylor & Francis Group/CRC Press).

Troitsky, V., Ghisellini, P., Pechkova, E. and Nicolini, C. (2002). DNASER II. Novel surface patterning for biomolecular microarray. *IEEE Transaction on Nanobiosciences*, 1, pp. 73–77.

Voinova, M. V., Rodahl, M., Jonson, M. and Kasemo, B. (1999). Viscoelastic acoustic responses of layered polymer films at fluid-solid interfaces: continuum mechanics approach. *Physica Scripta*, 59, pp. 391–396.

CHAPTER 3

LABEL FREE DETECTION OF NAPPA VIA MASS SPECTROMETRY

Rosanna Spera[1], Francesco Badino[1], Eugenie Hainsworth[2], Manuel Fuentes, Sanjeeva Srivastava[2], Joshua LaBaer[2] and Claudio Nicolini[1]

[1]*Nanoworld Institute, CIRSDNNOB-University of Genova and Fondazione EL.B.A., Italy*
[2]*Institute of Proteomics, Harvard University, Cambridge MA USA*

Our research relates to the implementation of Nucleic Acid Programmable Protein Array (NAPPA) analysis by matrix assisted laser desorption/ionization mass spectrometry (MALDI MS). Such analysis enables interrogation of protein–protein interactions in a label–free manner by desorption and ionization of analytes (*e.g.*, proteins synthesized from the immobilized cDNA and/or protein bounded to these). Mass spectrometry, in fact, has the unique advantage of being able to determine not only the presence but also establish the identity of a given ligand. However, the development of a MALDI MS-compatible protein microarray is complex since existing methods for forming protein microarrays do not transfer readily onto a MALDI target. Here we present a novel method for implementing a procedure to analyze protein contents on NAPPA protein microarrays using MALDI TOF mass spectrometry.

3.1 Introduction

The potential of label–free approaches to complement and even to improve other detection technologies has so far demonstrated only limited success. Protein microarrays and mass spectrometry present an innovative and versatile approach to study protein structure, abundance and function at an unprecedented scale. Protein microarrays have most commonly taken the form of collections of immobilized antibodies that can be used, for example, to monitor protein expression levels (Schweitzer *et al.*, 2002). Classically protein microarrays have been analyzed by fluorescent, radioactive labels or via antibody based detection systems, but not by using mass spectrometry. These methods of analyzing protein microarrays are therefore restricted by the availability of appropriately labeled ligands. Examples of labeled ligands that have been used successfully include fluorescent-labeled antibodies and radio- or fluorescently–labeled small molecule ligands.

A label-free method to detect interactions in a microarray format would be a major advancement, and it would greatly broaden the range of applications to areas where labeled compounds are not available or where labeling would alter the properties of the ligand. Amongst the label-free detection methods mass spectrometry has the unique advantage of being able to determine not only the presence but also the identity of a given ligand. However, the development of a MALDI MS-compatible protein microarray is complex since existing methods for forming protein microarrays do not transfer readily onto to a MALDI target. In addition MALDI procedure typically require the co-crystallization of the aqueous analyte with the "matrix" (an acidic energy absorbing molecules), to promote ionization of the analytes (Karas and Hillenkamp 1988). The method of co-crystallizing analyte and matrix for MALDI typically results in a heterogeneous crystallization process and yields discrete, spatially separated crystals that each contain differing amounts of matrix and analyte. As a consequence a significant lower limit is imposed to the size of individual target areas that can be analyzed by MALDI MS methods: the target area generally has as area of at least 500 μm^2.

In literature there are different examples of capture of analytes for mass spectrometric analysis (Brockman and Orlando 1995, Wang *et al.*, 2001). However, all of these methods suffer from one or more limitations and, in synthesis, do not enable the immobilization of large numbers of different purified proteins in the form of a MALDI MS-compatible microarray platform. We are implementing a procedure to analyze Nucleic Acid Programmable Protein Array (NAPPA; U.S. Patent No. 6,800,453) (Ramachandran *et al.*, 2004; Ramachandran *et al.*, 2005, Ramachandran *et al.*, 2008) by MALDI TOF mass spectrometry (Spera and Nicolini 2008). To achieve this goal the NAPPA protein array has been realized on a metal surface to allow the mass spectrometric analysis.

3.2 Materials and Methods

3.2.1 *The experimental layout*

A probe, for use with a laser desorption/ionization mass spectrometer, comprises a support having an electro conductive target surface; this surface is essential where the probe is to be subjected to MALDI MS analysis. Whilst the support could be made wholly of an electro conductive material (which term is used herein to include semi-conductive materials) it is preferred to coat a rigid support, *e.g.*, a glass, with an electro conductive material such as gold. To analyze NAPPA it can be mounted on an adapter, which carries it into a mass spectrometer. Such an adapter was realized modifying a standard MALDI target (Figure 3.1): lodging was realized on a standard Bruker Scout Target (Bruker Daltonics, Leipzig, Germany).

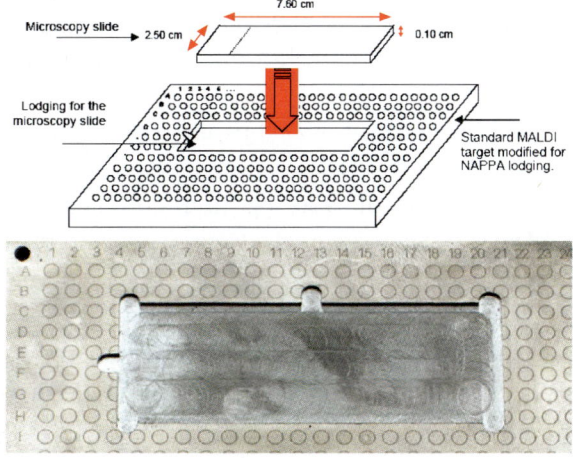

Figure 3.1. Adaptation of MALDI standard target to carry NAPPA (above), and MALDI standard target adapted to carry NAPPA (below).

The NAPPA microarray was then produced on a standard size microscope glass slide covered with a thin layer of gold. The gold surface was then functionalized by amino-PEG-thiols generating a homogeneous self-assembled monolayer onto the gold surface.

The lodging dimension is exactly the same of the array, guaranteeing the immobility of the glass once put in the lodging.

3.2.2 Fabrication of NAPPA

The study is carried out on three kinds of NAPPA:
 (i) Human Kinase NAPPA spotted with 24 × 72 200 microns spots of kinase genes from HIP FLEXGene human kinase collection (pDNR-Dual, complete set); such geometry is the same adopted for fluorescence analysis.
 (ii) NAPPA of 12 boxes of 4 × 4 spots (each spot in a box is of the same gene) of 300 microns (Figure 3.2); such geometry guaranties the production of an amount of proteins suitable for protein identification *via* fingerprinting. The genes immobilized on the NAPPA are from HIP human plasmid collection and are the following (the corresponding expressed proteins are also reported):
  - CdKN1A_Human: gene of 495 bp and corresponding protein of 18,119 kDa;
  - CdK2_Human: gene of 897 bp and corresponding protein of 33,930 kDa;
  - JUN_Human: gene of 996 bp and corresponding protein of 35,676 kDa;
  - P53_Human: gene of 1182 bp and corresponding protein of 43,653 kDa.
 (iii) High-density NAPPA of the same geometry as described previously, using the same genes, from HIP human plasmid collection (Ramachandran *et al.*, 2008).

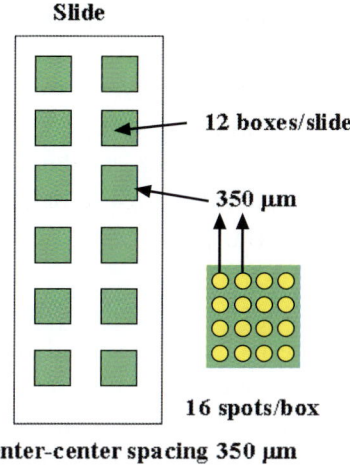

Figure 3.2. NAPPA geometry for MS analysis.

The genes spotted in the 12 box, each containing 16 spots (Figure 3.2), are: P53, JUN, CdK2, CdKN1A, A, B, C, D, Unknown 1, unknown 2 and vector (duplicate spots). This NAPPA has been projected in order to test step by step the possibility to identify the synthesized proteins: four known genes were immobilized (P53, JUN, CdK2, CdKN1A), the same genes have also been spotted in an unknown order (spots named A, B, C, D), and moreover there were two totally unknown genes (named Unknown1 and Unknown2) which were also immobilized. Such genes disposition allowed gradually testing the capability for protein identification: proteins A, B, C and D once identified them correctly, the unknown proteins (unknown1 and unknown2).

3.2.3 *NAPPA expression*

To express proteins we followed the protocol (modified from Ramachandran *et al.*, 2008):
- Blocking: 1 hr on rocking shaker at room temperature with SuperBlock. We used 30 mL in a pipette box for 3 slides.
- Rinsed with milli-Q water. Dried with nitrogen.

- Applied HybriWell (Grace Biolabs) gasket to the slide.
- Pre-heated the incubator to be used for IVT at 30 °C.
- Prepared IVT: 130 μL of IVT lysate mix for slide. Each tube after component addition contained 400 μL of lysate mix and the lysate tubes could not be re-frozen, *e.g.*, 1 tube = 3 slides = 400 μl → 16 μl TNT buffer, 8 μl T7 polymerase, 4 μl of –Met, 4 μl of –Leu or –Cys, 8 μl of RNaseOUT, 160 μl of DEPC water, 200 μl of reticulocyte lysate.
- Added IVT mix. Pipette the mix slowly. Gently massaged the HybriWell to get the IVT mix to spread out and cover all of the area of the array. Applied small round port seals to both ports.
- Placed the slides on a bioassay dish with divider on top of the levelling shelf inside the incubator. Incubated for 1.5 hr at 30 °C for protein expression, followed by 30 min at 15 °C for the query protein to bind to the immobilized protein.
- Removed the HybriWell and immersed each slide in water immediately; washed 3 times, 3 minutes each, in a pipette box. Dried with nitrogen.

3.2.4 *Autoflex analysis*

On each spot about 0.2 ng of protein is synthesised (an estimate from unpublished data of HIP), which is enough to analyze intact protein but not to perform fingerprinting. For this reason, for the kinase NAPPA we performed MS analysis of the intact proteins only, while for the other two kinds of NAPPA we performed proteolytic digestion of the synthesized proteins. We performed different tests to optimize both the placement of the matrix on the array in order to maximize the sensibility of the measurement and the digestion protocol of the NAPPA (before and after proteins expression).

For kinases NAPPA we spotted 0.3 μl of matrix per spot to analyze the expressed proteins. The matrix was made of energy absorbing molecules that were deposited onto the surface in a solvent, and the solvent evaporated off. Typical energy absorbing molecules for protein analysis included crystals of α-cyano-4-hydroxy-cinnamic acid (HCCA) and sinapinic acid (Bruker Daltonics, Leipzig, Germany). These molecules were detached in a saturated solution of of 2/3 of 0.1% TFA (trifluoroacetic acid, Sigma Aldrich) and 1/3 of acetonitrile (Sigma Aldrich), and 0.3 μl of this solution were spotted on each NAPPA spot and allowed it to dry. We utilized both the matrix on the basis of the molecular weight of the analyzed protein.

For protein fingerprint once the proteins were synthesized on the NAPPA we added 5 μl of 0.01 mg/ml trypsin (Trypsin Proteomics Grade, Sigma Aldrich) solution in 25 mM ammonium bicarbonate (Sigma Aldrich) per box (of 16 spot) and incubated the trypsin at 37°C for 4 hours (Koopmann and Blackburn, 2003; Finnskog et al., 2004). At the end of the digestion we spotted on each box 2.5 μl of HCCA matrix and let it dry. To calibrate the spectra, we spotted 1 μl of peptide standard solution and 1 μl of protein standard solution (Bruker Daltonics, Leipzig, Germany; both in 1 μl of HCCA matrix) on the gold surface next to every box of 16 spots. Once the array was dried it was placed on an opportune MALDI target and analyzed.

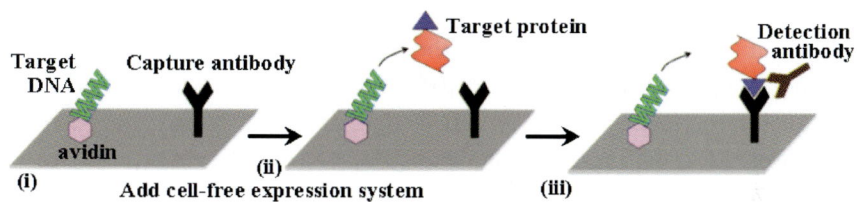

Figure 3.3. NAPPA technology (Spera and Nicolini 2008).

The analyses were performed on an Autoflex MALDI-TOF mass spectrometer (Bruker Daltonics, Leipzig, Germany) operated in linear and reflector mode. The resulting mass accuracy was < 300 ppm for intact proteins and < 20 ppm for peptides. MALDI TOF mass spectra were acquired with a pulsed nitrogen laser in positive ion mode. The mass lists obtained were submitted for a data bank search. We used a specific software for protein data interpretation, Biotools (Bruker Daltonics, Leipzig, Germany), that allowed an automated protein identification via library search with fully integrated MASCOT software (Matrix Sciences, Ltd. www.matrixscience.com).

In principle label free signals are due to all the molecules immobilized in the spot containing NAPPA chemistry (cDNA, Capture GST antibody, protein, *etc.*, see Figure 3.3). To identify the signals of protein it is necessary to acquire, for each spot, a spectrum before the protein synthesis. Once spectrum is acquired the array can not be utilized for protein synthesis (*i.e.*, being MS a destructive analytical method). For this reason it was necessary to have at least two identical gold NAPPA, one slide which can be analyzed before protein synthesis, and another slide after the synthesis. Once spectra of each spot was acquired, before and after the transcription (in the same conditions of digestion, matrix utilization and with the same MS method), it was possible to compare the different mass list obtained in order to identify the peaks of the protein.

Our first goal was then to analyze the fragmentation pattern of NAPPA reagents: GST, Streptavidin/BSA, Capture antibody and cDNA. For this reason we analyzed two identical NAPPA, one before and the other after protein expression in order to identify the peaks relative to the synthesized protein only. The protocol for the digestion and the analysis of the NAPPA before and after protein expression was same. The trypsin digestion (as described above) of the NAPPA before or after the protein synthesis was performed and 2.5 µl of HCCA matrix was spotted on the box of interest. The array was left to dry, and then it was placed on the opportune MALDI target for analysis.

3.3 Results

3.3.1 *Human kinase NAPPA*

As mentioned above we had only one human kinase gold NAPPA, for this reason we analyzed it only after protein synthesis. The spectra obtained for the analysis of three different spots are shown in Figure 3.4.

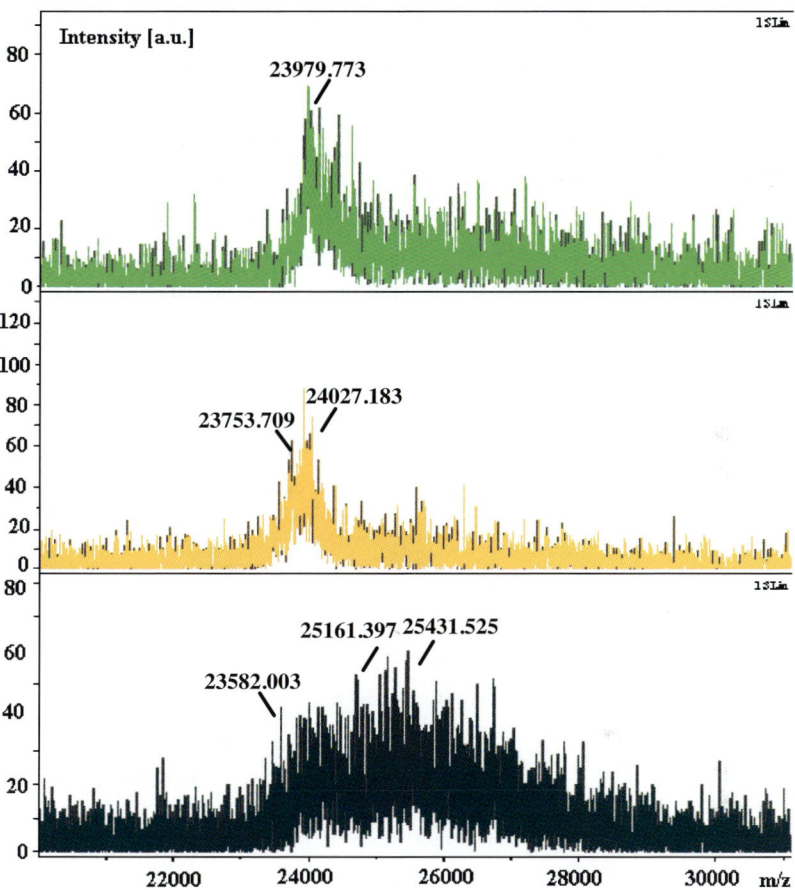

Figure 3.4. MALDI TOF MS Spectra of Human Kinase NAPPA after protein synthesis, high weights range (20–35 kDa) for 3 different spots/genes (from the top Gene NA7-A12, Gene NA7-E12 and Gene NA8-D12). The spectra were acquired in linear mode and the matrix utilized was sinapinic acid. The spectra were calibrated externally by Protein Standard Solution I (Bruker). (Reprinted with the permission from Spera and Nicolini, Nappa microarray and mass spectrometry: new trends and challenges, Essential in Nanoscience Booklet Series, © 2008, underwritten by Taylor & Francis Group, LLC).

The peaks shown in the Figure 3.4 are presumably due to proteins synthesized on the spots; however, the amount of protein synthesized on a single spot was insufficient to perform protein fingerprint; for this reason a new geometry for NAPPA was adopted.

3.3.2 NAPPA for MS

We have analyzed the new NAPPA, which we thought can be expressed for MS analysis. First of all we tested the possibility to digest the proteins synthesized on the array in order to perform fingerprint and to identify them.

Table 3.1. We reported in the 1st column the number of peaks identified in the known samples (P53, Jun, CdK2, CdKN1A) mass spectra; in the 2nd column the number of peaks identified in the unknown samples (A, B, C, D) mass spectra; in the 3rd column the number of matching peaks of the two experimental mass list (of known and unknown sample) without the background peaks.

Peak number of known sample experimental mass list	Peak number of unknown sample experimental mass list	Number of peaks matched in two samples (known and unknown) without the background peaks
JUN	C	JUN = C
56	53	17
CdKN1A	D	CdKN1A = D
46	49	7
P53	A	P53 = A
58	55	14
CdK2	B	CdK2 = B
58	61	19

We performed protein expression and, then, proceeded for the fingerprint of synthesized proteins (P53, JUN, CdK2, CdKN1A, A, B, C, D, unknown1, unknown2 and vector) following the previously described protocol. At the end of the digestion we spotted 2.5 µl of HCCA matrix on each box and allowed it to dry. We acquired spectra for each sample (P53, JUN, CdK2, CdKN1A, A, B, C, D, unknown1, unknown2 and vector). In order to identify the A, B, C and D samples; we matched, with the aid of a *"matching algorithm"* implemented by us and described below, their experimental mass lists with that of the known samples (P53, JUN, CdK2, CdKN1A).

The final identifications (those with the best scores) are reported in Tables 3.1 and 3.2 and have been confirmed in different experiments. In synthesis we identified the sample A as P53, B as CdK2, C as JUN and D as CdKN1A.

Table 3.2. We reported in the 2st column the percentage of matching peaks of the two experimental mass list (of known and unknown sample); in the 3nd column the percentage of matching peaks of the two experimental mass list, after the subtractions of the background peaks.

Sample identification	Matched peaks	Matched peaks (without background peaks)
C = JUN	92.5%	78.3%
D = CdkN1A	89.5%	60.0%
A = P53	80.0%	73.9%
B = CdK2	83.6%	76.7%

3.3.2.1 *Matching algorithm*

The "*matching algorithm*" has been written in C language, in Windows XP. It has been used the compiler Freecommand Linetools freely downloadable from Embarcadero Developer Network.

The first step of the algorithm is to compare the eight input mass lists (those of the four known samples with those of the four unknown samples) — within a given tolerance threshold (or error), identify the "background peaks" and save them in an output file deleting them from each mass list. The background peaks represent the background signal due to the molecules common to all sample spots (*e.g.*, antiGST, BSA/streptavidin, ...).

In the algorithm they are defined as the peaks present in at least n peak lists, with $n = 4, 5, 6, 7, 8$; in the results reported in this article we chose $n = 5$.

The second step is to carry out the proper matching between the known (P53, JUN, CdK2, CdKN1A) and unknown (A, B, C, D) four mass lists (purified from background peaks) namely:
- A mass list with P53, JUN, CdKN1A and CdK2 mass lists;
- B mass list with P53, JUN, CdKN1A and CdK2 mass lists;
- C mass list with P53, JUN, CdKN1A and CdK2 mass lists;
- D mass list with P53, JUN, CdKN1A and CdK2 mass lists;

in order to identify the unknown proteins with a certain probability. The "*matching algorithm*" flowchart is reported hereafter.

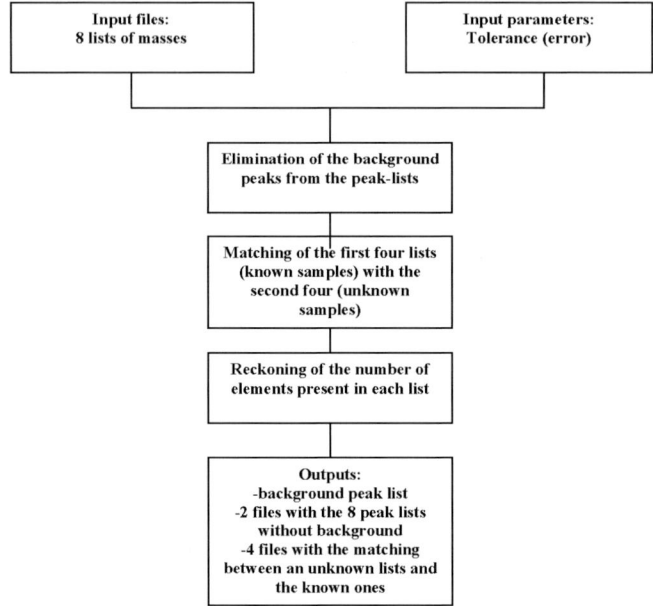

To compare two mass lists the algorithm calculates for each couple of masses the value:

$$\Delta = |m_{known} - m_{unknown}|$$

and assumes equal two masses for which $\Delta \leq 1$.

For each mass list — after the subtractions of the background peaks — the algorithm counts the number of peaks and starting from the number of mass values matching for each couple of mass lists it calculates the percentage of matched peaks as:

% of matched peaks for two mass lists = (number of matching masses/ number of peaks in the unknown sample mass list) × 100

The final identifications is made considering the match with the best score.

Hereafter we report the pictures of the spectra coupled as identified (Figures 3.5–3.8).

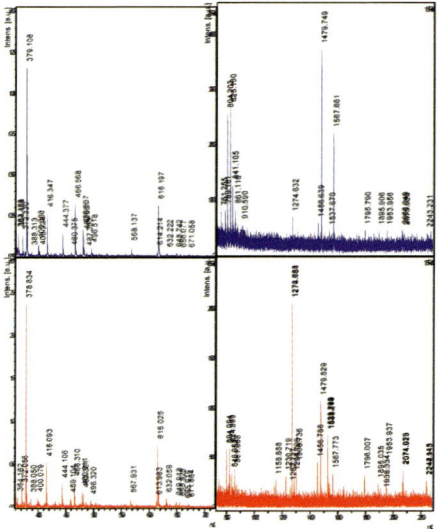

Figure 3.5. MALDI TOF Spectra of NAPPA after protein triptych digestion, 5–20 kDa range, for CdKN1A (upper) and D (bottom) samples.

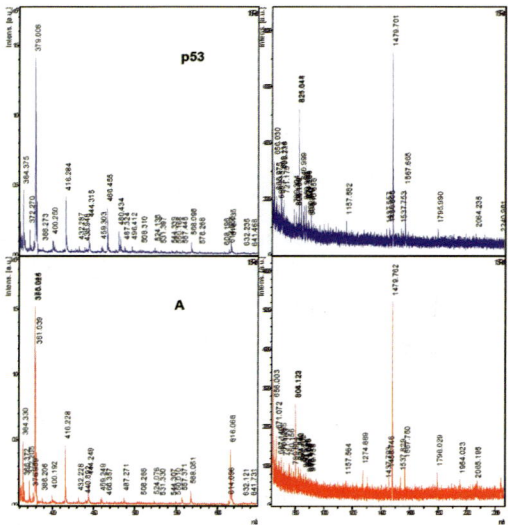

Figure 3.6. MALDI TOF Spectra of NAPPA after protein triptych digestion, 5–20 kDa range, for P53 (upper) and A (bottom) samples.

74 R. Spera et al.

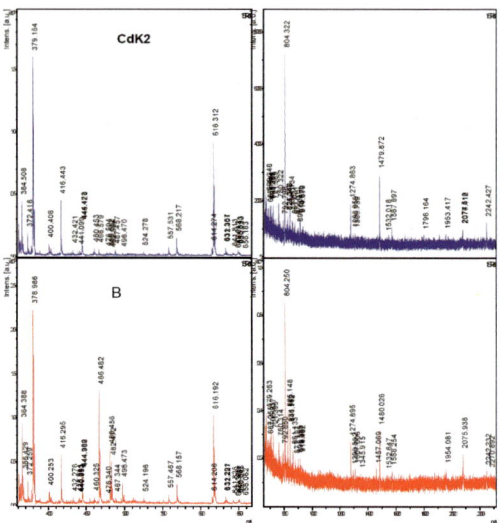

Figure 3.7. MALDI TOF Spectra of NAPPA after protein triptych digestion, 5–20 kDa range, for CdK2 (upper) and B (bottom) samples.

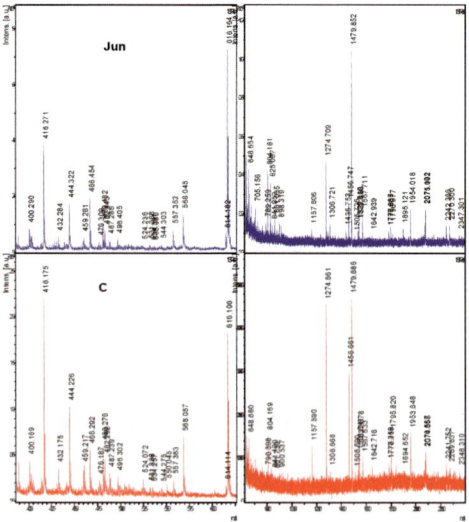

Figure 3.8. MALDI TOF Spectra of NAPPA after protein triptych digestion, 5–20 kDa range, for Jun (upper) and C (bottom) samples.

We submitted the known proteins experimental mass lists to a data bank search using Biotools software (Bruker Daltonics, Leipzig, Germany) in order to obtain their fingerprint.

Protein score is $-10*\text{Log}(P)$, where P is the probability that the observed match is a random event. Protein scores greater than 64 are significant ($p<0.05$). In our analysis, despite the fact that all the scores obtained were rather low to allow, an identification of the proteins was still possible.

Table 3.3. We reported: in the 1st column the scores obtained by a Mascot data bank search of the experimental mass lists. Protein score is $-10*\text{Log}(P)$, where P is the probability that the observed match is a random event. Protein scores greater than 64 are significant ($p<0.05$). In the 2nd column the scores obtained by a Mascot data bank search of the experimental mass lists including only the masses matching the theoretical values. In the 3rd column the number of experimental masses matching with the theoretical mass lists.

Score obtained with the experimental mass list	Score obtained with the experimental mass list (including only the masses matching the theoretical ones)	Number of experimental masses matching the theoretical values
JUN		
35	68	6
CdKN1A		
31	50	5
P53		
28	64	10
CdK2		
<22	36	5

Indeed, to gain more confidence in our data we used "*matching algorithm*" to match the experimental mass lists of the known samples with the theoretical mass lists of the same proteins calculated starting from the protein sequence (obtained from HIP, http://plasmid.med.harvard.edu/PLASMID, and ExPASy data bank, http://www.expasy.ch/) and performing a theoretical digestion of the sequence with 'Sequence Editor' a special tool of Biotools software

(Bruker Daltonics Inc.). We resubmitted the proteins experimental mass lists — now including only those masses matching the theoretical ones — to a data bank search obtaining scores significantly higher. In the Table 3.3, we have reported the scores and the number of matching between the experimental and the theoretical mass lists.

From these results we concluded that it is rather difficult to perform fingerprint of the NAPPA proteins directly. One problem could be that the protein is synthesized with the GST tag and it is attached to GST antibody (immobilized on the array).

Once performing tryptic digestion we obtain fragment of the entire protein complex (GST tag plus protein plus anti-GST). A possible solution that we are presently investigating is to cut the GST tag by a specific cleavage enzyme, and then collect and analyze only the synthesized proteins (cut from the GST tag — anti-GST complex).

3.4 Conclusions

We have presented the results obtained in our research to implement a procedure to analyze Nucleic Acid Programmable Protein Array (NAPPA) by MALDI TOF mass spectrometry. In summary, an identification of the proteins expressed on NAPPA appears possible although initial scores obtained are rather low.

We have analyzed two kinds of NAPPA, first a Human Kinase NAPPA (24 × 72 200 microns spots) and second a NAPPA designed specifically for MS analysis (12 boxes of 4 × 4 spots). From the analysis of the first NAPPA we realized that the amount of protein synthesized in each spot was enough to obtain the intact protein spectrum but not enough to perform its fingerprint. For these reason a new geometry for NAPPA to be analyzed with MS was thought and it was successfully achieved. Analysis of new NAPPA was very successful.

It resulted into identification of unknown proteins, knowing that they were the same yet spotted on the NAPPA but in another order. In synthesis we identified the sample A as P53, B as CdK2, C as JUN and D as CdKN1A. To perform these identifications we matched the experimental mass lists of the known and unknown proteins. So we identified those peaks due not to the protein but to the other molecules immobilized on the NAPPA, named 'background' peaks due to the molecules immobilized on the NAPPA before the protein synthesis.

We finally submitted the known proteins experimental mass lists — without the background peaks — to a data bank search engine and obtained their fingerprint with some degree of success, despite the problems due to the confusing presence of both GST tag synthesized with the protein and of GST antibody immobilized on the array.

In conclusion our results on NAPPA analysis by MALDI-TOF MS are very encouraging being the experiments and the results perfectly repeatable.

Acknowledgments

This project was supported by grants to Fondazione EL.B.A. by MIUR for "Funzionamento" and by a FIRB International Grant on Proteomics and Cell Cycle (RBIN04RXHS) from MIUR (Ministero dell'Istruzione, Università e Ricerca) to CIRSDNNOB-Nanoworld Institute of the University of Genova.

References

Brockman, A. H. and Orlando, R. (1995). Probe-immobilized affinity chromatography/mass spectrometry. *Analytical Chemistry*, 67, pp. 4581–4585.

Finnskog, D., Ressine, A., Laurell, T. and Marko-Varga, G. (2004). Integrated protein microchip assay with dual fluorescent and MALDI read-out. *Journal of Proteome Research*, 3, pp 988–994.

Karas, M. and Hillenkamp, F. (1988). Laser desorption ionization of proteins with molecular masses exceeding 10,000 Daltons. *Analytical Chemistry*, 60, pp. 2299–2301.

Koopmann, J. O. and Blackburn, J. (2003). High affinity capture surface for matrix-assisted laser desorption/ionisation compatible protein microarrays. *Rapid Communications in Mass Spectrometry*, 17, 455–462.

Ramachandran, N., Hainsworth, E.,Bhullar, B., Eisenstein, S., Rosen, B., Lau, A. Y., Walter, J. C. and LaBaer, J. (2004). Self-assembling protein microarrays. *Science*, 305, pp. 86–90.

Ramachandran, N., Larson, D. N., Stark, P. R., Hainsworth, E. and LaBaer, J. (2005). Emerging tools for real-time label-free detection of interactions on functional protein microarrays. *FEBS Journal*, 272, pp. 5412–5425.

Ramachandran, N., Raphael, J. V., Hainsworth, E., Demirkan, G., Fuentes, M. G., Rolfs, A., Hu, Y. and LaBaer, J. (2008). Next-generation high-density self-assembling functional protein arrays. *Nature Methods*, 11, pp. 535–538.

Schweitzer, B.Roberts, S., Grimwade, B., Shao, W. P., Wang, M. J., Fu, Q., Shu, Q. P., Laroche, I., Zhou, Z. M., Tchernev, V. T., Christiansen, J., Velleca, M. and Kingsmore, S. F. (2002). Multiplexed protein profiling on microarrays by rolling-circle amplification. *Nature Biotechnology*, 20, pp. 359–365.

Spera, R. and Nicolini, C. (2008). NAPPA microarrays and mass spectrometry: new trends and challenges. Nano Article - Taylor & Francis/CRC Press.

Wang, S., Diamond, D. L., Hass, M., Sokoloff, R., Vessella, R. L. (2001). Identification of prostate specific membrane antigen (PSMA) as the target of monoclonal antibody 107-1A4 by proteinchip; array, surface-enhanced laser desorption/ionization (SELDI) technology. *International Journal of Cancer*, 92, pp. 871–876.

CHAPTER 4

LABEL FREE NAPPA VIA NANOGRAVIMETRY

Manuela Adami[1], Sameh Sallam[1], Roberto Eggenhöffner[1], Marco Sartore[1], Eugenie Hainsworth[2], Joshua LaBaer[2] and Claudio Nicolini[1]

[1]*Nanoworld Institute, CIRSDNNOB-University of Genova and Fondazione EL.B.A., Italy*
[2]*Institute of Proteomics, Harvard University, Cambridge MA USA*

Nanogravimetry label-free applications to the Nucleic Acid Programmed Protein Array technology is reviewed. DNA micro-arrays constitute a major advance in functional proteomics for the *in situ* generation of proteins micro arrays. The use of two different nanogravimetric technologies allows to investigate the kinetics during the *in situ* proteins expression. The first nanogravimetric apparatus is capable to detect frequency variations and protein production kinetics from four independent transducers driven in parallel, with a miniature flow-cell to follow each step of the protein expression protocol directly on the quartz. A software tool able to drive the signal acquisition and to analyze the acquired data was developed. First results on the *in situ* expression of proteins are reported. The second nanogravimetric instrumentation has been developed since very recently to account for viscoelastic behavior of protein films smoothly connected to the quartz crystal in terms of the quality factor. The system can detect also the crystal oscillations variations at expressing proteins in a liquid environment. The complex nature of these experiments is briefly discussed and preliminary results reported.

4.1 Introduction

Nanogravimetry is a powerful technique to monitor adsorption processes at interfaces in different chemical and biological research areas (Ramachandran *et al.*, 2004, 2005; Spera and Nicolini, 2008;

Koopamann and Blackburn, 2003). The advantages of the quartz crystal microbalance (QCM) technique are that a label free detection of molecules is allowed, reagent preparation procedures are greatly simplified and safety concerns associated, for example, with radioactive materials are circumvented. In this review, we apply this label-free technology to the Nucleic Acid Programmed Protein Array (NAPPA) technology (Rodahl *et al.*, 1997; Voinova *et al.*, 1999), a step forwards in functional proteomics. In NAPPA, plasmids directing the expression of GST-fusion proteins are spotted along with an antibody against GST. The fusion proteins produced by use of a coupled transcription/translation reaction are captured and immobilised via their GST tag. Since these arrays are produced just before their use, problems of protein stability are avoided. The use of these two technologies allows mainly the analysis and the study of kinetics during the in-situ proteins expression. The potential of label-free approaches to complement and even to improve other detection technologies has never been higher as can be seen in references. Labelling techniques as the use of isotopes, of fluorescent labels, affinity tags and others were the bases in proteomics research. However, labels have some drawbacks: the accurate measurements of kinetic constants and binding equilibrium are prevented, beyond antibody cross-reactivity problems. Thus, the interest of researchers is moving clearly towards label-free proteomics techniques, looking at essentially naked proteins (Shaffer, 2007). Classical label-free techniques such as column chromatography, mass spectroscopy, and calorimetry are implemented and renewed by technological improvements. Indeed, heat transfer down to the nanowatt range can be detected and it turns out a non-destructive technique when used on proteins that denature reversibly (www.calorimetrysciences.com). Innovative label-free techniques make use of light and sound. Surface Plasmon Resonance (SPR) exploits the linear dependence between resonance energy and concentration of biomolecules as DNA or proteins (www.biacore.com). Sound waves can be generated by the vibration of a quartz crystal; sensor based on a quartz crystal coated with gold oscillating at a specific frequency was introduced (www.akubio.com). Binding events change the frequency at varying the amount of mass linked to the surface. With SPR, surface mass changes are detected, with QCM-based method, like the surface acoustic wave (SAW), both mass changes and changes in viscosity on the surface can be detected. These approaches are relevant since, sometimes, proteins change conformation and this changes viscosity as well. Also fibre optics technology can be suitable as label-free technique (www.corning.com), introduced in the 1970s along with microplates. These two lines of development have recently being merged together in an optically based label-free sensing system based on a 384-well microliter plate, the Corning Epic system. Label-free sensing technologies have a lot of advantages that are exploited in new developments. An electronic addressing system for assembling DNA microchips was developed (Höök *et al.*, 2002) whereas a new microchip system was announced (De Paul *et al.*, 2001) to identify clinically relevant proteins without the need for fluorescent tags.

4.2 Materials and Methods

4.2.1 *QCM-frequency technique*

Nanogravimetry exploits the piezoelectric quartz crystals properties to vary the resonance frequency (f) when a mass (Δm) is adsorbed to or desorbed from their surface according to well-known equation:

$$\Delta f / f_0 = -\Delta m / A \cdot \rho \cdot l$$

where A is the area covered by the adsorbate and ρ and l are the quartz density and thickness, respectively (Bartolucci *et al.*, 1997). As shown in Figure 4.1, the nanogravimetric instrumentation consists of three gold coated quartz crystals, each with circular shape and having printed one 300 microns NAPPA spot containing 400 genes, using the following genes and corresponding expressed proteins of the indicated molecular weight:
- CdK2 Human Gene of 897bp expressing 33, 93 kDa protein
- JUN_Human Gene of 996 bp expressing 35, 67 kDa protein
- P53_Human Gene of 1182 bp expressing 43, 65 protein

Parallel functional tests on regular NAPPA glasses confirmed the functioning of the resulting quartz samples. E-coil/K-coil chemistry as in the case of Surface Plasmon Resonance experiments have been used (Höök *et al.*, 2002). The nanogravimetric samples were embedded into glass-like structures for easy handling. The sample holder contained multiple quartz crystals samples 8 mm wide, as above indicated. Our instrument is able to detect frequency variations and kinetics from 4 quartzes, one control and three NAPPA, driven in parallel simply by a PC-USB port (as pictured in Figure 4.1).

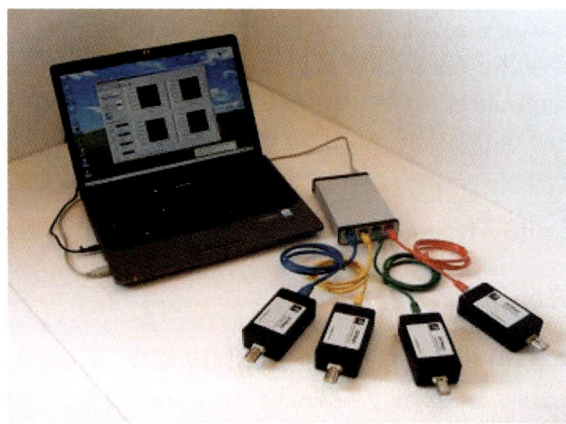

Figure 4.1. Four sensors enable four parallel measurements: one element can be used as reference and the remaining three as working transducer. In this manner, unspecific adsorption and/or uncontrolled parameters can be taken into account.

In order to realize a test system to detect protein interactions, different organic layers were deposited over a quartz crystal to make the adhesion with proteins possible. The first two layers (one dedicated to the bonding of the gold surface of quartz using gold-thiol reaction, another one to grant compatibility with proteins possible) were successfully applied, and this fact was confirmed by reduction of resonant frequency of the system. Streptavidin adhesion was also successful as proved by a further frequency reduction. The interaction with another organic macromolecule, namely with the natural streptavidin complementary molecule, biotin, is presently not detectable in terms of further frequency reduction being too small.

4.2.2 *QCM-D dissipation (quality) factor technique*

In recent years it became more and more evident that quartz resonators used in fluids are more than mere mass or thickness sensors. The sensor response is also influenced by viscoelastic properties of the adhered biomaterial, surface charges of adsorbed molecules and surface roughness. The combined measurement of frequency shifts (Δf) and dissipation factor (D) has been termed in literature as "QCM-D" (Q-Sense AB, Göteborg, SE) and it provides information about the viscoelastic properties, *i.e.*, shear modulus, viscosity of adsorbed adlayers and the surrounding bulk liquid environment. Starting from this scenario, we decided to implement QCM-D measurement in our 4 channels nanogravimetric instrumentation for NAPPA processes, although at present we are reporting results from prototype instruments only. In most situations the adsorbed film is not rigid and the Sauerbrey relation becomes invalid. A film that is "soft" (viscoelastic) will not fully couple to the oscillation of the crystal, hence the Sauerbrey relation will *underestimate* the mass at the surface. A soft film dampens the crystal's oscillation. The damping or, dissipation (D), of the crystal's oscillation reveals the film's softness (viscoelasticity). D is defined as:

$$D = \frac{E_{lost}}{2\pi E_{stored}}$$

where E_{lost} is the energy lost (dissipated) during one oscillation cycle and E_{stored} is the total energy stored in the oscillator.

The coupling of the crystal surface to a liquid drastically changes the frequency, because when a quartz crystal oscillates in contact with a liquid, a shear motion on the surface generates motion in the liquid near the interface. The oscillation surface generates plane-laminar flow in the liquid, which causes a decrease in the frequency, depending on the liquid density and viscosity: for a 10 MHz crystal with one face exposed to diluted solutions, near room temperature, Δf is about 2 KHz. The dissipation of the crystal is measured by recording the response of a freely oscillating crystal that has been vibrated at its resonance frequency.

D is also equal to $1/Q$, the quality factor of the quartz crystal. A typical 10 MHz, AT-cut crystal working in a vacuum or gaseous environment has a dissipation factor in the range of $10^{-6} \div 10^{-4}$.

A higher Q indicates a lower rate of energy dissipation relative to the oscillation frequency, so the oscillations die out more slowly.

If a substance slips on the electrode, frictional energy is dissipated and D can be used to interfere the coefficient of friction of the adsorbed film; if the film is viscous, energy is also dissipated due to the oscillatory motion induced in the film and D can be used to interfere the internal friction in the film.

We developed a first experimental set-up to measure D: the method is based on the principle that when the driving power to an oscillator is switched off at $t = 0$, the amplitude of oscillation A decays as an exponential damped sinusoid:

$$A(t) = A_0 \, e^{-t/\tau} \sin(\omega t + \varphi) + const$$

where τ is the decay time constant, φ is the phase and *const* is the dc offset. The decay constant is related to D by: $D = 1/(\pi f \tau)$.

Based on a mathematical model, the impedance real part is related to the D factor of the quartz while its imaginary part is related to the frequency shift. Up to now, we are able to reveal the real part variation and to find D.

Based on this concept, the quality factor analysis offers acoustic label-free sensors: by collecting both the dissipation and the resonance frequency of a quartz crystal, QCM-D can be used to study the formation of thin films of proteins onto surfaces in liquid. By measuring several frequencies and the dissipation, it is possible to judge if the adsorbed film is rigid or soft, and the kinetics of structural changes and mass changes can be obtained. Additional information is obtained when the QCM-D is operated in pulsed mode. By measuring the damping of the amplitude of the crystal vibration after the alternating electric field has been turned off, one can determine the amount of energy the adsorbed layer can dissipate. Since the signal from rigid layers is only weakly damped, it can be said to have a small dissipation factor. In other words, there are relatively few degrees of freedom into which energy can be lost. Conversely, soft layers have many degrees of motional freedom and a correspondingly large dissipation factor. The amplitude of the signal from a crystal with such a layer adsorbed on it will decay very rapidly.

With respect to the above approach, we changed completely the electronics scheme because we realized that, in real experiments, the method based on decay curves modelling is not so quick to follow the real kinetics. As shown in Figure 4.2, we have realized hence a second laboratory prototype able to acquire the D factor respect to the former prototype, already equipped with driving software and with the analysis software for the kinetic plots interpretation.

Figure 4.2. Electronic apparatus for the detection of the Q-factor.

Accordingly, we designed and implemented a new method, based on an impedance network analyzer: the quartz crystal is seen as an impedance whose real and imaginary part vary according to the surface chemistry of the quartz itself.

To show the effects just described, the Figure 4.3 shows the resonance curve acquired with three samples of glycerol in distilled water at different concentrations, *i.e.*, at different viscosity and at narrow temporal distance each other. In Figure 4.4 a real time acquisition of the D factor is reported *vs.* time for solutions of fructose in water of different viscosity. We point out that the resonance frequency peaks are almost unchanged irrespectively of the different concentration of glycerol, whereas the *D* factor measurement shows a strong dependence on fructose concentration.

The quartz crystal microbalance with dissipation technique is ideally suited for *in situ* characterization of polymer films grown from the surface because it is a label-free, in-line technique that directly reports on the amount and viscoelastic properties of polymer films, as they grow on the surface in real-time.

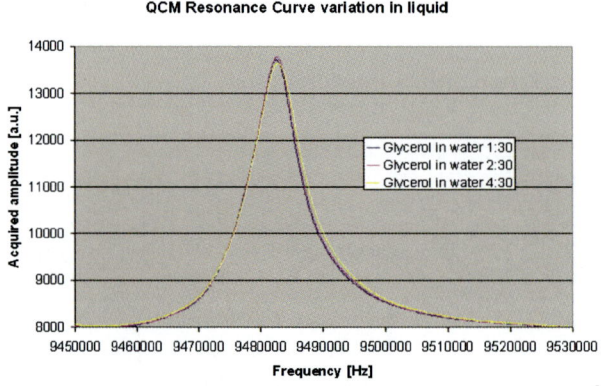

Figure 4.3. Example of a resonance curve set with solutions at different viscosity and density.

A particular feature of the QCM-D technique is that frequency changes reflect not only the adsorbed mass, but also solvent molecules trapped or mechanically coupled to the adsorbed film such as structural water in swollen polymer films.

Some authors' results suggest that QCM-D is likely to be an important quantitative spectroscopic tool to characterize the interfacial architecture and mechanical properties of soft-hydrated materials. In the mean time, some attempts to immobilize proteins on quartzes have been performed, while different protocols have to be tested yet in order to maximize the transducer sensitivity.

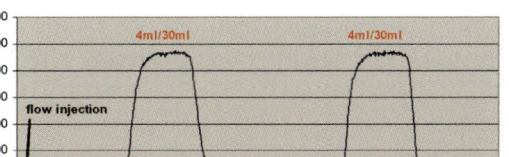
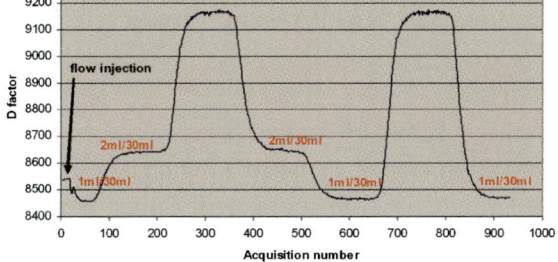

Figure 4.4. Example of a *D* factor recording vs. time with solutions at different viscosity.

The basic idea is to use one quartz as the working element, ad hoc modified in order to proper sense the interesting protein and the second one as reference element, in order to take into account the specific adsorption.

Figure 4.5. Flow-cell to follow each step of the protein expression.

In order to understand better what happens on the NAPPA spot, we have designed miniature flow-cell to follow each step of the protein expression protocol directly on the quartz as shown in Figure 4.5, where an element able to control temperature was also introduced.

4.3 Results

The quartz resonant frequencies and the D factor were measured at different phases for each NAPPA gene protocol to produce proteins, namely:

(i) in bare quartz;
(ii) after the printing of NAPPA chemistry;
(iii) after the blocking step, 1 hr on rocking shaker at room temperature or 4 °C, overnight in cold room with Super Block or milk. Rinse with Milli-Q water. Dry with filtered pure air. Apply HybriWell gasket to each slide. Use the wooden stick to rub the excess adhesive;
(iv) beginning of the protein expression through the heating protocol;
(v) after washing out the loosely bound materials and dry with pure air.

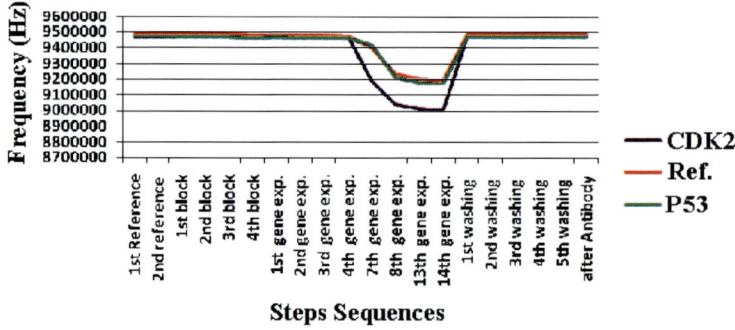

Figure 4.6. Behaviour of the resonance frequency as function of time, related to the various steps of the protein expression protocol.

As shown in Figure 4.6, our QCM apparatus is capable to measure the time variation of the resonance frequency entering Sauerbrey's equation. No frequency changes are detected in the various preparation steps, for both the reference and gene spots. It should be remarked that the reference spot contains the same biological materials as the other spots but the specific genes P53 and CDK2, separately spotted on different quartz crystals. Thus, as shown by the red curve in Figure 4.6, we observe mass density variations at the beginning of the protein expression procedure also for the reference spot. The P53 shows a time dependence slightly higher than in the reference, pointing out a limited protein production from this gene in the protocol conditions adopted. A much higher production of proteins is shown by CDK2 expression. All the three examined systems exhibit a saturation trend in the final stage of protein expressions and, at the end of this step, the recovery of the initial mass density after the crystal surface washing and drying. The expressed proteins are washed out almost completely, also in the reference system. More experimental details are reported (Eggenhöffner et al., 2009).

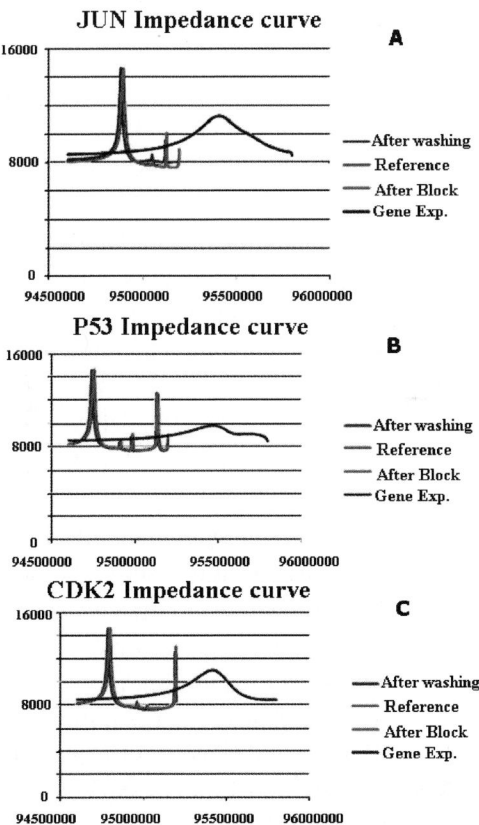

Figure 4.7. Dependence with frequency (Hz) of the impedance amplitude (a.u.) detected with the QCM-D in the D-factor measurement mode. Figures a,b,c are related to JUN, P53 and CDK2, respectively.

As discussed above, the Q quality factor is a dimensionless parameter that compares the time constant for decay of an oscillating physical system amplitude to its oscillation period. Since the Q factor the is inversely proportional to the dissipation D factor, the inspection of the resonance frequency variation and of the resonance shape curve provides qualitative insights on the formation of thin films of proteins onto surfaces in a liquid environment. The behaviour of the D factor obtained from impedance measurements for Reference, P53 and CdK2 samples during NAPPA experiment is reported in Figure 4.7 as function of the frequency. The NAPPA samples experience the same protocol as reported in the previous chapter. As in the conventional nanogravimetric experiments, significant frequency shifts are observed only in the gene expression step for all the three examined genes.

We observe in Figure 4.7a-c the resonance variation with frequency within the various stages of the treatment to promote JUN, P53 and CdK2 human gene expression. The initial resonance curve, the reference curve, after blocking curve and final curve obtained after washing expressed protein out of the quartz appear to coincide almost exactly. Very significant variation of both resonance frequencies and amplitude shape is obtained by our measurement at the end of gene expression lasting 20 minutes. The flattering of resonance curve signals the decrease of the quality factor of QCM resonance curve, resulting in a qualitative indication of the presence of proteins weakly linked to the quartz crystal. However, the positive shift observed systematically by the three protein expressions is clearly a very unexpected results. The dynamical system response is that at increasing the protein concentration due to the gene expressions the material would decrease the link to the quartz, resulting in an apparent loss of material. The weak link to the crystal is confirmed by the easily achieved removal of proteins from the crystal surface after washing. In these conditions, even biological material attached to the quartz before expression loose progressively the links and, at the end of the genes expression, the unexpected result is obtained.

The D factor results suggest that the protein expression process works efficiently for all proteins, whereas, on the bases of conventional nanogravimetry as shown in Figure 4.6 it would appear that CDK2 is produced in higher concentration, and that P53 is produced with lower efficiency. In another experiments on NAPPA we observed a similar behaviour of JUN and P53 on the bases of Sauerbrey's equation. On the contrary, JUN and CDK2 spots show similar results of the D factor in the expression process. P53 shows the larger dissipation effects as shown in Figure 4.7b.

5. Conclusions and Suggestions

Future activities are needed to complete the pilot study on Label Free NAPPA, namely to analyze samples printed with the same genes quartz each containing 16 spots 300 micron wide arranged as for MS. Considering that while Biacore (www.biacore.com) needs purified proteins to explore bindings, NAPPA might not need them if the following experiments will be carried out successfully with our label-free approaches.

Nanogravimetry can be employed to estimate the result with the known expected mass increase, knowing that the FOS binds interact with JUN but not with P53 and CdK2 gene:

(a) the kinetics of binding with our four frequency outlets (3 samples and 1 reference) for the three proteins expressed by the three single genes spotted;

(b) the kinetics of binding of FOS gene being first expressed in NAPPA and then transferred to the other three NAPPA genes along with lysate.

The (a) experiment will be carried out against a reference containing unexpressed NAPPA genes and the (b) experiment instead against the expressed FOS using "washed" lysate.

This label-free technique can be utilized as an "*a-priori* technique" useful to investigate the kinetic of an expressed protein for an optimization step before the NAPPA preparation: pH, temperature, reagent composition and concentration could be investigated before a massive production with NAPPA technology.

In conclusions, our review points out that conventional nanogravimetry is a well settled approach to study the protein expression in NAPPA, whereas the dissipation mechanisms in NAPPA at the stage of protein expressions or even in simpler gene systems deserve further investigations since our results have revealed essentially their very complex nature.

Acknowledgments

This project was supported by grants to Fondazione EL.B.A. by MIUR for "Funzionamento" and by a FIRB International Grant on Proteomics and Cell Cycle (RBIN04RXHS) from MIUR (Ministero dell'Istruzione, Università e Ricerca) to CIRSDNNOB-Nanoworld Institute of the University of Genova.

References

Bartolucci, S., Guagliardi, A., Pedone, E., Depascale, D., Cannio, R., Camardella, L., Rossi, M., Nicastro, G., Chiara, C., Facci, P., Mascetti, G. and Nicolini, C. (1997). Thioredoxin from Bacillus acidocaldarius: characterization, high-level expression in Escherichia coli and molecular modeling. *Biochemical Journal*, 328, pp. 277–285.

De Paul S. et al., (2001). *Measuring bound water in protein-resistant coatings*, Poster AVS Conference.

Eggenhöffner, R., Sallam, S., Bezerra, T., Adami, M., Pechkova, E. and Nicolini C. (2009). D factor and nanogravimetry on proteomics with a nanogravimeter of new design. *Proteins*, submitted.

Höök, F., Vörös, J., Rodahl, M., Kurrat, R., Böni, P., Ramsden, J. J., Textor, M., Spencer, N. D., Tengvall, P., Gold, J. and Kasemo, B. (2002). A comparative study of protein adsorption on titanium oxide surfaces using in situ ellipsometry, optical waveguide lightmode spectroscopy, and quartz crystal microbalance/dissipation. *Colloids and Surfaces B: Biointerfaces*, 24, pp. 155–170.

Koopmann, J. O. and Blackburn, J. (2003). *High* affinity capture surface for matrix-assisted laser desorption/ionisation compatible protein microarrays. *Rapid Communication in Mass Spectrometry*, 17, pp. 455–462.

Ramachandran, N., Hainsworth, E., Bhullar, B., Eisenstein, S., Rosen, B., Lau, A. Y., Walter, J. C. and LaBaer, J. (2004). Self-assembling protein microarrays. *Science*, 305, pp. 86–90.

Ramachandran, N., Larson, D. N., Stark, P. R., Hainsworth, E. and LaBaer, J. (2005). Emerging tools for real-time label-free detection of interactions on functionalprotein microarrays. *FEBS Journal*, 272, pp. 5412–5425.

Rodahl, M., Höök, F., Fredriksson, C., Keller, C. A., Krozer, A., Brzezinski, P., Voinova, M. and Kasemo B. (1997). Simultaneous frequency and dissipation factor QCM measurements of biomolecular adsorption and cell adhesion. *Faraday Discussions*, 107, pp. 229–246.

Shaffer, C., G & P Vol. 7, No. 10, October, 2007, pp. G6-G7.
Spera, R., Nicolini, C. (2008). *Nappa microarrays and mass spectrometry:new trends and challenges.* Nano Article - Taylor & Francis/CRC Press.
Voinova, M. V., Rodahl, M., Jonson, M. and Kasemo, B. (1999). Viscoelastic acoustic responses of layered polymer films at fluid-solid interfaces: continuum mechanics approach, *Physica Scripta*, 59, pp. 391–396

CHAPTER 5

LABEL-FREE NAPPA: ANODIC POROUS ALUMINA

Enrico Stura[1], Claudio Larosa[1], Tercio Bezerra Correia Terencio[1],
Eugenie Hainsworth[2], Niroshan Ramachandran[2],
Joshua LaBaer[2] and Claudio Nicolini[1]

[1]*Nanoworld Institute, CIRSDNNOB-University of Genova and Fondazione EL.B.A., Italy*
[2]*Institute of Proteomics, Harvard Medical School. Boston, MA*

Anodic porous alumina (APA) is a well known geometrically ordered inorganic oxide, honeycomb shaped, grown from ultra-pure aluminium by anodic oxidation. This material was used in this work as substrate for biological macromolecules hosting, specifically deposited to work using the NAPPA technology. After a surface characterization, intended to verify the compatibility of the micro-structured inorganic oxide with the NAPPA method, two detection approaches were tested, using fluorescence and electrochemically. In both cases the detection of the macromolecules were successful, even if a numerical evaluation still has to be performed.

5.1 Introduction

In order to optimize the properties of NAPPA microarray, Anodic porous alumina (APA) can be a useful and interesting technology. This insulating micro- or nanostructured material can be used in different ways, both for the traditional fluorescence approach to the protein array and also for a novel experimental method based on electrochemical measurements.

The first way of using APA takes advantage of its optical properties, in fact the high aspect ratio (depth/width ratio) of the pores makes this material a natural wave guide for any fluorescent molecule present on the bottom of the pores, avoiding crosstalk of many point–light sources too close. The second way, instead, uses the dielectric properties of Al_2O_3 as structure for the realization of an electrically anisotropic system; the electrochemical reactions occurring on the bottom of the well (caused by the interaction between the biological probe molecule and the test molecule) induce variation on the electrical response to alternating voltage signal, and they can be quantified by means of scanning electrodes moving on the surface of the array, placed in a proper solution (Stura *et al.*, 2007). The walls of the pore behave like insulator decoupling from the electrochemical events occurring few microns/millimeters away from the measurement place.

The second option is a label–free approach to the analysis of protein arrays, since no fluorescent/marked molecule is implied, while the first one still requires fluorescent molecules to spread the luminous signal through the APA wave guide. The potential of label–free approaches to complement and even to improve other detection technologies of proteins being expressed in NAPPA microarrays has never been higher (Ramachandran *et al.*, 2004; 2005, Spera *et al.*, 2008). In this manuscript we summarize the findings of an extended pilot study being carried out with APA. In the framework of the NAPPA project, APA experiments were performed in order to provide a structure able to host macromolecules in a better way than letting them free on a planar surface.

5.2 Materials and Methods

Until now, the synthesis process of Anodic Porous Alumina allows us to produce pores with diameter ranging from 300 nm to few microns, with a mean pore density of about 4×10^6 pores/mm^2. The work carried in our laboratories allowed us to define a preliminary protocol for the synthesis of APA, using a two step anodization process (an electropolishing step followed by the actual anodization). Here two extreme results are presented, showing AFM images of the anodized surface for the two different conditions of synthesis, keeping constant electrode distance D, acid concentration and aluminium purity. The porous form of alumina is now suitable for application to protein/nucleic acid deposition. Optimal pores width ranges between 300 and 700 nm, and depth, measurable using SEM or AFM, is estimated in the range of 300 nm to 200 μm according to the duration of the anodization process. The last step of our test consists in the spotting of a specific protein to verify the adhesion of the macromolecule to the nanostructured surface.

To obtain APA, different acids were tested, among which oxalic acid, phosphoric acid, and sulfuric acid, at different concentrations and voltage. After some preliminary tests, oxalic acid was chosen since the most regular structures were obtained using this reagent. The parameters that have to be monitored can be divided in two groups: The "primary" parameter group includes concentration of electrolyte, applied voltage, temperature, and the distance between the two electrodes inside the electrolytic cell. The secondary parameter group consists in current density, aluminium surface properties and cations present as impurity in the electrolyte. The latter group is made of values difficult to be measured, so for our procedures we only controlled the primary group.

The aim of this work is the production of highly ordered APA and pore diameters controlled optimizing the parameters of the synthesis. APA surface is intended for the immobilization of genes and proteins, so only a defined range in the geometry of the structured surface can be considered useful.

The set up used to synthesize APA consists in an electrolytic cell with an aluminium electrode and a platinum counter electrode. This cell is cooled by means of a spiral-shaped glass tube with cold water or refrigeration fluid continuously flowing across it, to keep the temperature of the solution as low as possible contrasting the raise of it due to the Joule dissipation phenomenon occurring during the anodization. Materials used for the synthesis were: high purity aluminium sheets (99.999%, 0.3 mm thick, Goodfellow), platinum sheets (99.9%, Goodfellow), oxalic acid and perchloric acid (Sigma), phosphoric acid (Fluka), ethanol (99.95%, Sigma), chromium trioxyde (Sigma) and high purity water (MilliQ system).

5.2.1 *Experimental details*

Aluminium sheets was used as initial material to grow APA, the sample surfaces were treated with a 1:4 (v:v) mixture of ethanol and perchloric acid for three minutes to roughly clean the surface from impurities. Then a first anodization process was run, in the cooling system and applying a current of 20 V using 1M oxalic acid as electrolytic solution. Aluminium sheet were connected as negative electrode and Platinum sheet as positive electrode. The reaction was continued for two hour with vigorous stirring maintained during anodization in order to keep temperature uniform on the electrode surface, avoiding preferential paths for the current due to temperature gradients.

The first anodization was applied for two hours, after this time the sample was rinsed with distilled water and treated with a mixture of phosphoric acid and chromium trioxyde (1:4 v:v) to remove the first aluminium oxide layer. After this removal, a second electrolysis process was run for 30 minutes using the same parameters described for the first one. The set up of the synthesis of APA was realized in a completely tunable way in order to study a numerical model allowing to control (at least roughly) the pore density, pore depth, wall thickness of APA. For this reason three series of experiments have been performed, to define the key parameters to be used in the reaction phase. At the end of each experiment series, a physical parameter was defined, chosen as the one that gave the best (most reliable, tunable and reproducible) result during all the current series of experiments. The following series used this fixed parameter, and varied the one that was to be found in the next series of experiments. Furthermore, to grow spots only on selected areas of the array, a surface photolithographic system was implemented. A mask has been printed on acetate transparent sheets and used to cover the aluminium layer protected by photoresist (poly(methyl methacrilate) (PMMA)). In this way, after the development in basic solution (NaOH 3M), a selectively unprotected layer of aluminium has been anodized, obtaining an ordered matrix of spots. Morphological parameters of anodic porous alumina, like mean pore diameter, pore density were determined by the image analysis of the AFM scanning, the porosity was obtained from the ratio:

$$\text{Porosity \%} = (S_{pore}/S_{wall}) \times 100$$

where S_{pore} is the surface area of pores and S_{wall} is the oxide geometric surface.

5.2.2 Mechanical tests for anodic porous alumina

Since APA provides a structure with very thin walls, its fragility must be taken in account handling the patterned surface.

Two tests were carried out to characterize mechanical properties of APA grown over aluminium: a grip test and a ball-crush test, both carried out in compliance with the DIN standards for mechanical evaluation (protocols DIN 4838 T 100, DIN 51097, DIN 51098 and DIN 51130).

5.2.2.1 *Grip test*

The grip test (Figure 5.1) performed on APA over aluminium consists in a measure of the mechanical force required to pull a pin, in contact with the analyzed surface forming an angle of 60°, across a linear path over the surface itself.

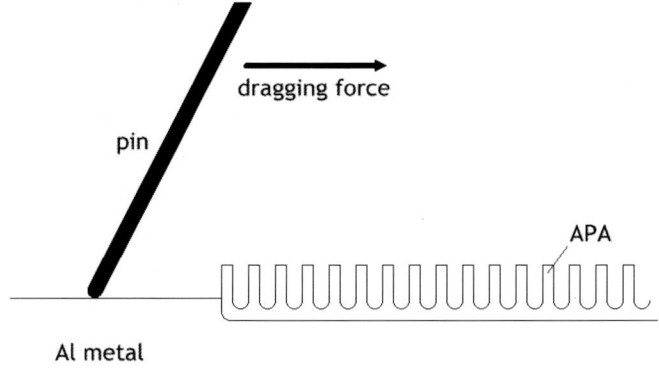

Figure 5.1. Setup for grip test.

The same test has been carried out with two test pins, with different tip diameters (10 and 60 µm). Results are shown in Figure 5.2. The drag path length was 15 mm, and exactly in the middle of the path the surface changed from aluminium metal to APA. It's clearly visible the increase of the friction factor, particularly for the thinner tip (this should be due to the more similar size of the tip and the APA pattern).

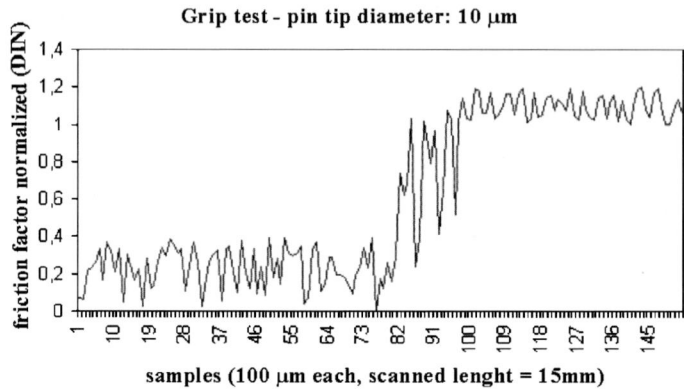

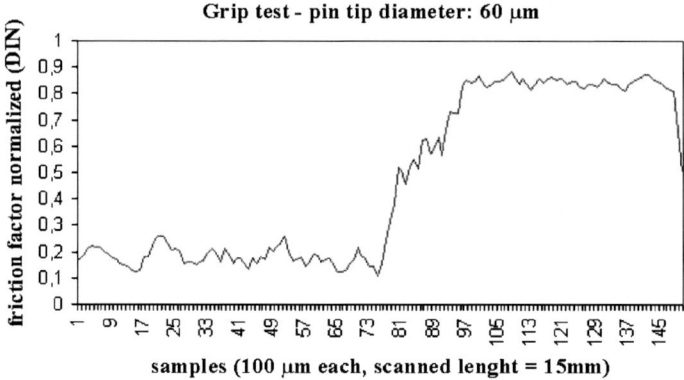

Figure 5.2. Result charts for grip test with different tip sizes.

This test proved the strongly increased grip of the APA surface with respect to the aluminium metal surface.

5.2.2.2 Ball-crush test

The second test that was performed to mechanically characterize APA surface was a ball-crush analysis. This test consists in throwing small steel balls, with controlled kinetic energy, with impact angles ranging in 135°, and then measure the loss in weight of the blasted sample. Two samples were analyzed: APA over aluminium and pure aluminium surface. Experiment data: energy = 2 J/ball, ball radius = 1 mm², 100 balls in 135° range, forced ventilation:

Mass loss of Al: 0.012%

Mass loss of APA over Al: 3%

Since the mass of the second sample (APA over aluminium) was estimated to be 95% Al and 5% APA, the result of this experiment proved that about 60% of APA was pulverized by impacting steel balls.

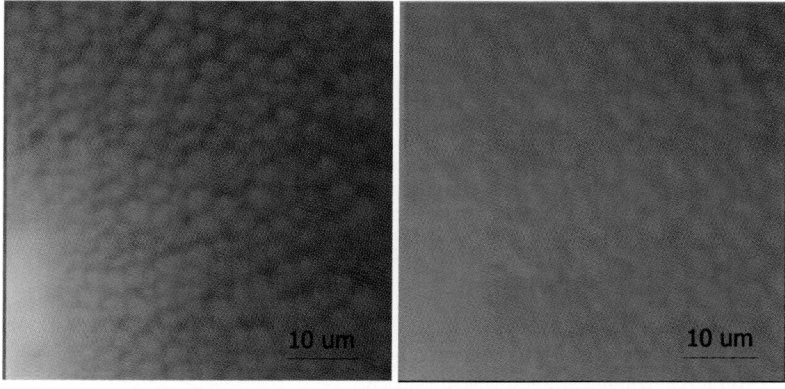

Figure 5.3. Fluorescent dye removal by forced rinsing.

After these two tests, the main consideration that can be done is that mechanically APA can grant a very good friction improvement (useful to attach over it any kind of compatible matter), but isn't applicable for direct mechanical stress (it must be handled with care).

5.2.2.3 *APA over glass slides and reversibility test*

The task of evaporating aluminium over glass has been accomplished avoiding its detachment during the anodization process, typical problem due to the incompatibility of cold borosilicate glass to the vapours of aluminium. This phenomenon was contrasted by means of a thin layer of chromium (deposited by sputtering) as medium element between glass and aluminium. Using the so obtained APA surface, a test was performed to verify the possibility to deposit a colored fluid in discrete spots (with diameter of 300 um).

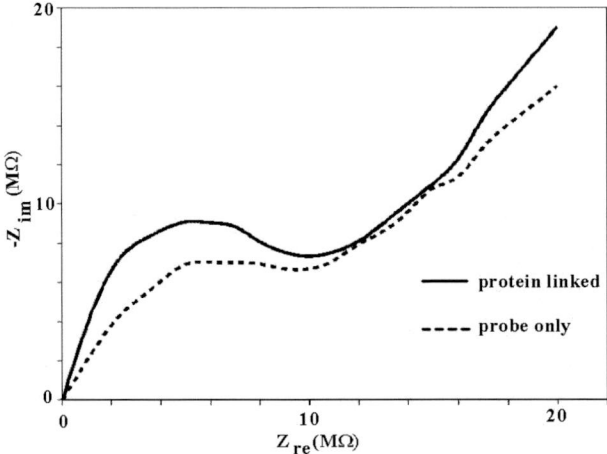

Figure 5.4. Impedance spectroscopy plots in two spots of APA surface, one with protein hybridized to the probe molecule and another with probe only. Frequency ranges from 1 Hz to 100 KHz, voltage applied was 400 mVpp.

This test was carried out using Cy3 fluorescent molecule on CCL20 protein in aqueous solution, spotting on APA surface by means of a Gilson micropipette, and after this operation the spots were able to be clearly detected, inspecting the sample using a fluorescence optical microscope.

The ability to rapidly exchange that fluid with a different fluid (*i.e.*, wash the first fluid away) was also demonstrated; after the experiment described above, the fluorescent molecules in solution were rinsed using MilliQ water, but part of them remained trapped (because of capillarity). Forcing the removal with stronger fluodinamic movements (pointing the flux of rinsing water to the surface of APA using a Hamilton syringe), it was possible to remove them. A series of CCD figures show and quantify what fraction of label remains in normal condition and how much after the optimal removal.

5.3 Results

5.3.1 *Electric impedance spectroscopy set up for NAPPA analysis*

A possible solution to take advantage of the APA technology for hosting biomolecules to have a label-free analysis is electric impedance spectroscopy (EIS). Similar approaches have been conducted for high density protein spotting on different surfaces (Koopmann and Blackburn, 2003), our task was to analyze the surface with relatively inexpensive instrumentation and methods. It's known that using EIS it's possible to detect different amounts of organic materials deposited even indirectly over conducing surfaces.

After the hybridization/expression experiment, using a scanning electrode controlled by a manipulator (MPC 200 by Sutter Technologies) via a PC, different EIS measures were performed in different spots on the surface of the array in phosphate buffered saline solution (PBS); the experimental setup is shown in Figure 5.5.

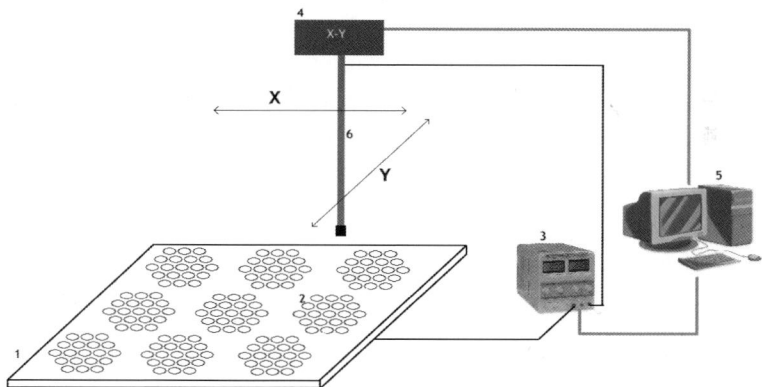

Figure 5.5. Set up (instrumentation) to analyze NAPPA elements using impedentiometric measurements: 1 – Aluminium substrate, serving also as counter electrode. 2 – APA spot, obtained by lithography, with biomolecules bound 3 – AC signal generator, controlled by PC. 4 – XY bidimensional actuator, controlled by PC, positioning the scanning electrode upon spots. 5 – PC, controlling bidimensional mover and AC signal generator. 6 – Scanning electrode, dipped in the solution containing NAPPA and buffer.

The walls of APA reduced the crosstalk effect, so the impedance parameters that have been measured were almost exclusively function of the single spot over which the scanning electrode was placed when the single measurement was carried out. The adsorption of the macromolecule to the surface was similar to what is obtained using specific substrates (Höök *et al.*, 2002; Rodahl *et al.*, 1997). In Figure 5.4 it is possible to compare EIS plots traced when the scanning electrodes were close (50 μm from the surface) to a test spot where the protein was expressed (straight line), and to another spot with the probe molecule only (dotted line). The two analysis spots were reciprocally distant 200 μm only, the experiments were carried out using frequencies ranging from 1 Hz to 100 KHz, applying 400 mV peak to peak voltage.

5.3.2 *Sponge effect and application to existing spotting systems*

The use of APA in biomolecular microarrays can be helpful for the features described above, but requires particular care in the positioning of the protein/nucleic acid probes. Since APA is an aluminium oxide structure, containing empty spaces and very thin walls, it's very easy to crush it impacting with spotting instrument. But thanks to the very high capillarity of this inorganic system, as soon as a wet object gets in contact, a "sponge effect" takes place, draining the solution inside the wells.

5.3.3 *AFM analysis of APA samples*

A topographical analysis of the samples obtained using different synthesis parameters was carried out by means of a customized AFM microscope (Elbatech Srl, Marciana, Italy). The results are shown in Figures 5.6–5.9, with synthesis details quoted.

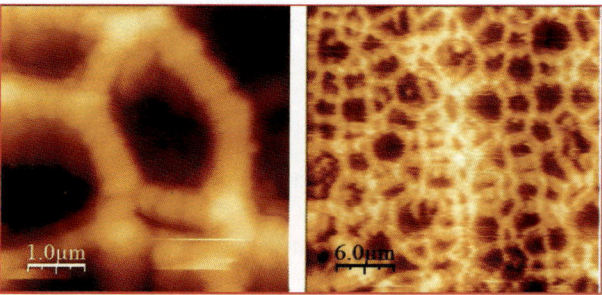

Figure 5.6. Aluminium purity 99.999%, acid concentration 0.5 M, reaction time 30', voltage 46 V, distance between two electrodes 1 cm.

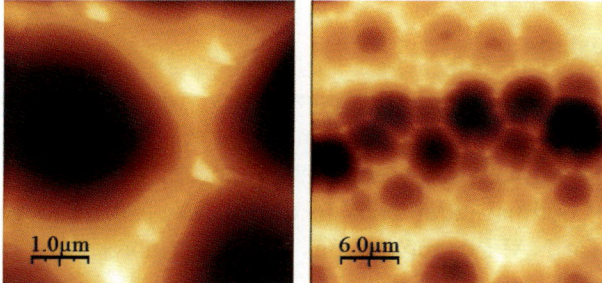

Figure 5.7. Aluminium purity 99.999%, acid concentration 0.5 M, reaction time 150', voltage 30 V, distance between two electrodes 1 cm.

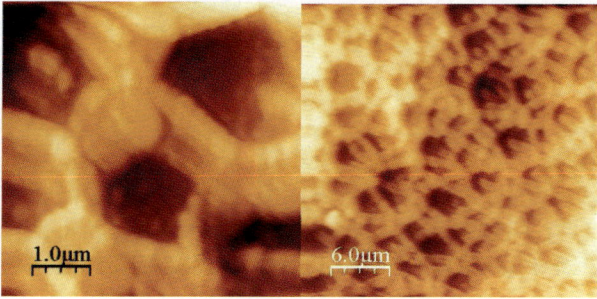

Figure 5.8. Aluminium purity 99.99%, acid concentration 1 M, reaction time 30', voltage 30V, distance between electrodes 2 cm.

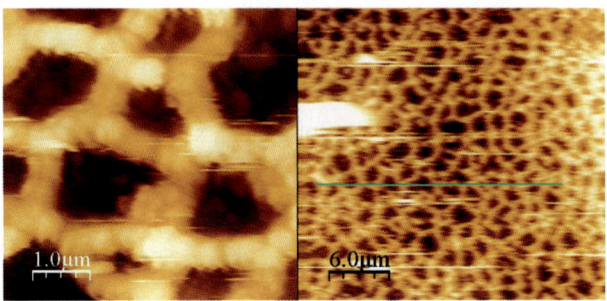

Figure 5.9. Aluminium purity 99.999%, acid concentration 1 M, reaction time 120', voltage 40 V, distance between electrodes 1 cm.

5.4 Conclusions

The influence of synthesis parameters (electrodes distance, acid concentration, time, voltage) in anodic porous alumina has here been defined and through a deterministic approach it appears possible to identify the optimal parameters to obtain pores of the desired diameter and depth for NAPPA applications either in fluorescence or Label Free using electrochemistry (Stura *et al.*, 2007).

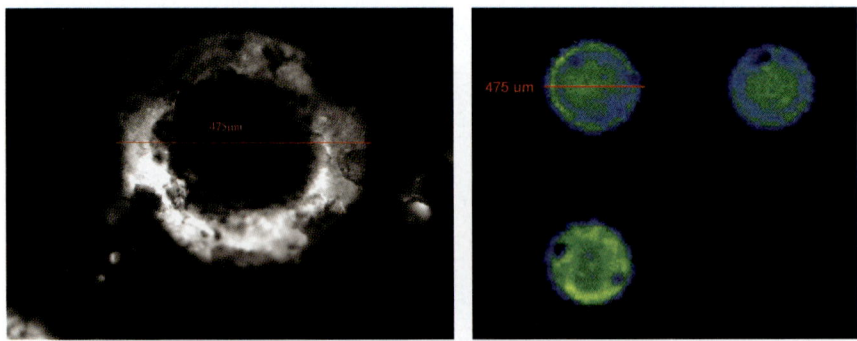

Figure 5.10. Single NAPPA fluorescence spot on APA (left) and on glass (Ramachandran *et al.*, 2005).

Preliminary NAPPA printing using APA patterned surface proves indeed encouraging with standard HIP configuration. The anodized aluminium surface, observed by CCD camera in fluorescence shows the occurrence of gene expression if treated with 150 ml of Cy3 and left to wait for 10 minutes (Figure 5.10).

The experiments show that the interface of the APA is resistant to the mechanical spotting process of the robot and the Cy3 fluorescence presents a diffuse pattern typical of NAPPA arrays. The conditions can be further optimized using piezoelectric technologies (Arrayjet, UK; Gesin, UK) or sound waves equipments (Labcyte, USA). These nanospotters should give optimal result, even if with less precision and requiring more process time with respect to the normally used mechanical spotters.

Acknowledgments

This project was supported by grants to Fondazione EL.B.A. by MIUR for "Funzionamento" and by a FIRB International Grant on Proteomics and Cell Cycle (RBIN04RXHS) from MIUR (Ministero dell'Istruzione, Università e Ricerca) to CIRSDNNOB-Nanoworld Institute of the University of Genova.

References

Höök, F., Vörös, J., Rodahl, M., Kurrat, R., Böni, P., Ramsden, J. J., Textor, M., Spencer, N. D., Tengvall, P., Gold, J. and Kasemo, B. (2002). A comparative study of protein adsorption on titanium oxide surfaces using in situ ellipsometry, optical waveguide lightmode spectroscopy, and quartz crystal microbalance/dissipation. *Colloids and Surfaces B: Biointerfaces*, 24, pp. 155–170.

Koopmann, J. O. and Blackburn J. (2003). High affinity capture surface for matrix-assisted laser desorption/ionisation compatible protein microarrays. *Rapid Communication in Mass Spectrometry*, 17, pp. 455–462.

Ramachandran, N., Hainsworth, E., Bhullar, B., Eisenstein, S., Rosen, B., Lau, A. Y., Walter, J. C. and LaBaer, J. (2004). Self-assembling protein microarrays. *Science*, 305, pp. 86–90.

Ramachandran, N., Larson, D. N., Stark, P. R., Hainsworth, E. and LaBaer, J. (2005). Emerging tools for real-time label-free detection of interactions on functionalprotein microarrays. *FEBS Journal*, 272, pp. 5412–5425.

Rodahl, M., Höök, F., Fredriksson, C., Keller, C. A., Krozer, A., Brzezinski, P., Voinova, M. and Kasemo, B. (1997). Simultaneous frequency and dissipation factor QCM measurements of biomolecular adsorption and cell adhesion. *Faraday Discussions*, 107, pp. 229–246.

Spera, R. and Nicolini, C. (2008). *Nappa microarrays and mass spectrometry: new trends and challenges*. Nano Article - Taylor & Francis/CRC Press.

Stura, E., Bruzzese, D., Valerio, F., Grasso, V., Perlo, P. and Nicolini, C. (2007). Anodic porous alumina as mechanical stability enhancer for LDL-cholesterol sensitive electrodes, *Biosensors and Bioelectronics*, 23, pp. 655–660.

CHAPTER 6

LABEL FREE DETECTION OF NAPPA VIA ATOMIC FORCE MICROSCOPY

Marco Sartore[1], Roberto Eggenhöffner[1], Tercio Bezerra Correia Terencio[1], Enrico Stura[1], Eugenie Hainsworth[2], Joshua LaBaer[2] and Claudio Nicolini[1]

[1]*CIRSDNNOB University of Genova and Fondazione EL.B.A., Nanoworld Institute, Genova (Italy)*
[2]*Harvard University, Institute of Proteomics, Cambridge USA*

Modern proteomics requires protein properties and functionality to be studied globally on arrays of a large number of elements from which high throughput data are accessible by a variety of biophysical and biochemical tools. Structures of organized molecules on arrays on suitable substrates allow the required fast characterization of their properties and interactions. Atomic force microscopy is a mature technology for the characterization of surface morphology with a potential high impact: applications to protein technology are reviewed to assess if such intrinsically nanoscopic technique can be regarded as a possible candidate to attain high-throughput protein production data from NAPPA arrays. The morphology and the profiles obtained at the highest resolution before and after the expression of three JUN, CdK2 and P53 proteins are reported and discussed in terms of these specific interests for future developments from the actual micro towards nano-scale ordered arrays. Further, by measuring interaction forces with functionalized AFM tips, the presence of specific final products can be detected among imaged objects.

6.1 Introduction

The potential of label-free approaches to complement and even to improve other detection technologies has never been higher for Nucleic Acid Programmable Protein Array (NAPPA) (Ramachandran *et al.*,

2004, 2005). Among other techniques, the Atomic Force Microscopy (AFM) appears a challenging tool (Spera and Nicolini 2008; Koopmann and Blackburn, 2003) to study in deeper details the NAPPA cover slip surface: a preliminary characterization was carried out within this framework, keeping in mind that, at present, optimal high-throughput measurements can be achieved by means of AFM with experimental complications as compared to other techniques.

Two approaches can be used: imaging and measuring interaction forces. Essentially, the former is useful to get images of the substrate at the various stages of the experiment as here successfully shown, whereas the latter can be used to detect the presence of final products among the "objects" found by imaging the surface.

The functionalized tips are produced attaching a capture molecule to the end of the cantilever in order to detect binding events. Both techniques are not new *per se*, as reported by many pioneering companies in cantilever technology, but the application to monitor label free protein expression by NAPPA is actually quite new, more effective and less invasive with respect to standard fluorescence technology (Rodahl *et al.*, 1997; Voinova *et al.*, 1999).

6.2 Materials and Methods

The extended study was carried out on NAPPA realized with 400 genes per 300 µm spot, using the following genes and corresponding expressed proteins of the indicated molecular weight:

(i) CdK2 human gene of 897 bp and 592 kDa protein;
(ii) JUN human gene of 996 bp and 657,36 kDa protein;
(iii) P53 human gene of 1182 bp and 7801,12 kDa protein.

The geometric characteristics and the molecular arrangements of the above proteins are given below in Figure 6.1.

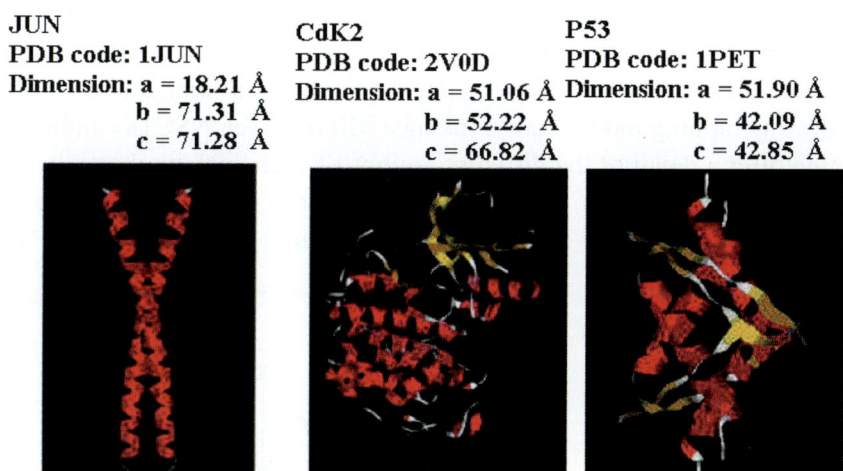

Figure 6.1. Structural details and PDB structures of the proteins investigated in the present work by AFM.

6.2.1 Technical details for NAPPA imaging by AFM

A suitable geometry onto cover slips was developed to host mica samples compatible with the NAPPA printing machine and in order to prepare samples to be imaged by AFM (atomic force microscopy).

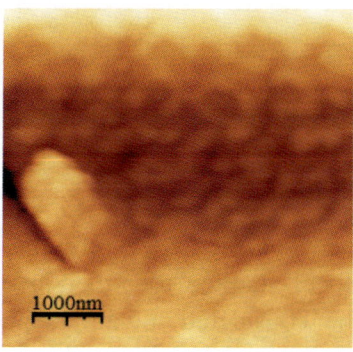

Figure 6.2. Typical topography of the vector spot.

These samples were printed with four 300 µm spots per sample each containing 0.2 ng, following a squared scheme placing each gene and the vector at each corner. These samples have been imaged by AFM, collecting tapping mode images (using a MikroMasch NSC18 cantilever) which offer a detail of the different regions of each spot, as shown by the vector spot imaged in Figure 6.2. A uniform deposition of aligned elements with length 1–2 µm and of intercalated rings of diameter 500 nm is observed.

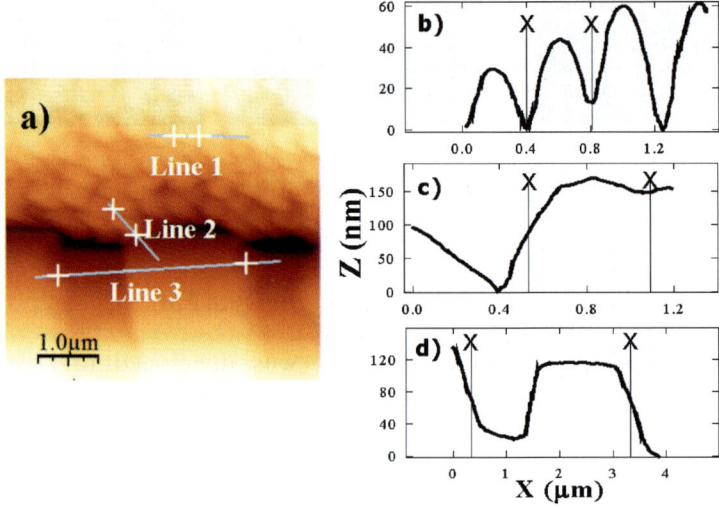

Figure 6.3a–d. 2D AFM image of microspheres in comparisons with TGZ02 grid and profiles along three directions as indicated in Figure a: among microspheres (line 1, Figure b), along the boundary between TGZ02 and microspheres (line 2, Figure c) and along TGZ02 (line 3, Figure d). The latter, 3 µm wide, gives the standard calibration step height of 100 nm.

The images obtained for the single proteins expressed by genes through the same protocol adopted for other investigations (Spera and Nicolini, 2008) can be considered a valid support to the chemistry involved in the sample preparation and processing. The AFM measurements were calibrated by polystyrene micro spheres of known size in comparisons with known TGZ02 grid size (Figure 6.3) and with the unknown proteins size (Figure 6.4). In the latter, the expressed CdK2 protein spot images by AFM in 2D (left) and 3D (right) are reported. To investigate the interplay between protein and micro spheres, profiles along lines 1 and 2 are reported for CdK2 and micro spheres regions, respectively. The profile along line 3 crossing CdK2 and the micro spheres regions is also reported.

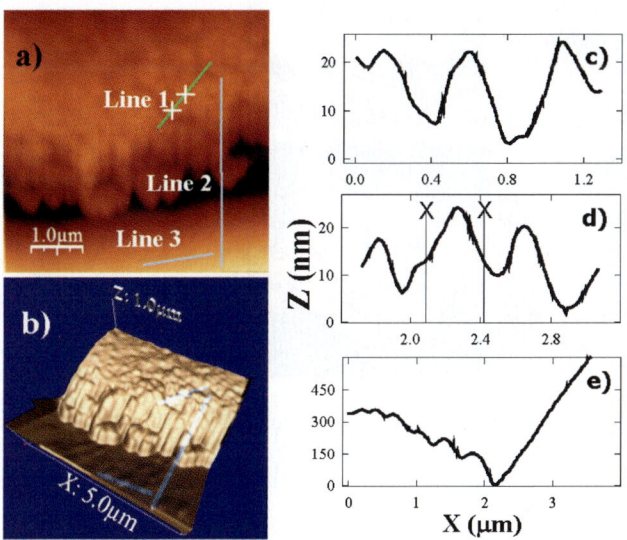

Figure 6.4. AFM images of microspheres deposited on NAPPA spot containing the expressed CdK2 protein in 2D (a) and 3D (b). The profile along line 1 in the CdK2 area is reported in (c), the profile along line 2 in the microsphere area in (d) and the profile along line 3 crossing the CdK2 and the microsphere area in (e). Profile along line 3 crosses CdK2 and microspheres regions.

6.3 Results

A thorough statistical analysis based on the AFM images of the NAPPA genes spots containing JUN, CdK2 and P53, respectively, either before (Figure 6.5–6.7 a, b) and after (Figure 6.5–6.7 c, d) protein expression was undertaken. Results are actually supported by the quantitative surface analysis from the calibrations shown in Figures 6.3 and 6.4 discussed above. 2D (Figure 6.5–6.7 a, c) and 3D (Figure 6.5–6.7 b, d)

AFM results are reported with profile analyses along thick lines shown in the 2D images.

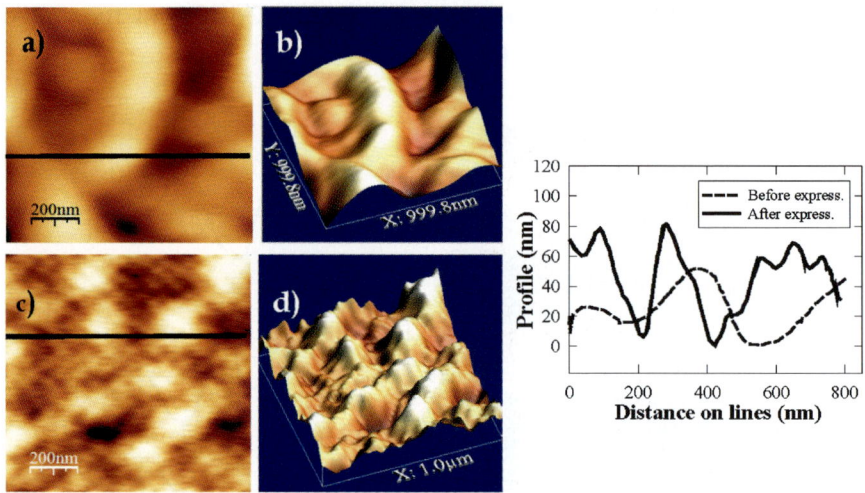

Figure 6.5. AFM topography of a $1 \times 1~\mu m^2$ before (a: in 2D and b: in 3D) and after JUN protein expression (c: in 2D and d: in 3D). In the graph, solid and dashed lines are obtained along the thick black lines shown on 2D images (a) and (c) after and before the protein expression, respectively.

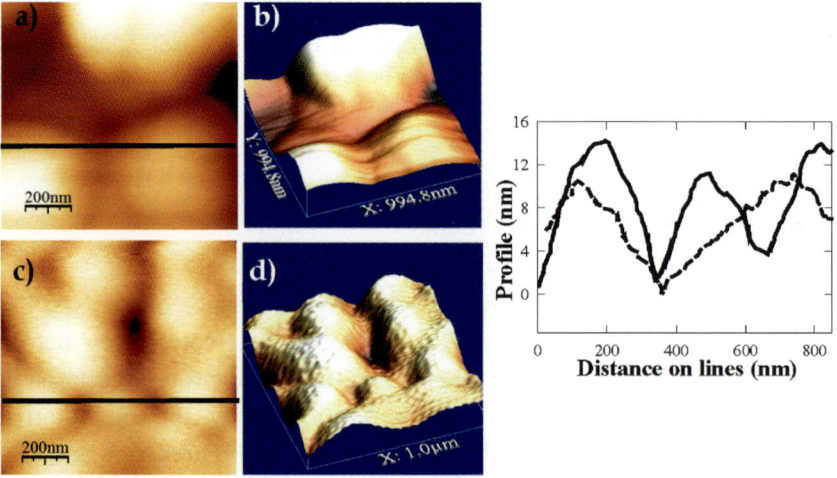

Figure 6.6. Same as in Figure 6.5, in the present figure for the CdK2 protein.

Each of the protein spots was analyzed with the highest resolution and with higher details at the highest resolution with their cross sections along the lines indicated in the Figures. Temperature was kept at 18 °C with cold room refrigerator to preserve protein function.

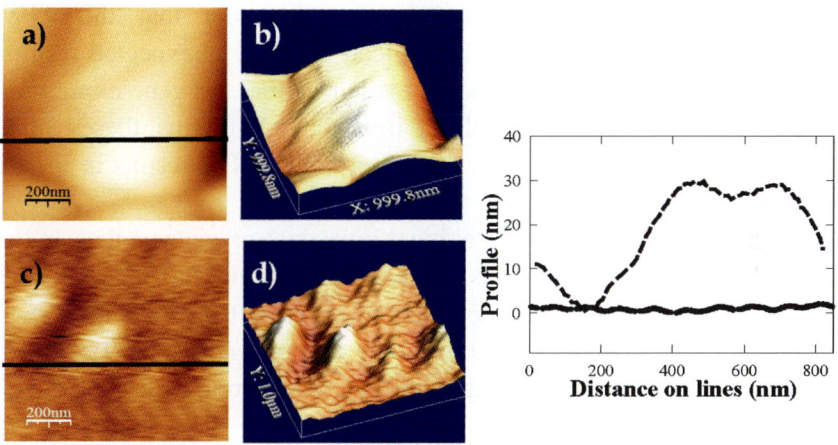

Figure 6.7. Same as Figure 6.5, in the present figure for the P53 protein.

As shown in Figure 6.5 for JUN protein, the topographical result of the protein film after the expression with respect to the spots before the expression protocol is traced by the decreased space period of the surface corrugation and the occurrence of chains longer than a few micron with narrow peaks along the lines that are of much higher amplitude after the expression.

Figure 6.6, related to CdK2, shows that the protein growth gives rise to corrugations with a periodic behavior similar to Figure 6.5 but with comparable profile amplitude before and after protein expression. The CdK2 expression appears of lower intensity with respect to JUN protein. Smoother peaks still develop on the protein film surface as in Figure 6.5 for JUN. As shown by the profiles in Figures 6.5 and 6.6 the space periods between two peaks are of the order of 200 nm.

In Figure 6.7c,d for the P53 protein at the end of the expression, the region in the AFM image shows a depleted profile amplitude with respect to the same surface observed before the expression (Figures 6.7a,b). For this protein, smooth peaks of small profile amplitude are obtained over lines of investigations: as shown in Figure 6.5d, the space period is close to 200 nm, similarly to JUN and CdK2, but the profile intensities are much lower.

The comparison of Figures 6.5÷6.7 points out that the AFM investigations at the highest resolution lead to a few unexpected results. Namely, we have strong independent evidences that the gene expression protocol can efficiently express all the investigated proteins: these results are confirmed by a lot of statistics of observations. The results in Figure 6.7 conflict with this expectation for the P53 protein.

These results from the topographical investigation suggest that, in the study of NAPPA protein surfaces, the high resolution AFM imaging over 1×1 μm^2 areas needs to be implemented by the AFM imaging over to the whole spot size, which is 300 μm wide (Bezerra *et al.*, 2009). A such wide area exploration can be achieved by AFM through the collage of 50×50 μm^2 areas.

The imaging over selected intermediate area of 5×5 μm^2, not reported in the present work, was performed for all our NAPPA samples without significant differences in the topographycal results with respect to Figures 6.5–6.7 above.

6.3.1 *NAPPA investigations with the functionalization of cantilever and the measurement of interaction forces*

In order to go a step over and start real label-free sensing by AFM, some preliminary experiments on the protein-protein binding has been performed with the AFM (Elbatech Srl, Marciana, Italy), in order to calibrate the system. In this case, a proper tip functionalization process for the proteins of interest must be carried out, with dedicated experiments to find out the best operating conditions both for the sample preparations and for the AFM-side optimization.

Cantilever bending is related to the change in surface stress, and the change in resonant frequency is related to the change in mass.

The readout is accomplished by optical method with the usual laser diodes. The detection of dual properties is an advantage of cantilever technology, because all the other label–free technologies are all a function more or less of a change in mass.

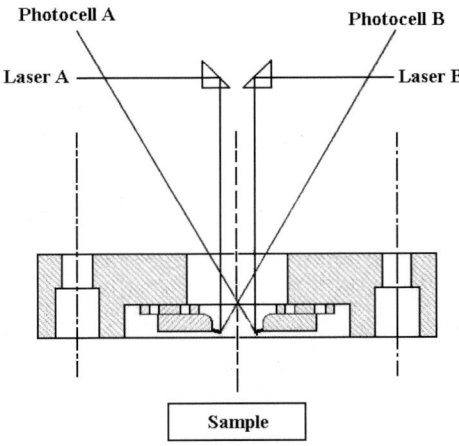

Figure 6.8. Assembly of a 2-cantilever AFM measuring head.

Cantilevers are capable to change surface stress related to surface energy conditions, a route prevented to other techniques. In a measurements of surface stress and resonant frequency variation, one is studying the change of energy per binding event.

At present, this is far to be achieved: however, a comprehensive approach should forecast to sustain the interaction at more than one cantilever, thus exploiting the benefit of parallel measurements.

A given binding test might be repeated at more sites, or to set differently–functionalized tips to detect presence of different proteins, or to build a working — and reference–pair. In any case, the first step to pursue is to design a measuring head capable of handling two or more cantilevers concurrently. This step was accomplished and Figure 6.8 shows a block assembly of the head designed to this purpose.

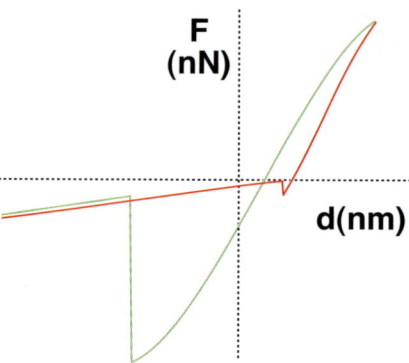

Figure 6.9. Typical force-distance curve acquired on a mica sample with an un-coated cantilever. The red line is the approach scan, the green line is the retract scan. The same measurement can be applied to protein samples.

One possible further development towards an AFM-based label-free detection scheme is the use of Kelvin-Probe, a variation of the AFM that measures the electrostatic forces (potential) between the probe tip and the surface of a material (Figure 6.9). The conducting tip and the sample are characterized by (in general) different work functions. This measurement is carried out by a two step technique. During the first step, a line scan, the sample topography is imaged by normal tapping mode. In the second step, the probe is raised above the sample by a fixed distance (10 nm for example) and the surface potential is measured, following the topography previously recorded to maintain a constant tip-sample distance. In order to achieve single protein resolution, this technique needs a very sharp tip.

Thus, to pursue this method a tip of particularly high aspect ratio must be developed, for example with the aid of carbon nanotubes.

6.4 Conclusions and Advancements

Results of the present work confirm the reliable performances of AFM in investigating properties of Label Free NAPPA, *i.e.*, to determine the surface morphology of the three expressed model proteins. Future activities as printing 16 spots 300 µm wide of the same genes on mica surface for improved AFM imaging are in progress. AFM might take advantage in future of the reduced spot dimensions that require to explore nowadays the full 300 µm wide spot size although at low resolution.

The approach of printing 16 spots of the same genes 300 µm wide on mica surface was exploited also for surface plasmon resonance experiments (Ramachandran et al., 2005 and mass spectrometry (Spera and Nicolini, 2008) to study protein interactions in the three different NAPPA genes and to determine the fraction of proteins appearing as monomer, dimer and tetramer. Knowing that JUN forms about 10% dimer, P53 about 50% tetramer and CK2 single monomer only, AFM could become a quite efficient and fast tool.

Further, considering that Biocore instrumentation (www.biocore.com) needs purified proteins to explore bindings NAPPA through plasmon resonance, in our NAPPA approach this stage might not be required and our label-free approach based on AFM experiments might be carried out successfully.

Acknowledgments

This project was supported by grants to Fondazione EL.B.A. by MIUR for "Funzionamento" and by a FIRB International Grant on Proteomics and Cell Cycle (RBIN04RXHS) from MIUR (Ministero dell'Istruzione, Università e Ricerca) to CIRSDNNOB-Nanoworld Institute of the University of Genova.

References

Bezerra Correia Terencio, T., Eggenhöffner, R., Sartore, M. and Nicolini, C., (2009). A combined experimental study of Nucleic Acid Programmable Protein array by Atomic force microscopy, Fluorescence and Nanogravimetry. *Nanotechnology*, submitted.

Koopmann, J. O. and Blackburn, J. (2003). High affinity capture surface for matrix-assisted laser desorption/ionisation compatible protein microarrays. *Rapid Communication in Mass Spectrometry*, 17, pp. 455–462.

Rodahl, M., Höök, F., Fredriksson, C., Keller, C. A., Krozer, A., Brzezinski, P., Voinova, M. and Kasemo, B. (1997). Simultaneous frequency and dissipation factor QCM measurements of biomolecular adsorption and cell adhesion. *Faraday Discussions*, 107, pp. 229–246.

Voinova, M. V., Rodahl, M., Jonson M. and Kasemo, B. (1999). Viscoelastic acoustic responses of layered polymer films at fluid-solid interfaces: continuum mechanics approach. *Physica Scripta*, 59, pp. 391–396.

Ramachandran, N., Hainsworth, E., Bhullar, B., Eisenstein, S., Rosen, B., Lau, A. Y., Walter, J. C. and LaBaer, J. (2004). Self-assembling protein microarrays. *Science*, 305, pp. 86–90.

Ramachandran, N., Larson, D. N., Stark, P. R., Hainsworth, E. and LaBaer, J. (2005). Emerging tools for real-time label-free detection of interactions on functional protein microarrays. *FEBS Journal*, 272, pp. 5412–5425.

Spera, R. and Nicolini, C. (2008). *Nappa microarrays and mass spectrometry:new trends and challenges.* Nano Article - Taylor & Francis/CRC Press.

CHAPTER 7

CELL FREE EXPRESSION AND APA FOR NAPPA AND PROTEIN NANOCRYSTALLOGRAPHY

Eugenia Pechkova[1], Shaorong Chong[2], Shailesh Tripathi[1] and Claudio Nicolini[1]

[1]*CIRSDNNOB University of Genova and Fondazione EL.B.A., Nanoworld Institute, Genova (Italy)*
[2]*New England BioLabs, Massachussetts, USA*

This review describes a study of nanoporous APA (anodic porous alumina) substrate combined with the cell-free protein expression system and explore its application to functional NAPPA-based proteomics and to protein nanocrystallography. Two different photolithographic microstructuring techniques are utilized for the ordered nanopore arrays. The first process uses a positive resist, the second uses a negative one and has hydrophobic properties which increase specificity to biomolecular linking. Nanoporous alumina is formed at Nanoworld Institute by anodic process and yields straight holes with high aspect ratio. Similarly the recent advances in cell free expression system developed in New England BioLabs will be introduced for its final implementation in NAPPA technology. The utilization of both new technologies offers several advantages over conventional supports, making them very attractive to utilize for biomolecular samples in functional proteomics and in protein nanocrystallography applications, as discussed in this review.

7.1 Introduction

The present review concerns the use of anodized porous alumina (hereafter APA) as a substrate for two distinct applications, namely DNA microarray and protein nanocrystallography. As shown in these two different biomedical applications, the APA material could offer several advantages over conventional supports: high surface area enlargement, improved microfluidic properties, easy and cheap manufacturability, flexibility in porous dimension and confinement effect as a mean for selectivity and maximization of light emission. The most common problems still remaining on planar microarray supports is spot concentration and aggregation, which leads to non-uniform intensity profiles; in that respect porous materials with improved microfluidic properties can be a solution. From the other hand, the high throughput protein crystallography, including this using nanovolumes, allows to significantly reduce amount of protein used for the crystallization trial. Generally, crystallization on the porous support requires even smaller amount of protein in comparison of classical in-volume crystallization (Chayen et al., 2003).This principle can be applied for the APA-cell free combined method, but with much more influence on the crystallization process itself.

7.2 APA and NAPPA Microarray

Oligonucleotide and protein microarrays are currently intensively investigated for a broad range of applications in biomedical diagnostics, to simultaneously determine several parameters in individual samples from a limited amount of material (Templin et al., 2002). On the other hand, the fabrication of nanochannel-array materials has attracted considerable scientific and commercial attention.

Moreover, the application of ordered nanochannel arrays as two-dimensional photonic crystals has generated increasing interest in recent years. Among the applications for these materials are the inhibition or enhancement of spontaneous emission and the fabrication of tunable optical filters and microcavities (Foresi et al., 1997) — waveguides (Yablonovitch, 1987; See et al., 1995) — catalytic combustion (Suzuki et al., 2003). Anodic porous alumina, which has been studied extensively over the last five decades (See et al., 1953; Czokan 1980; Thompson and Wood 1983), has recently been reported to be a typical self-ordered nanochannel material (Masuda and Fukuda, 1995; Masuda et al., 1997). Self-organization during pore growth, leading to a densely packed hexagonal pore structure, has been reported in oxalic, sulfuric, and phosphoric acid solution (Jessensky et al., 1998).

The interpore distances of the regularly ordered pore arrangement have been extended to the large range of 50–420 nm (Li *et al.*, 1998, 1999). Masuda and Satoh first reported a two-step fabrication method for straight nanoholes in a thin membrane of alumina (Masuda and Satoh, 1996).

The structural characteristics of the ordered porous alumina make it not only a perfect template material for the fabrication of nanoscale structures, but also an outstanding candidate material for two-dimensional (2D) photonic crystals, which may show photonic band gaps that are adjustable in the visible to ultra-violet spectral region.

7.3 Background of Cell-Free Protein Synthesis

Since the early pioneering work of Nirenberg and Matthaei in 1961 (Nirenberg and Matthaei, 1961), which demonstrated *in vitro* protein translation using cell extracts, cell-free protein synthesis has become an important tool for molecular biologists and played a central role in a wide variety of applications (Katzen *et al.*, 2005). In the post-genomic era, cell–free protein synthesis is becoming one of the most important high throughput technologies for functional genomics and proteomics.

The biggest advantage, compared to protein production in living cells, is that cell-free protein synthesis is the quickest way to obtain an expressed phenotype (protein) from a genotype (gene). From a PCR or plasmid template, *in vitro* protein synthesis and functional assays can be carried out in a few hours. Moreover, it is independent of host cells. Proteins that are toxic or prone to proteolytic degradation can be readily prepared *in vitro*.

Commercially available cell-free protein synthesis systems are typically derived from cell extracts of *Escherichia coli* S30, rabbit reticulocytes or wheat germ. The drawback of any extract-based systems is that they often contain nonspecific nucleases and proteases that adversely affect protein synthesis. In addition, the cell extract is like a "black box" in which numerous uncharacterized activities may modify or interfere with the downstream assays.

7.4 APA-Cell Free and Protein Nanocrystallography

An other application being here pursued is protein crystallization, which has been and is being intensively studied (Pechkova and Nicolini, 2004). One of the main reasons here, making these studies so rewarding, is the fact that proteins lie in root of the live. Nature has been particularly kind in allowing discovering structure to biological function relationship for many proteins. Relatively big and highly perfect crystals are indispensable in performing structure determinations. Recently, the art and science in growing protein crystals profits substantially from multidisciplinary research teams involved in the investigations. Despite the remarkable progress in many aspects, still there is a lack of complete understanding. By changing the solution conditions, the crystallographer tries to exceed the solubility limit of the protein so as to produce crystals (McPherson, 1999). This plan rarely runs smoothly. After changing the solution conditions, one of several difficulties is usually encountered: (i) nothing happens, *i.e.*, the protein solution remains homogeneous; (ii) a new phase appears, but it is not a crystal. Instead, it is an aggregate or a liquid; or (iii) crystals do form, but they are unsuitable for structure determination because they give a poor X-ray diffraction pattern, been easily destroyed by radiation. It is often possible to overcome these difficulties by trial and error — repeated crystallization attempts with many different conditions — but this strategy does not always work. Even when it is successful, the lessons learned cannot be easily generalized; the conditions which work with one protein are not necessarily optimal for a different protein (Asherie 2004). For this reason, the discover of general rules and universal method for protein crystallization is needed. Indeed in our hands the nanotechnology-based protein crystallization method demonstrates that use the homologous protein nanotemplate consisting of protein LB monolayer(s) can stimulate protein nucleation and accelerate crystallization of any protein (Pechkova and Nicolini, 2003; 2004). Moreover, the protein crystals obtained by this method appear to be more stable to synchrotron radiation (Pechkova *et al.*, 2004).

In the experiments here reported in order to influence the nucleation and crystallization process for obtaining reproducible and stable results, we use the APA template. The hen-egg white lysozyme (HEWL) has been chosen as a model protein. The main idea of the innovative crystallization method here described is demonstrated in Figure 7.1.

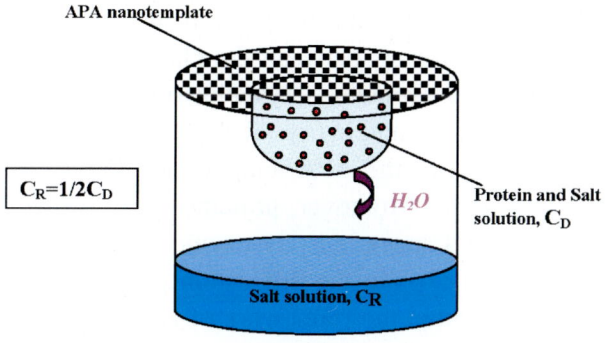

Figure 7.1. Protein crystallization method based on APA nanotemplate.

APA template created by means above technology (namely process II) is deposited on a solid glass support to be placed in the appropriate vapour diffusion apparatus. APA assumes the role of the template for protein nucleation and crystal growth. The details connected with the heterogeneous nucleation of protein crystals are still unclear (Nanev and Tsekova, 2000) despite many materials including mica, different inorganic minerals, polymeric film and amino acid-modified polymer surfaces (McPherson and Schlichta, 1988, Fermani *et al.*, 2001; Tang *et al.*, 2005) have been used for investigation of this phenomenon. The APA structure can act in protein crystallization also as porous glass (Rong *et al.*, 2004) or porous silicon surface (Chayen *et al.*, 2001), but being much more ordered (in fact, nanostructured) can produce more reliable results. In our hands, the protein LB protein film, deposited on APA surface appear to become more ordered and can supply better organized aggregates to protein solution, trigging the protein nucleation already in under saturated solution. During the screening procedure the following parameters can be varied: protein monolayer surface pressure, precipitant nature and concentration, number of the protein thin films monolayers, APA pore diameter. Temperature variation can be also used in order to control protein crystal nucleation and growth. To develop the above method we have elected the traditional hanging drop vapour diffusion method, because this protein crystallization technique is easy to perform, require quite small amounts of samples and permits easy variation of physical parameters during crystallization.

Interestingly the differences between the classical handing drop and APA-template modified protein crystallization method point to differences in nuclei number and in crystal shape (Pechkova *et al.*, 2009), pointing on the influence of the APA template to the overall nucleation and crystallization process. Moreover, proteins appear to be crystallized differently in constrained space, *e.g.*, lysozyme (Tsekova 2009). In the present review, we also explore the diffraction properties of the crystal grown in nanovolume within the framework of EMBO-NWI cooperation (Pechkova, Marquis, Cipriani and Nicolini, in preparation).

The possibility to modify the other crystallization techniques cannot be excluded, for example the sitting drop or sandwich drop vapour diffusion, micro-dialysis or crystallization under-oil.

7.5 Materials and Methods

7.5.1 *The PURExpress system and its advantages*

A complete *in vitro* reconstitution of protein translation from *E. coli* was accomplished in 2001 in Dr. T. Ueda's lab: the "PURE" system, for "Protein synthesis Using Recombinant Elements" (Shimizu *et al.*, 2001). Except for the ribosomes and tRNAs, which are highly purified from *E. coli*, the PURE system reconstitutes the *E. coli* translation machinery with fully recombinant proteins. These include 10 translation factors (IF1, IF2, IF3, EF-Tu, EF-Ts, EF-G, RF1, RF2, RF3, RRF), 20 aminoacyl-tRNA synthetases and several enzymes for energy regeneration. In addition, recombinant T7 RNA polymerase is used to couple transcription to translation. The PURE system represents an important step towards a totally defined *in vitro* transcription/translation system, thus avoiding the "black box" nature of the cell extract. The immediate advantage is the significantly reduced level of all contaminating activities. The PURE system has the capacity for a yield of more than 100 µg/ml, was first commercialised by the Post Genome Institute (PGI) (Tokyo, Japan), and then exclusively licensed to New England Biolabs (Ipswich, MA, USA) under the trade-name "PURExpress".

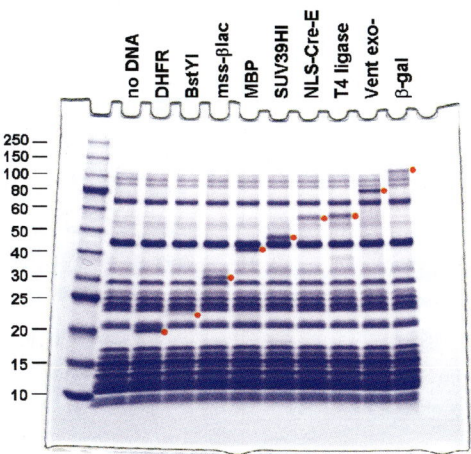

Figure 7.2. *In vitro* translation of a diverse set of target proteins using PURExpress system. The reactions were assembled for each of the target proteins in a final volume of 25 µl. The reactions contained 250 ng of plasmid DNA template, 20 units RNase Inhibitor and were incubated at 37 °C for 2 hours, then placed on ice. Aliquots (2.5 µl) were diluted 5-fold with H$_2$O, mixed with 3X SDS loading buffer, boiled for 3 minutes, resolved by SDS-PAGE with protein ladder (kDa, on the left) on a 10–20% Tris-glycine gel, and stained with Coomassie blue.

The PURExpress system is more robust and convenient than most extract-based systems for many *in vitro* applications. The activity of the synthesized protein can often be directly assayed without purification due to the low background activity of the translation mixture. The high purity of the PURExpress system even allows it to withstand more than 5 freeze-thaw cycles without losing its efficiency, further extending its shelf life (E. Cantor, unpublished observation). Figure 7.2 shows the *in vitro* expressions of a diverse set of protein targets using the PURExpress system, with molecular masses ranging from 20 kDa to more than

100 kDa. All recombinant protein factors inside the PURExpress system are His-tagged, allowing the synthesized protein to be "reverse-purified", Figure 7.3 (Shimizu *et al.*, 2001).

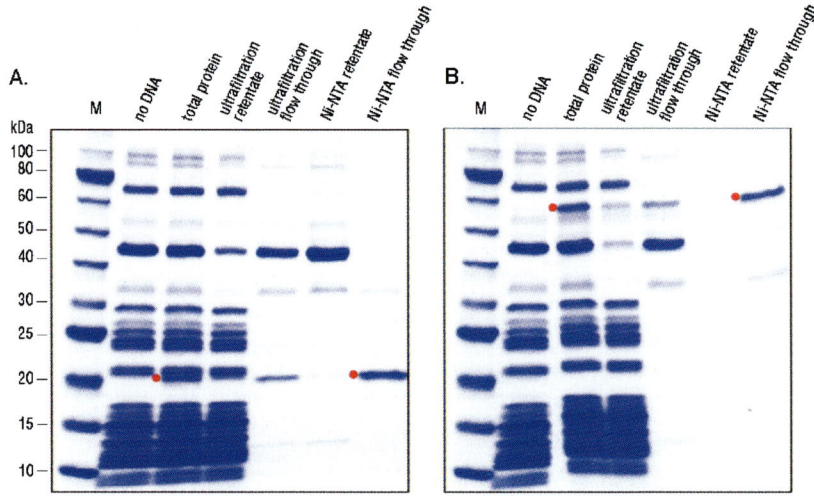

Figure 7.3. Proteins (DHFR (A), T4 DNA ligase (B)) expressed in the PURExpress system can be "reverse" purified by resin that binds His-tagged proteins (Ni-NTA). Compared to no template DNA added (lane 1), the expressed target proteins can be visualized on a 10–20% Tris-glycine gel stained with Commassie Blue (lane 2, red dots). After ultra-filtration, all ribosomal proteins are retained (lane 3) and the target proteins and recombinant factors are in the flow-through (lane 4). Since all recombinant factors are his-tagged, a subsequent purification by Ni-NTA removes these factors (lane 5), leaving the untagged target proteins in the flow-through with a relatively high purity (lane 6).

A comparative study has indicated that the reconstituted PURE system is sufficient but not efficient in protein translation when compared to the extract-based system (Hillebrecht and Chong, 2008) but compared to the eukaryotic system, the reconstituted system has a similar efficiency (Figure 7.4). Since then, PURExpess has been greatly improved and will reach a similar level as the extracted based system.

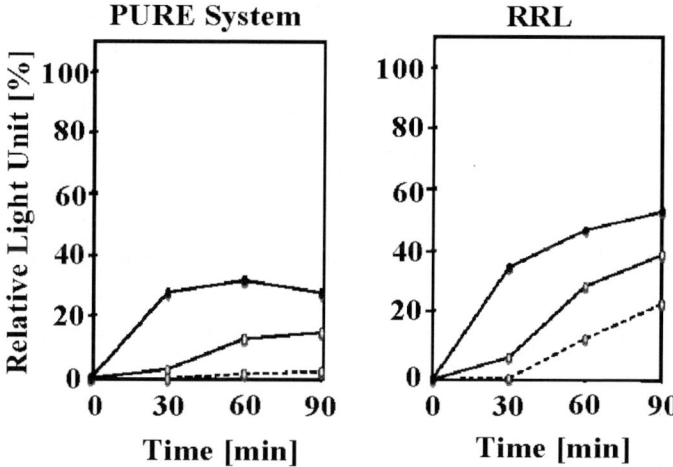

Figure 7.4. A time course of *in vitro* synthesis of Fluc in the reconstituted PURE system and an eukaryotic system (rabbit reticulocyte lysate, RRL) and at three temperatures (37 °C, filled ovals with solid line; 30 °C, open ovals with solid line; 20 °C, open ovals with dash line). The activities are averages of duplicate experiments and presented as the percentage of the relative light unit determined at 37 °C and 60 min.

7.5.2 *APA template preparation*

In this paper, two alternative processes are described for the fabrication of parallel, regularly arranged nanopore arrays with a high aspect ratio in anodic alumina. All chemicals ($HClO_4$, C_2H_5OH, H_3PO_4, oxalic acid, nitric acid, isopropyl alcohol) were purchased from Carlo Erba Reagents (Milan, Italy). For pholythographic process we used: positive resist AR-U4040 (Allresist, GmbH); negative resist SU-8 2 (Microchem, Inc., USA); SU-8 Developer (Microchem, Inc., USA). High–purity aluminum foils (Alfa Aesar, 0.25 mm÷0.50 mm thick, Puratronic, 99.998%). AFM measurements were done in tapping mode, in air by using a high aspect ratio silicon cantilever (NANOSENSORS, resonant frequency 350 KHz, radius of curvature of the tip less than 10 nm mounted on a SPM head (VEECO). SEM (Philips M525) measurement was done in standard secondary electron mode, HV 30. FIB (FEI) measurement was done with gallium ions (37 pA) both for cutting and imaging.

7.5.2.1 *APA process*

Briefly, ordered nanopore arrays are prepared by means of a two-step anodization process. Subsequently, a lithographic technique is applied to microstructure the nanopore arrays. High purity (99.999%) aluminum foils were electropolished in a mixture of $HClO_4$ and C_2H_5OH in order to smooth the surface. Anodization was conducted under constant cell potential in three types of aqueous solutions, sulfuric, oxalic, and phosphoric acids, which were used as electrolytes. Hexagonally-ordered pore domains can be prepared by a self-organization process under specific anodization conditions.

The pores originate at almost random positions and order thereafter by self-adjusting during their growth in a long-time anodization process. Because of that the pore arrangement on the surface is disordered and shows a broad distribution of size. Ordered pore domains can only be obtained at the bottom of the layers. Therefore, pores produced in the first anodization step are not parallel to each other. In order to obtain ordered nanopore arrays with straight holes, regularly arranged throughout the film, a two-step anodization process was used. After the first anodization step, the porous alumina was removed by a wet chemical etching in a mixture of phosphoric acid (6 wt.-%) and chromic acid (1.8 wt.-%) at 60 °C. At the bottom of the porous film there is a thin oxide layer, called barrier layer, covering the pore bottoms. The barrier layer is not flat but consists of periodically arranged crests and troughs. After the removal of the porous film, the remaining periodic concave patterns acts as a self-assembled template for the second anodization process. An ordered nanopore array is obtained after the second anodization when the same parameters are used as per the first anodization step, with the two-step anodization process being conducted in oxalic acid under 40 V and at 5 °C.

The anodization times of the first and second processes were 2 and 12 h, respectively. The period of the pore arrangement, or interpore distance, is about 350 nm. In order to observe the morphology of the bottom of the nanopore array, the remaining aluminum substrate was removed in a saturated $HgCl_2$ solution. A positive resist (AR-U4040, Allresist GmbH) was used. A wet chemical etch based on phosphoric and nitric acids with an additional small amount of wetting agent was used to transfer the patterned resist to the Al layer. The Al layer not covered by the resist was completely removed after 30 min of etching. Finally, the porous alumina in the areas of the Al windows was etched down with a chemical etching in aqueous H_3PO_4 solution (5 wt.-%). In the opened area the etching solution can penetrate into the holes and therefore the etching process takes place quite rapidly at the whole oxide/solution interface. On the other hand, the Al mask is not attacked by the etching solution, therefore protecting the areas that form the structure to be obtained. As a result, a highly selective etch output can be expected although the chemical etch is a microscopically isotropic process under our conditions.

7.5.2.2 APA process

In the second process of ordered porous alumina fabrication and microstructuring, ordered nanopore arrays are prepared by using a photolitographic technique that microstructures the nanopore arrays. A negative resist (SU-8, Micro-Chem) was used, that is a high contrast, epoxy based photoresist designed for micromachining and other microelectronic applications.

SU-8 shows very high optical transparency above 360 nm, which makes it ideally suited for imaging near vertical sidewalls in very thick films. SU-8 is best suited for permanent applications since, when imaged, cured and left in place it gives hydrophobic properties to final nanostructured surfaces. An aluminum sheet (250 μm thick) was cleaned to obtain maximum process reliability followed by isopropyl alcohol cleaning and deionized water rinse. After the resist has been applied to the substrate, it was soft baked in order to evaporate the solvent and thicken the film. The resist used is optimized for near UV (350–400 nm) exposure and is virtually transparent and insensitive above 400 nm.

After the exposure, a post expose bake (PEB) was performed to selectively cross-link the exposed portions of the film. The bake was performed either on a hot plate or in a convection oven. For present application, where the imaged resist is to be left as part of the final device, the resist was ramp/step hard baked at 150 °C for 15 minutes in a convection oven to further cross link the material. Finally, the surface not covered by resist has been anodized under constant cell potential in phosphoric acids 1.2 M, which was used as electrolyte. Hexagonally ordered pore domains can be prepared by a self-organization process under some specific anodization conditions. In alternative to aluminum sheet a glass surface can be used, where evaporation of an aluminum layer with a thickness desired is possible.

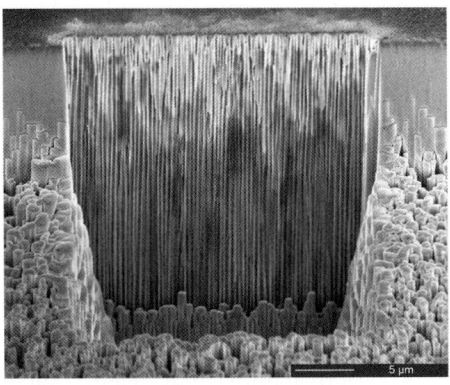

Figure 7.5. FIB system images of cross-sectional morphologies of the microarray spot, resulting at the end of photolithographic micro-structuring technique and 2 step anodization process. FIB-FEI measurement were done with gallium ions - 37 pA - both for cutting and imaging

Hexagonally ordered nanopore arrays with high aspect ratios based on a self-organization process in anodic alumina was fabricated. A two-step anodization technique was used in order to oxidize aluminum in phosphoric acid solution. The ordered pore arrays are straight and parallel and with polycrystalline structure.

The pore distance can be controlled by changing the anodic electrolyte and the applied voltage. Two different lithographic microstructuring techniques for the ordered nanopore arrays were reported, using a positive resist and a negative resist leaving surface hydrophobic. Spot of porous alumina 120–500 μm wide, 20–100 μm high were prepared. These spot were ready to be functionalized for the following Oligonucleotide and protein immobilization. A typical sample obtained by using the second microstructuring process is shown in Figure 7.5, where the whole microstructuring process is highly anisotropic and leads to a sharp edge. The anisotropy of the process can be seen with FIB system.

The side walls of the structures are very steep, and their roughness is determined by the quality of the mask and the aluminum transfer layer. Hexagonally ordered pore domains were prepared by a self-organization process under specific anodization conditions. This nanopatterning technique results in the APA template with hydrophobic properties, which increase specificity to biological sample linking.

7.5.3 *In vitro applications of the PURExpress system*

7.5.3.1 *High throughput functional genomics and proteomics*

The PURExpress system has a simple format that allows it to be easily integrated into high throughput platforms for functional genomics and proteomics studies. The absence of nuclease activities inside PURExpress system ensures the stability of linear DNA templates during protein synthesis. Individual DNA templates for *in vitro* expression can be generated by PCR, thus getting by the time-consuming cloning process. This feature will be particularly useful for the high throughput screenings at the whole genome scale, either for novel activities or for protein-protein interactions. For structural genomics projects, the PURExpress system can be an alternative route to acquire the "tough" protein targets which resist the traditional cellular expression (Graslund *et al.*, 2008).

7.5.3.2 *Protein engineering*

Directed evolution of proteins *in vitro* is a powerful tool for improving and creating biocatalysts. A number of *in vitro* evolution methodologies, such as mRNA display (Roberts and Szostak, 1997), ribosome display (Hanes and Pluckthun, 1997) and *in vitro* compartmentalization (Tawfik and Griffiths, 1998), depend on *in vitro* translation. The PURExpress system has been demonstrated to be uniquely suited for the *in vitro* selection of restriction endonucleases using the *in vitro* compartmentalization method (Zheng and Roberts, 2007), as it is free of non-specific nuclease activity. In this study, the PURExpress system and the DNA library were dispersed into more than 10^9 aqueous droplets in a water-in-oil emulsion. The droplet encapsulation provides a linkage between the phenotype (expressed protein) and the genotype (DNA), which sets the stage for the specific selection of restriction enzyme genes. Other researchers have reported that using the PURExpress system greatly improves the efficiency of ribosome display (Villemagne *et al.*, 2006).

Systematically mutagenized protein-coding libraries can be used as well to test if a specific mutation(s) affects protein function. Again, only a few PCR steps are needed to obtain the mutant protein, providing a quick experimental verification of hypotheses.

7.5.3.3 *Study of protein expression, translation and folding*

The PURExpress system contains the minimal set of factors necessary for *in vitro* protein translation. It is largely free of chaperones and other cellular factors for post-translational modifications, thus providing a starting point to study the involvement of these factors in transcription/translation regulation and nascent chain folding (Kaiser *et al.*, 2006). It can also be used to produce "clean" proteins, which, if purified from traditional cellular hosts, may come with undesired modifications or bound co-factors. A number of research labs studying translation routinely use home-made reconstituted systems to study different aspects of translation. Recently small RNA-mediated post-transcription regulation has been demonstrated in the reconstituted PURE system (Maki *et al.*, 2008).

7.5.3.4 *Incorporation of unnatural amino acids*

Another important advantage of the PURExpress system is the ability to control its composition. For example, omission of the release factor 1 (RF1) in the system allows unnatural amino acids to be efficiently incorporated at specific amber codon sites via chemically mis-acylated suppressor tRNA (Shimizu *et al.*, 2001; Noren *et al.*, 1989). It was recently reported that the translation apparatus of *E. coli* can tolerate a wide range of amino acid derivatives, revealing even greater potential for the ribosomal synthesis of unnatural peptides using reconstituted systems (Hartman *et al.*, 2007). In summary, future generations of cell-free protein synthesis systems should contain defined components, providing a clean background and at the same time be able to produce correctly folded proteins of any type in high yields. The PURExpress system clearly represents a breakthrough towards this goal. The improved reconstituted systems will be customized for expression of complex proteins such as large membrane proteins, multi-subunit assemblies and specifically modified proteins, *etc*.

The customization will also permit diversity in the protein synthesis conditions, such as elevated or decreased temperature, ionic strength and redox environment *etc*. Protein translation is one of the core processes in living organisms that serve as a "central node" to network other biological processes. There is no doubt that creative applications of the reconstituted systems will go beyond producing proteins and into such diverse fields as synthetic biology, systems biology and medical diagnostics.

7.5.4 *Crystal growth on APA template and in nanovolume*

Lysozyme from chicken egg white (cat. L6876) and thaumatin from T. Danieli (cat. T7638) were purchased by Sigma Aldrich (Milan, Italy). The proteins were filtered through a 0.45 µm mesh size filter (Millipore, Carrigtwoill, Co. Cork, Irleand) and reservoirs were filtered through a 0.20 µm mesh size filter (Albet-Jacs) before the experiments were set up.

For lysozyme crystallization, the APA template, prepared on the glass cover slide, was used in the common protein hanging drop crystallization well in the way to put the protein solution droplet in the contact with the nanotemplate.

In alternative, the sandwich drop wells were prepared with contemporary use of two APA nanotemplate. For both classical and APA method the lysozyme crystallization conditions were following: 4 ml of

40 HEWL solution in the 50 mM sodium acetate buffer was mixed with 2 ml of 0.9 M NaCl solution and equilibrated on the reservoir, containing the 0.9 NaCl. The classical hanging and sandwich drop experiment were carried out in parallel, using the same crystallization condition

In parallel, the thaumatin crystals were grown in nanovolumes by hanging drop method using the high throughput robotic system at EMBL, Grenoble. Two crystallization method were used: classical hanging drop and LB nanotemplate method (Pechkova and Nicolini, 2003, 2004). In the last method the protein LB nanotemplate was fist prepared by spreading the protein molecules on the air-water interface of the LB trough and then compressing them to a surface pressure of 20 mN/m. The protein monolayer was further transferred on the solid support — in this case — on the glass cover slide by LS (horizontal lift) deposition and dried into the nitrogen flux. This template was used as the template for accelerate the protein nucleation and growth.

The crystallization conditions used for thaumatine crystallization (both for LB and classical method) were a 10 nl drop containing 15 mg/ml of thaumatin in 100 mM ADA buffer pH 6.5 was mixed with the reservoir solution (1:1) containing 1 M sodium/potassium tartrate in 100 mM sodium/potassium tartrate buffer pH 6.5. The crystal have grown in about 24 hours, which is much faster then in the classical vapour diffusion with 5–10 microliter volume drops. Using the LB method, the crystals grown faster, than un classical hanging drop nanovolume method (Pechkova and Nicolini, 2001).

7.5.4.1 *APA protein crystals X-ray diffraction*

The crystals was mounted in the cryoloops, frozen in the liquid nitrogen flux up to 140 K and analysed by ID13 beamline at ESRF, Grenoble (France). The beam size was chosen in such a way to cover the similar fraction of the crystal to be compared. Number of analysed images was 98. Dataset was integrated in two sets (10 to 60 images) and (61 to 98 images).

7.5.4.2 Nanovolume protein crystals X-ray diffraction

X-Ray diffraction data for these crystals were collected at ID14-1 beamline. Exposure time for each image was 0.5 sec and wavelength of X-Rays used was 0.934 Å. Total 50 well diffracting images were obtained for both LB and classical crystals using ADSC CCD detector. Crystal to detector distance 136.94 in all cases.

7.5.4.3 Data processing and reduction

In crystallization trials, the axial ratios show a definite increasing trend in the presence of APA nanotemplate (Figure 7.6), while the number of crystals in the drops was decreased.

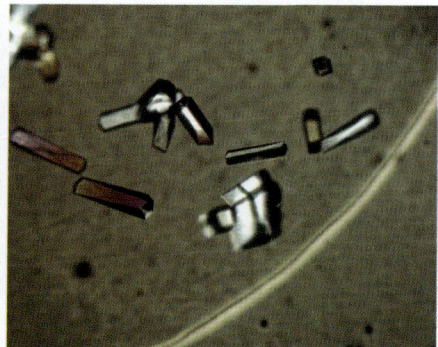

Figure 7.6. Crystal formed in presence (right) and absence (left) of APA layer (Pechkova *et al.*, 2009).

This effect confirmed that APA surface appear to influence the protein nucleation and growth. Since APA surface is negatively charged, it could attract the lysozyme molecules, which have the mostly positive charges on its surface. Thus, highly ordered pattern of APA can produce the same highly ordered lysozyme pattern which can favor the nucleation and crystallization in the specific way. Next, being porous material, APA can trig the nucleation inside the pores. In this case, the space restriction factor cam take place, causing the elongation of the c-axis of the crystal.

Data reduction of X-ray synchrotron radiation diffraction experiment the was carried out for classical and APA lysozyme crystal. On refining the final model the R factor obtained for the APA-grown crystal structure is less compared to the R factor of classical lysozyme. R factor indicates the quality of the final structure obtained and in case of APA it is much better compared to classical.

Table 7.1. The data collection statistics and analysis of APA-grown vs classical hanging drop crystal structure.

Parameter	APA	Classical
Resolution	1.9–39.0	1.7–39.0
Unit cell parameters		
a, c (where $a=b$)	78.808, 37.529	78.839, 37.256
$\alpha=\beta=\gamma$	90.00	90.00
group symmetry	P 43 21 2	P 43 21 2
Completeness	99.52	91.25
R factor	0.16822	0.17071
R free	0.22347	0.25038
R_{mrg}	0.241	0.061
I/Sigma	1.5	6.5
Water	215	227
Number of unique reflections	9169	11172
Redundancy	0.262	0.064
Cell Volume	233081,393	231567,944
RMS bonds	0,011	0,012
Alpha helix content	43	43
3_{10} helix	16	12
Coil	14	13
Strands	8	8
Turns	44	49

This means that agreement between the crystallographic model and the X-ray diffraction data is more in case of the APA compared to classical. Also the R free values are more in case of classical structure compared to APA. In many test calculations R free correlates very well with phase accuracy of the atomic model. The statistic and data collection of that the proteins crystals grown by APA template and those grown by classical hanging drop vapour diffusion method is shown in the Table 7.1. The quality of this data can be compared with the nanovolume protein crystal X-ray analysis (see below).

In the final structures obtained from the two methods another thing which is required to observe is the content of 3_{10} helix in the two forms. In case of classical it is less, *i.e*, 12 compared to 16 in case of APA. The structure of the 3_{10} helix is similar to the alpha helix accept instead of $i - i+4$ hydrogen bonding we have $i - i+3$ bonding. Alpha helix sometimes begin or end with the single turn of this 3_{10} helix.

Content of 3_{10} helix is similar in APA and 2cgi, which the structure used as the template for the refinement process. Content of alpha helix and the beta strands are same in both the cases, but the number of turns in case of classical is more in APA crystals, because this part is occupied by 3_{10} helix in case of APA. This helix structure is considered to be a reasonably stable structure in a molecule like alpha helix.

7.5.4.4 *Nanovolume-grown protein crystal structure: LB versus classical*

Various processing and structure parameters were compared in the Table 7.2. Number of water molecules is less for thaumatin LB based crystals compared to classical crystals.

Table 7.2. Comparison of crystal parameters and electron density of thaumatin crystals grown in the nanodroplets using robotic technique for LB (one crystal) and classical method (two crystals).

Parameter	LB	Classical	Classical 2
Resolution	1.6–45.23	1.6–45.79	1.6–45.83
a,c (where a=b)	58.099, 150.409	57.782, 150.219	57.797, 150.271
Completeness	99.52	91.25	98.93
R-factor	0.20571	0.18695	0.18349
Rmrg	0.128	0.060	0.061
I/Sigma	4.2	8.4	7.7
Water	235	264	265
Number unique reflections	33059	29913	32470
Redundancy	0.138	0.068	0.070
Cell volume	507704.64	501545.11	501979.255

These water molecules were added at 1σ for both classical and LB crystals. Less number of water molecules for LB based crystals, are in agreement as previously reported for lysozyme crystals (Pechkova *et al.*,

2005). Secondary structure obtained from two types of crystals is same. Electron density of different residues was compared at same contour level. Figure 7.7 shows the comparison of electron density map for classical and LB of carboxylic acid group of aspartic acid at 55 position.

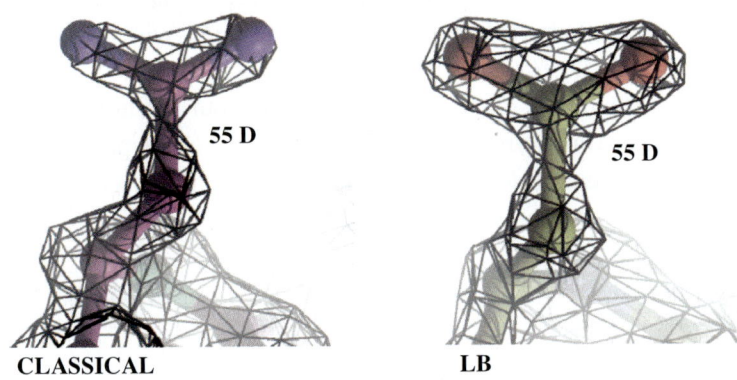

Figure 7.7. Comparison of electron density map for carboxilic acid group of aspartic acid contoured at 2.01 sigma for thaumatin crystal grown by classical and LB method.

It is clear that while the electron density for LB structure in this site is perfect, the classical structure has suffered from the radiation damage, loosing the part of electron density. This confirmed that the LB crystals are more radiation stable, due to less amount of water molecules in their structure. We have compared also the electron density map classical versus LB crystal structure for disulphide bond region of cystein at 71 position, disulphide bond formed by 149 cystein residue and proline residue at 135 position (not shown) with similar results.

7.6 Conclusions and Future Prospectives

The integration of the PURExpress system into APA and NAPPA is being presently exploited (in preparation).

Cell Free Expression and APA for NAPPA and Protein Nanocrystallography 141

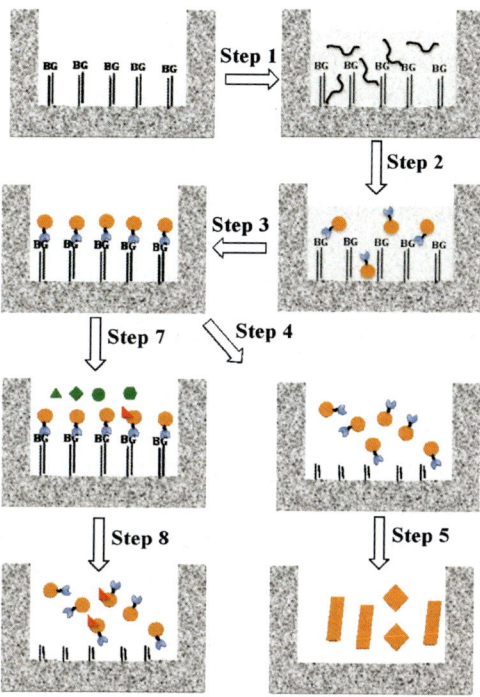

Figure 7.8. A scheme illustrating the application of APA-PURExpress for high-throughput nanocrystallography and label-free proteomics of protein-protein interaction. A template double-stranded DNA containing the gene of interest fused to a SNAP tag and an upstream T7 promoter is immobilized by APA technology. The free 5' end of the template DNA contains a benzylguanine derivative (BG), which is a substrate for the SNAP tag, forming a covalent linkage upon binding. By adding the PURExpress reconstituted cell-free translation system, the template DNA will be transcribed into mRNA (step 1), and then translated into a fusion protein containing the N-terminal SNAP tag and the C-terminal target protein (step 2). In the same APA pore, the SNAP tag allows the synthesized protein to bind to its own template DNA via the BG linkage, thus immobilized the target protein. The rest of the reaction mixture can be washed away (step 3). For nanocrystallography (step 4), adding a small amount of the DNA endunuclease will cleave the template DNA and release the fusion protein. Alternatively, a site-specific protease can also be used to cleave between the SNAP tag and target protein, releasing the untagged target protein. The target protein is then allowed to form nanocrystal *in situ* (step 5). For label-free high-throughput analysis of protein-protein interaction, the immobilized target protein (from step 3) is allowed to interact with a mixture of query proteins (step 7). After the binding reaction, the unbound proteins are washed away and the target protein complex is released by cleaving the template DNA (step 8). The protein complex can be characterized by subsequent MS analysis.

The single most important advantage of the PURExpress system is that it contains defined components with minimal contamination of unknown factors, making it the ideal tool for the nanocrystallography and label–free functional proteomics. Other cell-free systems all contain a large amount of factors derived from the cell extract that often create non-specific binding and high background. Using the PURExpress to express a fusion protein contains a SNAP tag (Gautier et al., 2008; Johnsson, 2009; Johnsson and Johnsson, 2007) allows immobilization of the protein. Therefore, application of the PURExpress in APA will create a more powerful NAPPA technology (Ramachandran et al., 2008) for both nanocrystallography and label-free functional proteomics applications, as illustrated in Figure 7.8.

This communication describes the construction of nanoporous APA (anodic porous alumina) substrate and its application to DNA chip technology (Grasso et al., 2005; Troitsky et al., 2002) and to protein nanocrystallography (Pechkova and Nicolini, 2003, 2004). Two different photolithographic microstructuring techniques for the ordered nanopore arrays are utilized. While the first process uses a positive resist, the second uses a negative one and has hydrophobic properties to increase specificity to biomolecular linking. Nanoporous alumina formed by anodic process yields straight holes with high aspect ratio and has been used as substrates for genomics and protein nanocrystallography. This offers several advantages over conventional supports making APA very attractive to use as supports for both applications here discussed.

The lithographic microstructuring technique described above is a valid alternative to process performed by Masuda et al. (1996) for the microstructuring of alumina. Using this technique we obtained a microarray of anodic porous alumina spots. Subsequently the test of functionalization and oligonucleotide immobilization demonstrated that the spots are suitable as matrix for DNA microarray application (Troitsky et al., 2002). Besides, the resist giving hydrophobic properties to spot boundary surface avoids possible liquid spread outside the spot and increases specificity to biomolecules linking.

The bi-dimensional photonic–crystal like behavior of the high aspect ratio ordered hollowed structure is such that the embedded dots or macromolecules emit light both more selectively and with a controlled pattern distribution (Gaponenko *et al.*, 2000) thus reducing cross-interference between adjacent points, typical in traditional planar microarrays having photoactive molecules emitting light in a Lambertian distribution.

This characteristic of APA allows the preparation of very dense microarrays, reducing the distance between the spots. Future efforts should concentrate on testing porous alumina matrices for protein–chip application in order to exploit the advantages offered by this material over planar conventional supports.

The experiments now in progress with protein nanocrystallography points to the possibility to influence *ab-initio* the protein crystallization rate and the crystal shape which can in turn lead to the generalization and to the control of the entire procedure.

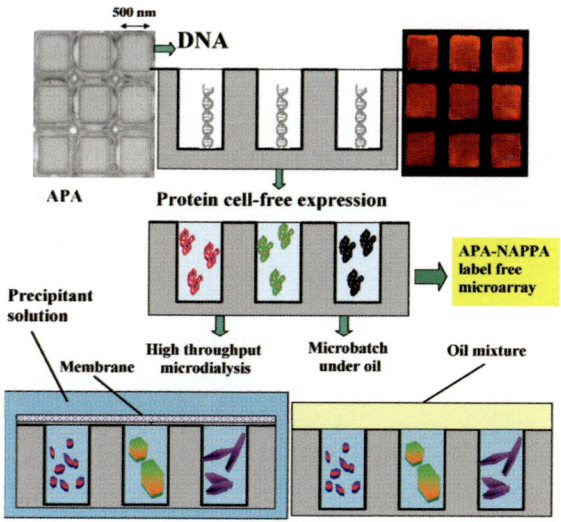

Figure 7.9. The future APA-NAPPA-cell free based nanobiocrystallography.

This means that the utilization of APA in controlling the shape of the biocrystals being formed may lead to interesting new applications in the field of 3D nanoelectronics (Pechkova and Nicolini, 2003) and of 3D nanomaterials (Nicolini and Pechkova, 2006). We also propose here a possible future development of the discussed APA-nanocrystallography application, whereby the APA channels can be used as the very small crystallization wells for the protein. In this case, the protein amount can be further drastically reduced in comparison of the conventional robotic system. The overall system (APA template with the channels filled with Protein and Precipitate solution) can be covered with the layer of the silicon oils mixture for the crystallization under oil. In the alternative, the protein solution-filled channels can be covered with the micro dialysis membrane and overall system can be immersed in the precipitant solution for the micro dialysis crystallization (Figure 7.9). In this manner, the advantages of high throughput protein crystallization method could be optimally combined with the Cell free expression system and with the APA nanotemplate heterogeneous crystallization.

Acknowledgments

This project was supported by grants to Fondazione EL.B.A. by MIUR for "Funzionamento" and by a FIRB International Grant on Proteomics and Cell Cycle (RBIN04RXHS) from MIUR (Ministero dell'Istruzione, Università e Ricerca) to CIRSDNNOB-Nanoworld Institute of the University of Genova.

References

Asherie, N. (2004). Protein crystallization and phase diagrams. *Methods*, 34, pp. 266–272.

Chayen, N. E., Saridakis, E., El-Bahar, R. and Nemirovsky, Y. (2001). Porous silicon: an effective nucleation-inducing material for protein crystallization. *Journal of Molecular Biology* 312, pp. 591–595.

Czokan, P. (1980). *Advances in Corrosion Science and Technology*, Vol. 7 (Ed: M. G. Fontana, R. W. Staehle) Plenum New York p. 239.

Fermani, S., Falini, G., Minnucci, M. and Ripamonti, A. (2001). Protein crystallization on polymeric film surfaces. *Journal of Crystal Growth*, 224, pp. 327–334.

Foresi, J. S., Villeneuve, P. R., Ferrera, J., Thoen, E. R., Steinmeyer, G., Fan, S., Joannopoulos, J. D., Kimerling, L. C., Smith, H. I. and Ippen, E. P. (1997). Photonic-bandgap microcavities in optical waveguides. *Nature*, 390, pp. 143–145.

Gaponenko, N. V., Davidson, J .A., Hamilton, B., Skeldon, P., Thompson, G. E., Zhou, X. and Pivin, J. C. (2000). Strongly enhanced Tb luminescence from titania xerogel solids mesoscopically confined in porous anodic alumina. *Applied Physics Letters*, 76, pp. 1006–1008.

Gautier, A., Juillerat, A., Heinis, C., Correa, I. R., Jr., Kindermann, M., Beaufils, F. and Johnsson, K. (2008). An engineered protein tag for multiprotein labeling in living cells. *Chemistry & Biology*, 15, pp. 128–136.

Graslund, S., Nordlund, P., Weigelt, J., Bray, J., Gileadi, O., Knapp, S., Oppermann, U., Arrowsmith, C., Hui, R., Ming, J. *et al.* (2008). Protein production and purification. *Nature Methods*, 5, pp. 135–146.

Grasso V., Lambertini V., Ghisellini P., Valerio F., Stura E. Perlo P. and Nicolini, C. (2006). Nanostructuring of a porous alumina matrix for a biomolecular microarray. *Nanotechnology*, 17, pp. 795–798.

Hanes, J. and Pluckthun, A. (1997). In vitro selection and evolution of functional proteins by using ribosome display. *Proceedings of the National Academy of Sciences USA*, 94, pp. 4937–4942.

Hartman, M. C., Josephson, K., Lin, C. W. and Szostak, J. W. (2007). An expanded set of amino Acid analogs for the ribosomal translation of unnatural peptides. *PLoS ONE*, 2, pp. e972.

Hillebrecht, J. R. and Chong, S. R. (2008) A comparative study of protein synthesis in in vitro systems: from the prokaryotic reconstituted to the eukaryotic extract-based. *BMC Biotechnology*, 8, 58.

Jessensky, O., Müller, F. and Gösele, U. (1998). Self-organized formation of hexagonal pore arrays in anodic alumina. *Applied Physics Letters*, 72, pp. 1173–1175.

Johnsson, K. (2009). Visualizing biochemical activities in living cells. *Nature Chemical Biology*, 5, pp. 63–65.

Johnsson, N. and Johnsson, K. (2007). Chemical tools for biomolecular imaging. *ACS Chemical Biology*, 2, pp. 31–38.

Kaiser, C. M., Chang, H. C., Agashe, V. R., Lakshmipathy, S. K., Etchells, S. A., Hayer-Hartl, M., Hartl, F. U. and Barral, J. M. (2006). Real-time observation of trigger factor function on translating ribosomes. *Nature*, 444, pp. 455–460.

Katzen, F., Chang, G. and Kudlicki, W. (2005). The past, present and future of cell-free protein synthesis. *Trends in Biotechnology*, 23, pp. 150–156.

Li A. P., Müller, F., Birner, A., Nielsch, K. and Gösele, U. (1999). Fabrication and microstructuring of hexagonally ordered two-dimensional nanopore arrays in anodic alumina. *Advanced Materials*, 6, pp. 483+.

Li, A. P., Müller, F., Birner, A., Nielsch, K. and Gösele U. (1998). Hexagonal pore arrays with a 50–420 nm interpore distance formed by self-organization in anodic alumina. *Journal of Applied Physics*, 84, pp. 6023–6026.

MacDonald, G. M., Steenhuis, J. J. and Barry, B. A. (1995). A difference Fourier-transform infrared spectroscopy study of chlorophyll oxidation in hydroxylamine-treated photosystem-II. *Journal of Biological Chemistry*, 270, pp. 8420–8428.

Maki, K., Uno, K., Morita, T. and Aiba, H. (2008). RNA, but not protein partners, is directly responsible for translational silencing by a bacterial Hfq-binding small RNA. *Proceedings of the National Academy of Sciences USA*, 105, pp. 10332–10337.

Masuda, H. and Fukuda, K. (1995). Ordered metal nanohole arrays made by a two-step replication of honeycomb structures of anodic alumina. *Science*, 268, pp. 1466–1470.

Masuda, H. and Satoh, M. (1996). Fabrication of gold nanodot array using anodic porous alumina as an evaporation mask. *Japanese Journal of Applied Physics*, 35, pp. L126–L129.

Masuda, H., Hasegawa, F. and Ono, S. (1997). Self-ordering of cell arrangement of anodic porous alumina formed in sulfuric acid solution. *Journal of Electrochemistry Society*, 144, pp. L127–L130.

McPherson, A. and Schlichta, P. (1988). Heterogeneous and epitaxial nucleation of protein crystals on mineral surfaces. *Science*, 239, pp. 385–387.

McPherson, A., (1999) CSHL Press, Cold Spring Harbor.

Murshudov, G. N., Vagin, A. A., Lebedev, A.,Wilson, K. S. and Dodson, E. J. (1999). Efficient anisotropic refinement of macromolecular structures using FFT. *Acta Crystallographica D*, 55, pp. 247–255.

Nanev, C. and Tsekova, D. (2000). Heterogeneous nucleation of hen-egg-white lysozyme - Molecular approach. *Crystal Research and Technology*, 35, pp. 189–195.

Nicolini, C. and Pechkova, E. (2006). Nanostructured biofilms and biocrystal. *Journal of Nanoscience and Nanotechnology*, 6, pp. 2209–2236.

Nirenberg, M. W. and Matthaei, J. H. (1961). The dependence of cell-free protein synthesis in E. coli upon naturally occurring or synthetic polyribonucleotides. *Proceedings of the National Academy of Sciences USA*, 47, pp. 1588–1602.

Noren, C. J., Anthony-Cahill, S. J., Griffith, M. C. and Schultz, P. G. (1989). A general method for site-specific incorporation of unnatural amino acids into proteins. *Science*, 244, pp. 182–188.

Pechkova, E. and Nicolini, C. (2001). Accelerated protein crystal growth onto the protein thin film. *Journal of Crystal Growth*, 231, 599–602.

Pechkova, E. and Nicolini, C. (2003). *Proteomics and Nanocrystallography*, Kluwer Academic Publishers, pp. 1–212.

Pechkova, E. and Nicolini, C. (2004). Protein nanocrystallography: a new approach to structural proteomics. *Trends in Biotechnology*, 22, pp. 117–122.

Pechkova, E. and Nicolini, C. (2009). APA induced crystal growth and design. *Crystal Growth and Design*, submitted.

Pechkova, E., Sivozhelezov, V., Tropiano, G., Fiordoro S. and Nicolini, C. (2005). Comparison of lysozyme structures derived from thin-film-based and classical crystals. *Acta Crystallographica D* 61, pp. 803–808.

Pechkova, E., Tropiano, G., Riekel, C. and Nicolini, C. (2004). Radiation stability of protein crystal grown by nanostructured templates synchrotron. *Spectrochimica Acta B*, 59, pp. 1687–1693.

Ramachandran, N., Raphael, J. V., Hainsworth, E., Demirkan, G., Fuentes, M. G., Rolfs, A., Hu, Y. and LaBaer, J. (2008). Next-generation high-density self-assembling functional protein arrays. *Nature Methods*, 5, pp. 535–538.

Roberts, R. W. and Szostak, J. W. (1997). RNA-peptide fusions for the in vitro selection of peptides and proteins. *Proceedings of the National Academy of Sciences USA*, 94, pp. 12297–12302.

Rong, L., Komatsu, H., Yoshizaki, I., Kadowaki, A. and Yoda, S. (2004). Protein crystallization by using porous glass substrate. *Journal of Synchrotron Radiation*, 11, pp. 27–29.

See, E. G., Joannopoulos, J. D., Meade, R. D. and Winn, J. N. (1995). *Photonic crystals: molding the flow of light*, Princeton University Press, Princeton, NJ.

See, E. G., Keller, F., Hunter, M. S. and Robinson, D. L. (1953) *Journal of the Electrochemical Society*, 100, pp. 411.

Shimizu, Y., Inoue, A., Tomari, Y., Suzuki, T., Yokogawa, T., Nishikawa, K. and Ueda, T. (2001). Cell-free translation reconstituted with purified components. Nature Biotechnology, 19, pp. 751–755.

Stura E., Bruzzese D., Valerio F., Grasso F., Perlo P. and Nicolini C. (2007). Anodic porous alumina as mechanical stability enhancer for LDL-cholesterol sensitive electrodes, *Biosensors & Bioelectronics*, 23, pp. 655–660.

Suzuki, Y., Saito, J., Horii, Y. and Kasagi, N. (2003). *The International Symposium on Micro-Mechanical Engineering*.

Tang, L., Huang, Y. B., Liu, D. Q., Li, J. L., Mao, K., Liu, L., Cheng, Z. J, Gong, W. M., Hu, J. and He, J. H. (2005). Effects of the silanized mica surface on protein crystallization. *Acta Crystallographica D* 61, pp. 53–59.

Tawfik, D. S. and Griffiths, A. D. (1998). Man-made cell-like compartments for molecular evolution. *Nature Biotechnology*, 16, pp. 652–656.

Templin, M. F, Stoll, D., Schrenk, M., Traub, P. C., Vohringer C. F. and Joos, T. O. (2002). Protein microarray technology. *Trends in Biotechnology*, 20, pp. 160–166.

Thompson, G. E., and Wood, G. C. (1983) *Treatise on Materials Science and Technology*, Vol. 23 (Ed: J. C. Scully) Academic New York p. 205.

Troitsky, V., Ghisellini, P., Pechkova, E. and Nicolini, C. (2002). DNASER II. Novel surface patterning for biomolecular microarray. *IEEE Transactions on Nanobiosciences*, 1, pp. 73–77.

Tsekova, D. S. (2009). Formation and growth of tetragonal lysozyme crystal at some boundary conditions. *Crystal Growth & Design*, 9, pp. 1312–1317.

Villemagne, D., Jackson, R. and Douthwaite, J. A. (2006). Highly efficient ribosome display selection by use of purified components for in vitro translation. *Journal of Immunological Methods*, 313, pp. 140–148.

Yablonovitch, E. (1987). Inhibited spontaneous emission in solid-state physics and electronics. *Physics Review Letters*, 58, pp. 2059–2062.

Zheng, Y. and Roberts, R. J. (2007). Selection of restriction endonucleases using artificial cells. *Nucleic Acids Research*, 35, pp. e83.

CHAPTER 8

STRUCTURAL AND FUNCTIONAL STUDIES ON THE *HELICOBACTER PYLORI* PROTEOME: THE STATE OF THE ART

Giuseppe Zanotti and Laura Cendron

Department of Biological Chemistry, University of Padua, Viale G. Colombo 3, 35121 Padua, Italy, and Venetian Institute of Molecular Medicine (VIMM), Via Orus 2, 35129 Padua, Italy
E-mail: giuseppe.zanotti@unipd.it

Helicobacter pylori, a major human pathogen identified less than 30 years ago, is nowadays among the most studied gram-negative bacteria. The complete genome of four different strains has been sequenced and a large amount of experimental data, both *in solution* and *in vivo*, has accumulated on it. In this chapter we review the structural studies till now performed, presenting an overview of the about 80 three-dimensional proteins structures available.

8.1 Introduction

H. pylori is a microaerophilic gram-negative bacterium that chronically infects the gastric mucosa of millions of people worldwide: it has been estimated that over 50% of the world population carries this infection (Covacci *et al.*, 1999). *H. pylori* colonization is often acquired during childhood and, in the absence of an appropriate therapy, it persists for life. It is associated with the development of several diseases, such as chronic gastritis, gastric and duodenal ulcer, gastric adenocarcinoma, and mucosa-associated lymphoma (Parsonnet *et al.*, 1991, Parsonnet *et al.*, 1994, Moss and Sood, 2003). Fortunately, only 10% of infected individuals will develop peptic ulcer disease and about 1% of them gastric neoplasia (Pisani *et al.*, 1999).

In order to survive in the difficult gastric environment, the bacterium has developed a series of appropriate features and tools (Cover and Blaser, 2009):

a) resistance to acidity, thanks to the presence of the enzyme urease that hydrolyzes urea to yield carbon dioxide and ammonia, which in turn buffers the pH in proximity of the bacterium (Sidebotham and Baron, 1990);
b) a set of flagella, which allow the bacterium to penetrate the gastric mucus layer and reach the epithelial cells (Appelmelk *et al.*, 2000);
c) a set of outer membrane proteins, some of which are able to bind to receptors on the surface of gastric cells and that diminish the rate of bacterial wash-out as a result of peristalsis (Mahdavi *et al.*, 2002);
d) the presence of transport systems for the uptake of nutrients, coupled with the ability to induce a local inflammation;
e) multiple systems to counteract the effects of reactive oxygen and nitrogen species (Wang *et al.*, 2006);
f) secreted proteins that limit the innate and adaptive immune response from the host (Wilson and Crabtree, 2007).

The ability of the bacterium to live in a very peculiar environment, thanks to the adaptive features developed during the thousands of years of cohabitation with humans, along with the pathogenicity of some of its strains, render *H. pylori* a very interesting subject, both from the point of view of the fundamental research and of the applied science. Its genome codes for about 1500 ORFs and several complete genomes of different strains are now available (Tomb *et al.*, 1997, Oh *et al.*, 2006, Alm *et al.*, 1999, Alm and Trust 1999). On the contrary, structural data available are quite limited: 171 files of atomic coordinates are present in the Protein Data Bank (http://www.pdb.org, Berman *et al.*, 2000) at June 2009, and 27 are waiting for release. In this chapter we will try to summarize our present structural knowledge about the bacterial proteome. Structures present in the protein data bank have been grouped in five sets: members of the *cag*-PAI, enzymes, proteins interacting with nucleic acids, toxins, unknown function proteins.

8.2 The *cag* Pathogenicity Islands

Strains isolated from patients with severe gastric diseases generally include in their genome the so-called "*cag* Pathogenicity Island" (cytotoxin-associated gene, *cag*-PAI) and are defined as Type I strains (Censini *et al.*, 1996). The *H. pylori cag*-PAI is a 40-kb chromosomal region which includes, depending on the clinical strain, between 27 and 30 open reading frames. One of them encodes for the immunodominant antigen CagA, which is localized at the 3′ end of the island (Tummuru *et al.*, 1994, Covacci *et al.*, 1993).

CagA, one of the major virulence factors of the bacterium, (Parsonnet *et al.*, 1997), is so far the only *H. pylori* macromolecule known to be translocated into the gastric epithelial cells by a Type IV Secretion System (TFSS) codified by the *cag*-PAI (Odenbreit *et al.*, 2000). Inside the host cell, CagA toxin induces multiple changes which cause the development of ulcers and cancer (Bourzac and Guillemin, 2005). A direct correlation of the presence of the *cagA* gene with diseases has been demonstrated (Covacci *et al.*, 1997; Blaser 1998).

Infection of cultured gastric epithelial cells with *cag*-PAI positive strains results also in interleukin-8 (IL-8) induction and a drastic cellular elongation. The discovery of mutations that can block CagA delivery but not IL-8 induction has led to the hypothesis that a second secreted effector molecule could be responsible for IL-8 induction. Recent experiments have implicated peptidoglycan, acting through the cytoplasmic receptor Nod1, as the secreted factor that induces IL-8 (Viala *et al.*, 2004), while others have hypothesized that in some strain backgrounds CagA itself is responsible for some IL-8 induction (Brandt *et al.*, 2005).

The *cag*-PAI of *H. pylori* encodes proteins with homologies to structural and functional components of TFSS of other gram-negative bacteria (Akopyants *et al.*, 1998, Christie *et al.*, 2005), such as *Agrobacterium tumefaciens*, *Brucella suis*, *Bartonella henselae* and *Legionella pneumophila*. From the genes encoded by the *cag*-PAI, eighteen have been proved to be required for translocation of CagA into gastric epithelial cells, but only four have sequence homologies to genes encoding components of the VirB/D4 type IV secretion system of

A. tumefaciens, the best-studied model of the TFSS. Other five Cag proteins have some features in common with *A. tumefaciens* TFSS components, and have been proposed to be functional homologues of these latter. The remaining genes are unique to the *H. pylori cag*-PAI, and very little is known about their localization and function.

To date, the three dimensional structure of five proteins belonging or related to the *H. pylori* cag-PAI have been published (Table 8.1) and the results are summarized below.

Table 8.1. Proteins belonging to the *cag* pathogenicity island.

Protein	Hypothetical function	PDB ID code and reference
VirB11	ATPase	1G6O, 1NLY, 1NLZ, 1OPX
HP1451 + VirB11	Inhibitor of VirB11	2PT7
ComB10	VirB10 homolog	2BHV
CagD	Unknown	3CWX, 3CWY
CagS	Unknown	3G3V
CagZ	Unknown	1S2X, 2A9E

VirB11 ATPase (HP0525). This protein functions as a gating molecule at the inner membrane, coupling the ATP binding/hydrolysis to macromolecular secretion. The crystal structure of a binary complex of HP0525 bound to ADP (Yeo *et al.*, 2000) shows that the protein is an examer, each monomer consisting of two domains formed by the N- and C-terminal halves of the sequence. ADP is bound at the interface between the two domains. In the hexamer, the N- and C-terminal domains form two rings, which define an open chamber open on one side and a closed one on the other. The closure and opening of the examer is regulated by ATP binding and ADP release (Savvides *et al.*, 2003).

The crystal structure of VirB11 in complex with a fragment of HP1451, a protein of previously unknown function, shows how the latter binds to the HP0525 examer, locking it in the closed form (Hare *et al.*, 2007). From the structure, it appears that HP1451 acts as an inhibitory factor of HP0525 to regulate *cag*-mediated secretion.

The other structures of the *cag*-PAI components refer to proteins whose exact functions are not known, but that do not belong to the pilus and to the structural core assembly. They are essential or important for CagA translocation or maximal induction of IL-8 secretion. **CagD** (HP0545) is a covalent dimer in which each monomer folds as a single domain, composed of five β-strands and three α-helices (Cendron *et al.*, 2009) (Figure 8.1A). It is involved in CagA translocation into the host epithelial cells, but it does not seem absolutely necessary for pilus assembly.

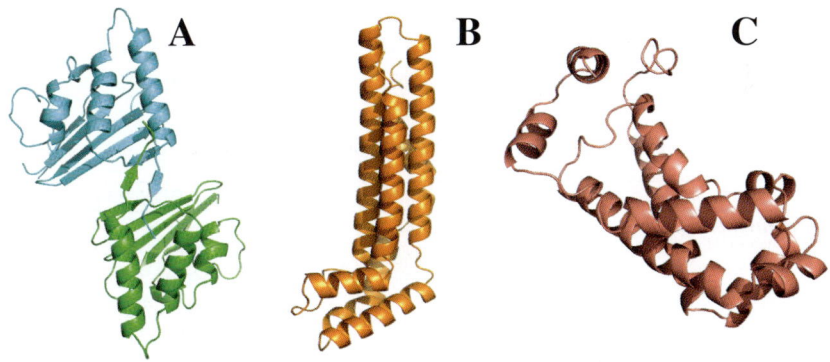

Figure 8.1. Cartoon view of CagD dimer (A), and the monomers of CagS (B) and CagZ (C).

It may serve as a unique multifunctional component of the T4SS which may be involved in CagA secretion at the inner membrane and is realized outside the bacteria most likely to promote additional effects on the host cell. **CagZ** (HP0526) is a 23 kDa protein essential fro the translocation of the major toxin CagA. Its structure consists of a single compact L-shaped domain, composed of seven α-helices including about 70% of the total residues (Cendron *et al.*, 2004) (Figure 8.1C). Three-dimensional homology searches did not reveal structural homologues, and CagZ can be considered representative of a new protein fold. The crystal structure of **CagS** (HP0534) demonstrates it is an all-α structure, composed of ten α-helices arranged in a single, compact "peanut-shaped" domain (Cendron *et al.*, 2007) (Figure 8.1B). Specific characteristics of the protein structure include the presence of a methionine cluster and a highly charged external surface.

Finally, **ComB10** is a VirB10 homolog of *H. pylori*, belonging to another type IV apparatus conferring competence to the bacteria for the release of DNA plasmids. ComB10 crystallizes as a dimer, providing detailed predictions about its self association (Terradot *et al.*, 2005).

8.3 Enzymes

The crystal structures of 46 enzymes of the bacterium are known (Table 8.2). Most of them are considered possible pharmacological targets and, for some, inhibitors have already been developed. They are grouped and classified according to their function and briefly summarized below.

Table 8.2. Enzymes. (continue on the next page).

Protein	Function	PDB ID code
Lipid metabolism		
FabZ	(3R)-Hydroxyacyl-acyl carrier protein Dehydratase	2GLL, 2GLM, 2GLP, 2GLV, 3B7J, 3CF8, 3CF9, 3DOY, 3DOZ, 3DP0, 3DP1, 3DP2, 3DP3
MCAT	Malonyl-CoA:acyl carrier protein transacylase	2H1Y
Enoyl-acyl carrier protein reductase	Reduces the double bond between C2 and C3 positions of a fatty acyl chain bound to the acyl carrier protein.	2PD3, 2PD4
Shikimate patway		
3-dehydroquinate synthase (DHQS)	Converts 3-deoxy-D-arabino-heptulosonate 7-phosphate (DAHP) into 3-dehydroquinate (DHQ)	3CLH
Chorismate synthase	Conversion of 5-enolpyruvylshikimate 3-phosphate to chorismate	1UMF
Shikimate kinase	Phosphorylation of 3-hydroxyl group of shikimic acid in the presence of ATP	1ZUH, 1ZUI
Type II 3-dehydroquinase	Interconversion of 3-dehydroquinate and 3-dehydroshikimate	1J2Y, 2C4W, 2C57, 2C4V

Table 8.2. Continuous from previous page.

Protein	Function	PDB-ID code
Enzymes with antioxidant properties		
Alkyl hydroperoxide-reductase (AhpC)	$ROOH + 2e^- \rightarrow ROH + H_2O$	1ZOF
Thioredoxin reductase	Reduction of Trx	2Q0K, 2Q0L
SOD, superoxide dismutase	Dismutation of superoxide to oxygen and hydrogen peroxide	3CEI
Catalase - HP0875	Conversion of hydrogen peroxide to water and oxygen	1QWL, 1QWM, 2IQF
Sugar metabolism		
Class II fructose-biphosphate aldolase	Cleavage of fructose-1,6-bisphosphate to dihydroxyacetone phosphate and glyceraldehyde-3-phosphate.	3C4U, 3C52, 3C56
Triosephosphate isomerase (TIM)	Interconversion between dihydroxyacetone phosphate and D-glyceraldehyde-3-phosphate	2JGQ
FlaA1 – HP0840	UDP-GlcNAc inverting 4,6-dehydratase	2GN4, 2GN6, 2GN8, 2GN9, 2GNA
Enzymes involved in cell wall biosynthesis		
Glutamate racemase	Stereoinversion of D- and L-glutamate	2JFX, 2JFY, 2JFZ2W4I
D-alanine-D-alanine ligase	Condensation reaction	2PVP
Undecaprenyl pyrophosphate synthase	Consecutive condensation reactions of farnesyl pyrophosphate with eight isopentenyl pyrophosphate	2D2R, 2DTN
UDP-N-acetylglucosamine acytransferase	Transfer of an R-3-hydroxyacyl chain from R-3-hydroxy-acyl carrier protein (ACP) to UDP-N-acetylglucosamine (UDP-GlcNAc) at the glucosamine 3-OH position	1J2Z
YbgC thioesterase (HP0496)	acyl-CoA thioesterase	2PZH
Mannose-6-phosphate isomerase	Isomerisation of of aldose substrates	2QH5

Table 8.2. Continuous from previous page.

Protein	Function	PDF Id code
Vitamin precursors biosynthesis		
Aspartate decarboxylase	conversion of L-aspartate into β-alanine	1UHD, 1UHE
TenA (HP1287)	Thiamin Biosynthesis	2RD3
Dethiobiotin synthetase (bioD)	Biotin biosynthesis	2QMO
NAD biosynthesis		
NAD$^+$ synthetase	Conversion of deamido-NAD$^+$ into NAD$^+$	1XNG, 1XNH
Quinolonic acid phosphoribosyltransferase - HP1355	Transfer of a phosphoribosyl moiety from 5-phosphoribosyl-1-pyrophosphate to quinolinic acid	2B7N, 2B7P, 2B7Q
Miscellanea		
Ferritin	Oxidoreductase, Fe storage	3BVE, 3BVI, 3BVK, 3BVL
Flavodoxin	Electron transfer	1FUE, 2BMV
Inorganic Pyrophosphatase	hydrolysis of inorganic pyrophosphate (PPi) to orthophosphate (Pi)	1YGZ, 2BQY, 2BQX
4-Oxalocrotonate tautomerase homolgue DmpI	Isomerization of b,g-unsaturated enones to their α,β-isomers	2ORM
a1,3-fucosyltransferase	LPS fucosylation	2NZW, 2NZX, 2NZY
ATP-dependent protease FtsH	Degrades misassembled membrane proteins for quality control	2R62, 2R65
Spermidine synthase	Transfer of the aminopropyl group from decarboxylated S-adenosylmethionine to putrescine	2CMG, 2CMH
NAD(P)H-flavin oxidoreductase	Superoxide radicals production	2H0U
hcpB (cysteine rich protein)	Hydrolysis of penicillinic acid	1KLX

Table 8.2. Continuous from previous page.

Protein	Function	PDB-ID code
Enzymes involved in peptides biosynthesis or degradation		
Peptide deformylase	Hydrolytic removal of the N-terminal formyl group from nascent ribosome-synthesised polypeptides	2EW7, 2EW5, 2EW6
Peptidyl-arginine deiminase	Conversion of arginine to citrulline	2CMU
PseC aminotransferase	Pse biosynthetic pathway	2FN6, 2FNI, 2FNU
apo-CopP	Metal transport	1YG0
Aminotransferase AspB	Transamination of branched-chain amino acids to ketoglutarate?	3EZS
Diaminopimelate decarboxylase	Lysine biosynthesis	(2QHG), 3C5Q
g-glutamyl-transpeptidase	Hydrolysis of the gamma-glutamyl bond in glutathione and its conjugates	3FNM, 2QM6, 2QMC, 2NQO
ClpP	Involved in removing of misfolded proteins	2ZL0, 2ZL2, 2ZL3, 2ZL4
ClpX	Involved in removing of misfolded proteins	1UM8
Urea metabolism		
Urease	Hydrolysis of urea to ammonia and carbon dioxide	1E9Y, 1E9Z
UREF	Urease accessory protein, chaperone	2WGL, 3CXN
AmiF formamidase	Hydrolysis of formamide to formic acid and ammonia	2DYU, 2DYV, 2E2K, 2E2L

8.3.1 Enzymes involved in fatty acids metabolism

The structures of three enzymes of *H. pylori* involved in fatty acid metabolism are known. All of them are possible pharmacological targets and *in vitro* inhibitors have been already designed.

- **β-Hydroxyacyl-acyl carrier protein dehydratase** (FabZ, EC 4.2.1.-) is an important enzyme for the elongation cycles of both saturated and unsaturated fatty acids biosyntheses in the type II fatty acid biosynthesis system pathway. This is the *H. pylori* protein with the greatest number of structures deposited (13), corresponding to inhibitor/protein complexes or mutants (Zhang et al., 2008). The enzyme structure is that of an hexamer; differently from the structures of other bacterial FabZs, it contains an extra short two-turn α-helix between α-3 and β-3, which plays an important role in shaping the substrate-binding tunnel.

- **Malonyl-CoA: acyl carrier protein transacylase** (MCAT, EC 2.3.1.39) is a critical enzyme responsible for the transfer of the malonyl moiety to holo-acyl carrier protein (ACP) forming the malonyl-ACP intermediates in the initiation step of type II fatty acid synthesis in bacteria. The crystal structure of MCAT has been determined at 2.5 Å resolution (Zhang et al., 2007). The crystal structure reveals a fold composed of two domains, a large one similar to that of α/β hydrolases, and a smaller ferredoxin-like subdomain.

- **Enoyl-ACP reductase** (ENR; EC 1.3.1.9) is a key enzyme of the Type II fatty-acid biosynthetic system in prokaryotes and plants. It makes use of NADH or NADPH as the cofactor to reduce the double bond between C2 and C3 positions of a fatty acyl chain bound to the acyl carrier protein in the terminal step of the fatty acid chain elongation cycle. The crystal structures of *H. pylori* ENR as ternary complexes with NAD^+ and triclosan has been determined (Lee et al., 2007).

8.3.2 Enzymes of the shikimate pathway

The shikimate pathway for aromatic amino acid biosynthesis is present in bacteria, fungi, and plants, but not in mammals. In this respect, the enzymes of this pathway represent an attractive target for the development of new antimicrobial agents, herbicides, and antiparasitic agents. The structure of four enzymes of the pathway has been determined, but others can be predicted by homology modeling, since various other enzymes are known from different bacteria.

- **Dehydroquinate synthase** (DHQS, EC 4.2.3.4) is a nicotinamide adenine dinucleotide (NAD)-dependent enzyme that converts 3-deoxy-D-arabino-heptulosonate 7-phosphate into 3-dehydroquinate. It catalyzes the second step in the shikimate pathway. The crystal structure complexed with NAD has been determined at 2.4 Å

resolution (Liu *et al.*, 2008). It possesses an N-terminal Rossmann-fold domain and a C-terminal alpha-helical domain (Figure 8.2a).
- **Chorismate synthase** (EC 4.2.3.5) catalyzes the conversion of 5-enolpyruvylshikimate 3-phosphate to chorismate. It requires reduced FMN as a cofactor, but the catalyzed reaction involves no net redox change. Both apo- and holo-protein structures of chorismate synthase have been determined at 2.25 Å resolution (Ahn *et al.*, 2004). The quaternary assembly of the enzyme is tetrameric, each monomer being characterized by a `β–α–β sandwich' fold. The FMN-binding site is mainly formed by a single subunit, with a small contribution from a neighboring one (Figure 8.2b).
- **Shikimate kinase** (EC 2.7.1.71) catalyzes the specific phosphorylation of the 3-hydroxyl group of shikimic acid in the presence of ATP. It is the fifth step in the shikimate pathway. The enzyme structure, solved at 1.8 Å resolution (Cheng *et al.*, 2005), presents an α/β fold, consisting of a central sheet of five parallel β-strands flanked by seven α-helices (Figure 8.2c).
- Other enzymes include **type II 3-dehydroquinase** (EC 4.2.1.10), which catalyzes the third step of the pathway (Lee *et al.*, 2003, Robinson *et al.*, 2006).

From the structures of the previous enzymes, potent *in vitro* inhibitors have been already designed.

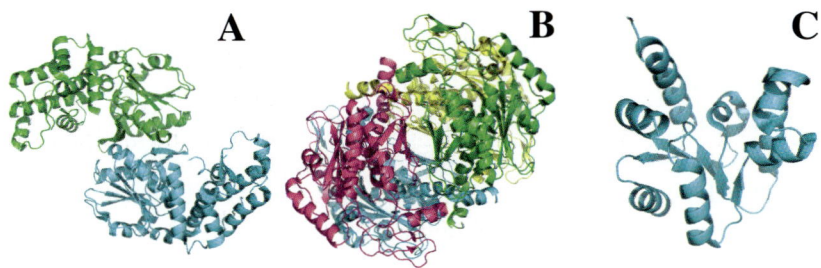

Figure 8.2. Cartoon view of Dehydroquinate synthase (A), Chorismate synthase (B) and Shikimate kinase (C) from *H. Pylori*.

8.3.3 *Enzymes with antioxidant properties*

The ability to respond to oxidative stress is of paramount importance for bacteria living in a difficult and hostile environment, like *H. pylori*. The inhibition of enzymes involved in the protection of the bacterium can be an efficient way to control the infection.

- AhpC is a thioredoxin (Trx)-dependent **alkyl hydroperoxide-reductase** (Figure 8.3A). It is a member of the ubiquitous 2-Cys peroxiredoxins family (2-Cys Prxs), a group of thiol-specific antioxidant enzymes. Prxs exert a protective antioxidant role in cells through their peroxidase activity, whereby hydrogen peroxide, peroxynitrite and a wide range of organic hydroperoxides (ROOH) are reduced and detoxified. Crystals of the recombinant protein diffract to 2.95 Å resolution (Papinutto *et al.*, 2005). The model shows an overall fold similar to that of the other 2-Cys Prx family members crystallized till now. These toroid-shaped complexes consist of a pentameric arrangement of homodimers [$(\alpha_2)_5$ decamer]. However, the model of AhpC from *H. pylori* presents peculiar differences with respect to other members of the family: apart from some loop regions, α5-helix and the C-terminus is shifted, preventing the C-terminal tail of the second subunit from extending toward this region of the molecule.
- Homodimeric **thioredoxin reductase** (TrxR). Thioredoxin in bacteria, thanks to the low redox potential, is the major dithiol reductant in the cytosol (Koharyova and Kolarova, 2008) (Figure 8.3B). The crystal structure of TrxR has been determined for both the oxidized and reduced form of the enzyme, at 1.7 and 1.45 Å resolution, respectively (Gustafsson *et al.*, 2007). The enzyme subunit is formed by two nucleotide-binding domains, FAD- and NAD(P)H, each of which is a variant of the classical Rossmann fold. A comparison of the corresponding proteins from *H. pylori* and *E. coli* suggests that the overall surface charge distribution in these proteins is conserved and that residue substitutions that change the shape of the binding surface may account for the species-specific recognition of thioredoxin by TrxR.

- **Superoxide dismutase** (SOD). Superoxide dismutases (SOD) are key enzymes for fighting oxidative stress. *H. pylori* produces a single SOD which contains iron. Its structure has been determined at 2.4 Å resolution (Esposito *et al.*, 2008) (Figure 8.3C). It is a dimer of two identical subunits with one iron ion per monomer. The protein shares 53% sequence identity with the corresponding enzyme from *E. coli*. The model, if compared with that of other dimeric Fe-containing SODs, shows a significant difference represented by an extended C-terminal tail, which makes HpSOD the biggest FeSOD reported to date. It is not clear which is the physiological function of this extended carboxyl terminus.
- Excessive hydrogen peroxide is harmful for almost all cell components, so its rapid and efficient removal is of essential importance for aerobically living organisms. Conversely, hydrogen peroxide acts as a second messenger in signal-transduction pathways. H_2O_2 is degraded by peroxidases and catalases, the latter being able both to reduce H_2O_2 to water and to oxidize it to molecular oxygen. *H. pylori* produces one monofunctional **catalase** (EC 1.11.1.6). The crystal structure, refined at 1.6 Å resolution (Loewen *et al.*, 2004), is closely related to those of other catalases already determined. Moreover, it appears to be the only class III catalase till now characterized that does not bind NADPH.

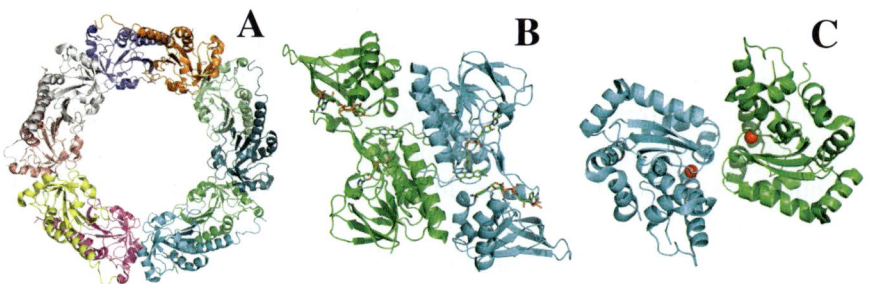

Figure 8.3. Cartoon view of AhpC (A), Thioredoxin reductase (B) and SOD (C). FAD and NAD(P)H cofactors are shown as ball-and-stick, Fe ions as red spheres.

8.3.4 *Enzymes involved in sugar metabolism*

The structure of two enzymes involved in sugar metabolism is known for this bacterium: **Triosephosphate isomerase** (TIM, EC 5.3.1.1, (Chu et al., 2008)) and Class II (zinc-dependent) **fructose bisphosphate aldolases** (Fonvielle et al., 2008) (EC 4.1.2.13). They both are classical enzymes and their overall fold does not differ significantly from that of the orthologue enzymes from other species.

In addition, the structure of **UDP-GlcNAc inverting 4,6-dehydratase** has been determined (Ishiyama et al. 2006). The *flaA1 (HP0840)* gene product has been shown to be involved in the synthesis of both flagella and LPS. FlaA1 is a 37-kDa protein whose sequence suggests that it is a member of the short-chain dehydrogenase/reductase superfamily.

8.3.5 *Enzymes involved in cell wall biosynthesis*

Enzymes essential to the bacterial cell wall biosynthesis pathway are considered as attractive targets for antibacterial drug discovery. In this respect two key enzymes of *H. pylori* are **Glutamate racemase** (Lundqvist et al., 2007) and **D-Alanine-D-alanine ligase** (Wu et al., 2008). The former presents a complex mechanism of regulation, including allosteric and kinetic activation, and uncompetitive inhibitors of the *H. pylori* enzyme have been used to probe the mechanistic and dynamic features of the enzyme. D-Alanine-D-alanine ligase is the second enzyme in the D-Ala branch of bacterial cell wall peptidoglycan assembly. Its structure has revealed important differences with the same enzymes of other gram-negative bacteria.

Other important enzymes for maintaining the integrity of the cell wall are:
- **Undecaprenyl pyrophosphate synthase** (UPPS, E.C. 2.5.1.31; (Kuo et al., 2008), which catalyzes consecutive condensation reactions of farnesyl pyrophosphate with eight isopentenyl pyrophosphate to form lipid carrier for bacterial peptidoglycan biosynthesis.

- **UDP-N-acetylglucosamine acyltransferase** (LpxA), the first enzyme of the lipid A biosynthetic pathway. It catalyzes the transfer of an R-3-hydroxyacyl chain from R-3-hydroxy-acyl carrier protein to UDP-N-acetylglucosamine at the glucosamine 3-OH position. Lipid A is the hydrophobic anchor of lipopolysaccharide in Gram-negative bacteria and is required for growth of most Gram-negative bacteria and for maintaining the integrity of the outer membrane as a barrier to toxic chemicals (Nikaido 1996). The crystal structure of LpxA from *H. pylori* has been solved (Lee and Suh, 2003) using 2.1 Å data and a model for the complex between LpxA and ACP has been proposed.
- **YbgC** proteins are bacterial acyl-CoA thioesterases associated with the Tol-Pal system. This system is important for cell envelope integrity and is part of the cell division machinery. In *E. coli*, YbgC associates with the cell membrane and is part of a protein network involved in lipid biogenesis. The putative homologue of YbgC in *H. pylori*, named HP0496, was found to interact with the cytotoxin CagA. The structure of HP0496 shows that it is a member of the hot-dog family of proteins, with an $\varepsilon-\gamma$ tetrameric arrangement (Angelini *et al.*, 2008). Enzymatic assays performed with the purified protein showed that HP0496 is a thioesterase enzyme more active against long-chain acyl-CoA substrates.

8.3.6 *Vitamins precursors*

L-Aspartate α-decarboxylase (ADC) catalyzes the conversion of L-aspartate into β-alanine, which is required for the synthesis of pantothenate (vitamin B_5), which is the precursor of 4'-phosphopantetheine and coenzyme A. The crystal structure shows that *H. pylori* ADC is a tetrameric enzyme (Lee and Suh 2004). Each subunit contains nine β-strands and three α-helices, folded into a double-ψ β-barrel structure. The isoasparagine complex structure shows the substrate analog covalently attached to the pyruvoyl group of the enzyme, so representing the enzyme-substrate Schiff base intermediate.

Other enzymes involved in the synthesis of vitamins are **TenA** (HP1287), which catalyzes a step in the synthesis of Thiamin (Vitamin

B_1; Barison *et al.*, in press) and **dethiobitin synthetase** (Chruszcz *et al.*, unpublished, PDB ID 2qmo) involved in the synthetic pathway of Biotin (Vitamin B_7 or H).

8.3.7 *Enzymes involved in NAD biosynthesis*

Nicotinamide adenine dinucleotides (NAD^+ and $NADP^+$) play a central role in all living systems. They are essential and ubiquitous coenzymes, and are involved in biochemical processes ranging from redox reactions to DNA repair, DNA recombination, and protein-adenosine diphosphate (ADP) ribosylation (Foster and Moat 1980). The structure of two enzymes has been solved in this group: **type II quinolinic acid phosphoribosyltransferase** (Hp-QAPRTase) and **NAD^+ synthetase**. Hp-QAPRTase catalyzes the transfer of a phosphoribosyl moiety from 5-phosphoribosyl-1-pyrophosphate (PRPP) to quinolinic acid (QA), yielding nicotinic acid mononucleotide (NAMN), pyrophosphate and CO_2, the last resulting from decarboxylation at position 2 of the quinolinate ring. This step is central and indispensable to NAD biosynthesis and cell survival in prokaryotes. The crystal structure is a trimer, each monomer being arranged in two domains (Kim *et al.*, 2006). The second enzyme, NAD^+ synthetase [NadE; EC 6.3.5.1], catalyzes a key reaction in NAD^+ biosynthesis by converting deamido-NAD^+ into the final product NAD^+ by a two-steps reaction. The crystal structures of the *H. pylori* enzyme in apo- and complex forms with NaAD and ATP were determined to a resolution of 2.3 and 1.7 Å, respectively (Kang *et al.*, 2005).

8.3.8 *Enzymes involved in peptides biosynthesis or degradation*

H. pylori **γ-glutamyltranspeptidase** (HpGT) is a general γ-glutamyl hydrolase and a demonstrated virulence factor. The enzyme confers a growth advantage to the bacterium, providing essential amino acid precursors by initiating the degradation of extracellular glutathione and glutamine. HpGT is a member of the N-terminal nucleophile (Ntn) hydrolase superfamily and undergoes autoprocessing to generate the active form of the enzyme. Its structure has been determined in complex with glutamate (1.6 Å), in an inactive HpGT mutant, T380A, in complex with S-(nitrobenzyl)glutathione (1.55 Å) and in complex with acivicin at 1.7 Å resolution. (Morrow *et al.*, 2007, Williams *et al.*, 2009).

Peptide deformylase (EC 3.5.1.27) catalyses the hydrolytic removal of the N-terminal formyl group from the nascent ribosome-synthesized polypeptide. Its activity is essential and it is present in all eubacterias and in the organelles of some eukaryotes. It utilizes a Fe^{2+} ion to catalyze the hydrolysis of an amide bond. The crystal structure has been solved in complex with different inhibitors (Cai *et al.*, 2006).

Peptidyl-arginine deiminase catalyzes the post-translational modification of proteins through the conversion of arginine to citrulline in the presence of calcium ions. The structure of a putative enzyme from *H. pylori* has been determined (Rajashankar et al., unpublished, PDB ID 2cmu).

Meso-diaminopimelate decarboxylase (DAPDC, EC 4.1.1.20) catalyzes the final step of L-lysine biosynthesis in bacteria. In its 2.3 Å crystal structure, the product L-lysine forms a Schiff base with the cofactor pyridoxal 5'-phosphate (Hu *et al.*, 2008).

PseC aminotransferase. Flagellins in *H. pylori* are heavily glycosylated with the novel sialic acid-like nonulosonate, pseudaminic acid (Pse). The glycosylation process is fundamental for the assembly of functional flagellar filaments and consequent bacterial motility. The amino transfer from UDP-2-acetamido-2,6-dideoxy-beta-L-arabino-4-hexulose is catalyzed by the aminotransferase PseC (HP0366). The crystal structures of the native protein, its complexes with pyridoxal phosphate alone and in combination with the UDP-4-amino-4,6-dideoxy-beta-L-AltNAc product, have been determined (Schoenhofen *et al.*, 2006).

ClpP and its ATPase compartment, ClpX or ClpA, remove misfolded proteins in cells and are of utmost importance in protein quality control. The ring hexamers of ClpA or ClpX recognize, unfold, and translocate target substrates into the degradation chamber of the double-ring tetradecamer of ClpP. The crystal structures of ClpP (Kim and Kim 2008) and ClpX have been determined, in complex with product peptides bound to the active site as well as in the apo state. The structures provide structural explanations for the broad substrate specificity and the product inhibition. The structures also suggest that substrate binding causes local conformational changes around the active site that ultimately induce the active conformation of ClpP. The overall structure of **ClpX** (Kim and Kim, 2003) is similar to that of heat shock locus U (HslU), consisting of two subdomains, with ADP bound at the subdomain interface. The nucleotide binding environment provides the structural explanation for the hexameric assembly and the modulation of ATPase activity.

8.3.9 *Urea metabolism*

H. pylori, a neutralophilic bacterium, is able to survive to the acid stomach environment mainly thanks to the presence of the urease system (Voland *et al.*, 2003). Regulation of the activity and assembly of urease, a Ni^{2+}-containing oligomeric heterodimer, is important for gastric habitation. The gene complex of the urease system encodes catalytic subunits (ureA/B), an acid-gated urea channel (ureI), and accessory assembly proteins (ureE-H). The urease enzyme neutralizes gastric acid by producing ammonia for the survival of the bacteria. Up to 30% of the enzyme associates with the surface of intact cells upon lysis of neighboring bacteria.

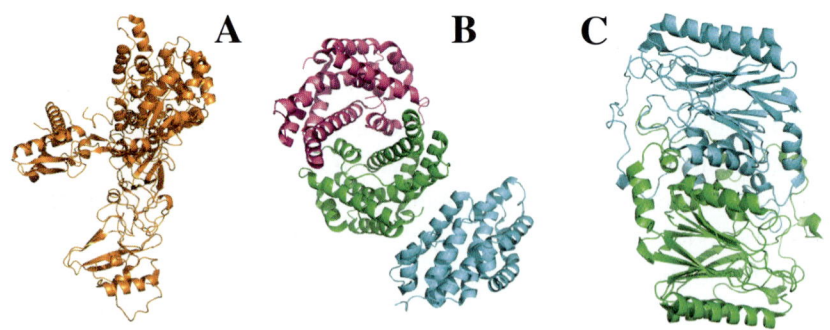

Figure 8.4. Cartoon view of the asymmetric unit of urease (A), PDB ID 1e9y), UREF (B, PDB ID 3cxn) and AmiF (C, PDB ID 2dyu).

The crystal structure of *H. pylori* **urease** has been solved at a resolution of 2.9 Å (Figure 8.4A). It comprises twelve catalytic units assembled in a 1.1 MDa spherical assembly, with an outer diameter of approximately 160 Å (Ha *et al.*, 2001). Under physiologically relevant conditions, the activity of the enzyme remains unaffected down to pH 3.

The crystal structure of other two accessory proteins of the urease system has been determined: **UREF** (Lam *et al.*, unpublished, PDB ID 3cxn) and **AmiF formamidase** (Hung *et al.*, 2007) (Figure 8.4B,C). The latter hydrolyzes formamide to produce formic acid and ammonia. It belongs to the nitrilase superfamily. The crystal structure of AmiF was solved to 1.75 Å resolution. The structure consists of a homohexamer, each subunit being related to the others by a 3-fold symmetry. Each subunit presents an α–β–β–α four-layer architecture, characteristic of the nitrilase superfamily.

8.3.10 *Miscellanea*

The crystal structure of HP **ferritin**, determined in its apo, low-iron-bound, intermediate, and high-iron-bound states (Cho *et al.*, 2009), is similar to other members of the ferritin family: it assembles as a spherical protein shell of 24 subunits, each of which folds into a four-helix bundle. The imidazole rings at the channel entrance are directly coupled to axial translocation of Fe ions through the 4-fold channel. These structures provide a structural evidence of the transition from a protein-dominated process to a mineral-surface-dominated process during biomineralization.

Flavodoxins are electron-transfer proteins involved in a variety of photosynthetic and non-photosynthetic reactions in bacteria (in eukaryotes, flavodoxins are subunit of multidomain proteins) (Sancho 2006). The redox activity of flavodoxin is due to its bound flavin mononucleotide cofactor (FMN). Flavodoxin has been shown to be reduced by the pyruvate-oxidoreductase (POR) enzyme complex of *H. pylori*. It was also suggested to be involved in the pathogenesis of gastric mucosa-associated lymphoid-tissue lymphoma (MALToma) and, being essential for the survival of the bacterium, the inhibition of the interaction between the POR-enzyme complex and flavodoxin could be of pharmacological relevance. Its X-ray structure is similar to flavodoxins of other bacteria (Freigang *et al.*, 2002). The structure of the apoprotein (Martinez-Julvez *et al.*, 2007) has allowed to explain dynamics features of this protein.

Inorganic pyrophosphatase is a ubiquitous cytosolic enzyme which catalyzes the hydrolysis of inorganic pyrophosphate to orthophosphate. The crystal structure of inorganic pyrophosphatase from *H. pylori* has been solved at 2.6 Å resolution (Wu *et al.*, 2005). The crystallographic asymmetric unit contains a homohexameric enzyme, arranged as a dimer of trimers. While most of the structural elements of the molecular model are highly conserved in the other enzymes of the family, some unique structural features are localized in the flexible loops near the active site, suggesting that the structural flexibility of these loops is required for the catalytic efficiency of PPase

4-Oxalocrotonate tautomerase (4-OT) catalyzes the isomerization of β,γ-unsaturated enones to their α,β-isomers. The enzyme is part of a plasmid-encoded pathway, which enables bacteria harboring the plasmid to use various aromatic hydrocarbons as their sole sources of carbon and energy (Hackert *et al.*, unpublished, PDB ID 2orm).

From the host defense aspect, innate immune recognition of *H. pylori* by surfactant protein D is influenced by the extent of LPS **fucosylation**. Furthermore, Le antigen expression affects both the inflammatory response and T-cell polarization that develops after infection. Although controversial, Le antigen expression and fucosylation in *H. pylori* have multiple biological effects on pathogenesis. $\alpha 1,3$-fucosyltransferase structure has been determined for the apo form and in complex with GDP, GDP-fucose (Sun *et al.*, unpublished, PDB ID 2nzw).

The **ATP-dependent protease**, FtsH, degrades misassembled membrane proteins for quality control. The crystal structures of FtsH ATPase domain in the nucleotide-free state and complexed with ADP, were determined (Kim *et al.*, 2008). The FtsH protein has an N-terminal transmembrane segment and a large cytosolic region that consists of two domains, an ATPase and a protease domain.

Spermidine synthase (putrescine aminopropyltransferase, PAPT,) catalyzes the transfer of the aminopropyl group from decarboxylated S-adenosylmethionine to putrescine during spermidine biosynthesis.

H. pylori PAPT (HpPAPT) has a low sequence identity with other PAPTs and a unique binding pocket between the two-domains structure (Lu *et al.*, 2007).

The structure of *H. pylori* **cysteine-rich protein** (HcpB) possesses a modular architecture consisting of four α/α-motifs that are cross-linked by disulfide bridges (Luthy *et al.*, 2002, Luthy *et al.*, 2004). It is capable of binding and hydrolyzing 6-amino penicillinic acid and 7-amino cephalosporanic acid derivatives and presents a fold distinct from that of any known penicillin-binding protein. The putative penicillin binding site is located in an amphipathic groove on the concave side of the molecule.

Apo-CopP is a hypothetical metal binding protein, possibly involved in Copper homeostasis (Smith *et al.*, 2009).

8.4 Proteins Interacting with DNA and Regulators of Gene Expression

The structure of five proteins interacting with nucleic acids are available. One is a **single-stranded DNA-binding protein** (SSB). SBB are known to play important roles in DNA replication, recombination, and repair. The crystal structure of a C-terminally truncated form has been determined in complex with a 35-mer ssDNA at a resolution of 2.3 Å (Chan *et al.*, 2009). SSB from *H. pylori* consists of an N-terminal ssDNA-binding domain with an oligonucleotide/oligosaccharide binding fold and a flexible C-terminal tail. The latter is possibly involved in protein-protein interactions. The second protein, *H. pylori* **IS608 transposase (TnpA)**, belongs to the IS200/IS605 family, which represents the smallest known DNA transposases (Barabas *et al.*, 2008). Its 3D structure explains how a TnpA-bound sub-terminal transposon DNA segment recognizes, by complementary base pairing, the specific DNA sequence to be cleaved, in a way reminiscent of the interdependence of protein and RNA in ribosomes.

DNA glycosylases catalyze the excision of chemically modified bases from DNA. The high-resolution crystal structures available are those of the m3A DNA glycosylase in the unliganded form and bound to alkylated bases 3,9-dimethyladenine and 1,N6-ethenoadenine (Eichman *et al.*, 2003). The enzyme achieves its specificity for positively-charged m3A not by direct interactions with purine or methyl substituent atoms, but rather by stacking the base between two aromatic side chains in a pocket that excludes 7-methylguanine.

Other proteins of this class are the C-terminal domain of a transcriptional regulatory protein, **ArsR DNA binding domain** (Gupta et al., unpublished, PDB ID 2K4J) and the**Hpy99I restriction endonuclease** (Sokolowska et al., unpublished, PDB ID 3fc3).

Other enzymes regulate gene expression indirectly, without interacting with nucleic acids. They are:

- **HobA** (HP1230), a protein that interacts with DnaA. In prokaryotes, the binding of DnaA to the oriC region of the chromosome DNA replication loads the primosome machinery and starts a new replication round. Among the proteins that control these events there is HobA, whose crystal structure has been solved at 1.7 Å. (Natrajan et al., 2007). The closest structural homologue of HobA is a sugar isomerase (SIS) domain containing protein, the phosphoheptose isomerase from *Pseudomonas aeruginosa* (Taylor et al., 2008). Whilst SIS proteins share strong sequence homology with DiaA from *E. coli*, HobA and DiaA have no sequence homology, despite they are structural homologues.

- **HP-RR** (HP1043). The three-dimensional structure of HP-RR by NMR and x-ray crystallography has been determined, revealing a symmetrical dimer with two functional domains (Hong et al., 2007). It is a new type of response regulator; it is essential for cell growth and does not require the well known phosphorelay scheme. HP-RR belongs to the prokaryotes two-components signal transduction systems, which regulate, through a phosphorylation-dependent process, cellular functions in response to environmental conditions. Its molecular topology resembles that of the OmpR/PhoB subfamily. It has been proposed that HP-RR is a new type of response regulator protein, which function as a cell growth-associated regulator in the absence of post-translational modification.

- **Luxs**. Quorum sensing is the mechanism used by bacteria to control gene expression in response to cell density. Characteristic signaling molecules, called AI-1 and AI-2 (autoinducer-1 and autoinducer-2), characterize the two major quorum-sensing systems, system 1 and system 2, respectively. The luxS gene is required for the AI-2 system of quorum sensing. LuxS and AI-2 are present in both Gram-negative and Gram-positive bacterial species and have been shown to be involved in the expression of virulence genes in several pathogens. Various structures of the LuxS protein from *H. pylori* have been reported with resolutions ranging from 1.8 Å to 2.4 Å resolution (Lewis *et al.*, 2001). Luxs monomer is characterized by an α/β fold, with a zinc ion bound to a Cys-His-His triad. Acting as a homodimer, the protein binds a methionine analog, S-ribosylhomocysteine (SRH). The zinc atom is in position to cleave the ribose ring in a step along the synthesis pathway of AI-2.

Table 8.3. List of the proteins that interact with nucleic acids or regulate protein expression.

Protein or gene	Hypothetical function	PDB ID code and reference
ISHO608 transposase	IS2000/IS605 family DNA transposase	2VIC, 2VIH, 2VJU, 2VJV, 2A6M, 2ALO
SSB	Single stranded DNA binding protein	2VW9
ArsR DNA binding domain	Putative transcriptional regulator	2K4J
Hpy99I b-b-a-Me type II restriction endonuclease	Restriction endonulcease	3FC3
3-methyladenine DNA glycolsylase (magIII)	Excision of chemically modified bases from DNA	1PU7, 1PU8
HOBA (HP1230)	Interacts with DnaA	2UVP
HP1043	Orphan response regulator	2PLN, 2HQN, 2HQO, 2HQR
Luxs		1J6X
apo-NiKR	Putative nickel response regulator	2CA9, 2CAD, 2CAJ

Nickel is a necessary cofactor of urease, whose activity is fundamental to the survival of *H. pylori* in the human stomach. Nickel-responsive gene regulation is mediated by **HpNikR**, a protein belonging to the ribbon-helix-helix family of transcriptional regulators. It has the double function of repressor and activator within an acid adaptation cascade. The crystal structure of the full-length HpNikR in a nickel-free conformation and two nickel-bound structures have been reported (Dian et al., 2006). Despite the overall fold of the protein is similar to its bacterial homologues, the three structures have allowed the determination of the route for nickel within HpNikR. In addition, they have revealed the cooperativity between the tetramerization domain and the DNA-binding domain.

8.5 Toxins and Other Proteins Directly Involved in Pathogenicity

VacA. *H pylori* VacA is a pore-forming toxin secreted by an autotransporter pathway. It causes multiple alterations in human cells and contributes to the pathogenesis of peptic ulcer disease and gastric cancer. Owing to its antigenic properties, it is a candidate antigen for inclusion in an *H. pylori* vaccine. VacA p55 domain plays an important role in mediating VacA binding to host cells and its crystal structure has been solved at 2.4 Å resolution (Gangwer et al., 2007). The structure is predominantly a right-handed parallel β-helix, a feature that is characteristic of autotransporter passenger domains.

Table 8.4. Toxins and other proteins directly involved in pathogenicity.

Protein	Hypothetical function	PDB ID code and reference
Vacuolating toxin p55 domain	Pore forming toxin	2QV3
TIP-alpha N34 (HP0596)	Virulence factor	2WCQ, 2WCR
NAP-A	Fe storage; Neutrophil activating protein	1JI4
MotB (domain)	Component of the bacterial flagellar motor	3CYP, 3CYQ
HpaA	Neuraminillactose-binding hemagglutinin homolog	3BGH
FIC	Filamentation protein	2F6S

TIPalpha is a unique virulence factor secreted by the bacterium. It enters gastric cells and both stimulates the production of the TNF-alpha and activates the NF-kappaB pathway. The crystal structure of a truncated version of Tipalpha (TipalphaN34) is available in two crystal forms (Tosi *et al.*, 2009). Tipα adopts a novel β1α1α2β2β3α3α4 topology arranged in three domains, possibly stabilized during the secretion process by several disulphide bridges.

NAP-A. An *H. pylori* protein that is highly immunogenic in humans and mice has been called HP-NAP or NAP-A, due to its ability of activating neutrophils. Its quaternary structure (Zanotti *et al.*, 2002) is similar to that of the dodecameric bacterial ferritins, but it has a different surface potential charge distribution. The presence of a large number of positively charged residues on its surface could well account for its unique ability in activating human leukocytes.

MotB. Important components of the bacterial flagellar motor are MotA and MotB proteins: they act together to generate a torque force acting on the FliG ring of the rotor. The C-terminal domain of MotB (MotB-C) possibly anchor the MotA/MotB complex to peptidoglycan of the cell wall. The first crystal structures of MotB-C and its complex with N-acetylmuramic acid (NAM) have been determined to 1.6- and 2.3-Å resolution, respectively (Roujeinikova 2008).

Neuraminyllactose-binding hemagglutinin homolog. The ability of certain strains of *H. pylori* to cause sialic acid-sensitive agglutination of erythrocytes has been attributed to the HpaA protein (Evans *et al.*, 1993). The crystal structure of this protein has been determined (Bonanno *et al.*, unpublished, PDB ID 3bgh).

Proteins containing **FIC** (filamentation induced by cyclic adenosine monophosphate) domains are found in both prokaryotic and eukaryotic organisms, but their function is not clear yet. Several data indicate that bacterial **FIC** domain-containing proteins, after being delivered into eukaryotic host cells, disrupt host cell processes. The crystal structure of HP-FIC has been determined by a Structural Genomic consortium (Cuff *et al.*, unpublished, PDB ID 2f6s).

8.6 Unknown Function Proteins

There are at least 17 proteins (Table 8.5) of known three-dimensional structure whose function is totally unknown. Some of them are considered relevant for pathogenesis or for survival of the bacterium. It must be underlined that the crystal structure in itself is often not sufficient to discover the function of a new protein.

Table 8.5. Proteins of unknown function.

Protein	Hypothetical function	PDB ID code and reference
HP0035	Unknown	3F42
HP0062	Unknown	2GTS
HP0184	Unknown	2ATZ
HP0218	PEPB-like	2EVV
HP0222	Unknown	1X93
HP0242	Unknown	2BO3
HP0242	Unknown	2OUF
HP0492	Unknown	2I9I
HP0495	Unknown	2H9Z, 2JOQ
HP0564	Unknown	2KIO
HP0892	Unknown	2OTR
HP0894	Unknown	1Z8M
HP1203	Unknown	2KOZ
HP1242	Unknown	1ZHC
HP1531	Unknown	1ZKE
Y162	Unknown	1MW7
HP (2-20)	Microbial peptide	1OT0

8.7 Conclusions

The three-dimensional structure of less than 80 different proteins is presently available for *H. pylori*. This is apparently a quite small number, if compared to the total of the gene products predicted for this organism (about 1500). One has anyhow to consider that a large fraction of the protein of the bacterium are common to other prokaryotes and their structure can be predicted, with reasonable reliability, by homology modeling. The remaining proteins, those with a completely new fold or without a significant sequence similarity with proteins of known structure, can be estimated to be than 20–30% of the genome. We can perhaps predict that in a reasonable time, possibly in the next 10–20 years, a complete picture of the bacterium from the structural point of view will be available. This, coupled with the huge amount of data on the physiology and biochemistry of *H. pylori* accumulated, will possibly allow a complete description of its behavior from a molecular point of view.

References

Ahn, H. J., Yoon, H. J., Lee, B. I. and Suh, S. W. (2004). Crystal structure of chorismate synthase: a novel FMN-binding protein fold and functional insights. *Journal of Molecular Biology,* 336, pp. 903–915.

Akopyants, N. S., Clifton, S. W., Kersulyte, D., Crabtree, J. E., Youree, B. E., Reece, C. A., Bukanov, N. O., Drazek, E. S., Roe, B. A. and Berg, D. E. (1998). Analyses of the cag pathogenicity island of Helicobacter pylori. *Molecular Microbiology,* 28, pp. 37–53.

Alm, R. A., Ling, L. S., Moir, D. T., King, B. L., Brown, E. D., Doig, P. C., Smith, D. R., Noonan, B., Guild, B. C., deJonge, B. L., Carmel, G., Tummino, P. J., Caruso, A., Uria-Nickelsen, M., Mills, D. M., Ives, C., Gibson, R., Merberg, D., Mills, S. D., Jiang, Q., Taylor, D. E., Vovis, G. F. and Trust, T. J. (1999). Genomic-sequence comparison of two unrelated isolates of the human gastric pathogen Helicobacter pylori. *Nature,* 397, pp. 176–180.

Alm, R. A. and Trust, T. J. (1999). Analysis of the genetic diversity of Helicobacter pylori: the tale of two genomes. *Journal of Molecular Medicine,* 77, pp. 834–846.

Angelini, A., Cendron, L., Goncalves, S., Zanotti, G. and Terradot, L. (2008). Structural and enzymatic characterization of HP0496, a YbgC thioesterase from Helicobacter pylori. *Proteins,* 72, pp. 1212–1221.

Appelmelk, B. J., Monteiro, M. A., Martin, S. L., Moran, A. P. and Vandenbroucke-Grauls, C. M. (2000). Why Helicobacter pylori has Lewis antigens. *Trends in Microbiology,* 8, pp. 565–570.

Barabas, O., Ronning, D. R., Guynet, C., Hickman, A. B., Ton-Hoang, B., Chandler, M. and Dyda, F. (2008). Mechanism of IS200/IS605 family DNA transposases: activation and transposon-directed target site selection. *Cell,* 132, pp. 208–220.

Berman, H. M., Westbrook, J., Feng, Z., Gilliland, G., Bhat, T. N., Weissig, H., Shindyalov, I. N. and Bourne, P. E. (2000). The Protein Data Bank. *Nucleic Acids Research,* 28, pp. 235–242.

Blaser, M. J. (1998). Helicobacter pylori and gastric diseases. *Britsh Medical Journal,* 316, pp. 1507–1510.

Bourzac, K. M. and Guillemin, K. (2005). Helicobacter pylori-host cell interactions mediated by type IV secretion. *Cellular Microbiology,* 7, pp. 911–919.

Brandt, S., Kwok, T., Hartig, R., Konig, W. and Backert, S. (2005). NF-kappaB activation and potentiation of proinflammatory responses by the Helicobacter pylori CagA protein. *Proceedings of the National Academy of Sciences USA,* 102, pp. 9300–9305.

Cai, J., Han, C., Hu, T., Zhang, J., Wu, D., Wang, F., Liu, Y., Ding, J., Chen, K., Yue, J., Shen, X. and Jiang, H. (2006). Peptide deformylase is a potential target for anti-Helicobacter pylori drugs: reverse docking, enzymatic assay, and X-ray crystallography validation. *Protein Science,* 15, pp. 2071–2081.

Cendron, L., Couturier, M., Angelini, A., Barison, N., Stein, M. and Zanotti, G. (2009). The Helicobacter pylori CagD (HP0545, Cag24) protein is essential for CagA translocation and maximal induction of interleukin-8 secretion. *Journal of Molecular Biology*, 386, pp. 204–217.

Cendron, L., Seydel, A., Angelini, A., Battistutta, R. and Zanotti, G. (2004). Crystal structure of CagZ, a protein from the Helicobacter pylori pathogenicity island that encodes for a type IV secretion system. *Journal of Molecular Biology*, 340, pp. 881–889.

Cendron, L., Tasca, E., Seraglio, T., Seydel, A., Angelini, A., Battistutta, R., Montecucco, C. and Zanotti, G. (2007). The crystal structure of CagS from the Helicobacter pylori pathogenicity island. *Proteins*, 69, pp. 440–443.

Censini, S., Lange, C., Xiang, Z., Crabtree, J. E., Ghiara, P., Borodovsky, M., Rappuoli, R. and Covacci, A. (1996. cag, a pathogenicity island of Helicobacter pylori, encodes type I-specific and disease-associated virulence factors. *Proceedings of the National Academy of the Sciences USA*, 93, pp. 14648–14653.

Chan, K. W., Lee, Y. J., Wang, C. H., Huang, H. and Sun, Y. J. (2009). Single-stranded DNA-binding protein complex from Helicobacter pylori suggests an ssDNA-binding surface. *Journal of Molecular Biology*, 388, pp. 508–519.

Cheng, W. C., Chang, Y. N. and Wang, W. C. (2005). Structural basis for shikimate-binding specificity of Helicobacter pylori shikimate kinase. *Journal of Bacteriology*, 187, pp. 8156–8163.

Cho, K. J., Shin, H. J., Lee, J. H., Kim, K. J., Park, S. S., Lee, Y., Lee, C., Park, S. S. and Kim, K. H. (2009). The crystal structure of ferritin from Helicobacter pylori reveals unusual conformational changes for iron uptake. *Journal of Molecular Biology*, 390, pp. 83–98.

Christie, P. J., Atmakuri, K., Krishnamoorthy, V., Jakubowski, S. and Cascales, E. (2005). Biogenesis, architecture, and function of bacterial type IV secretion systems. *Annual Review of Microbiology*, 59, pp. 451–485.

Chu, C. H., Lai, Y. J., Huang, H. and Sun, Y. J. (2008). Kinetic and structural properties of triosephosphate isomerase from Helicobacter pylori. *Proteins*, 71, pp. 396–406.

Covacci, A., Censini, S., Bugnoli, M., Petracca, R., Burroni, D., Macchia, G., Massone, A., Papini, E., Xiang, Z. and Figura, N. (1993). Molecular characterization of the 128-kDa immunodominant antigen of Helicobacter pylori associated with cytotoxicity and duodenal ulcer. *Proceedings of the National Academy of the Sciences USA*, 90, pp. 5791–5795.

Covacci, A., Falkow, S., Berg, D. E. and Rappuoli, R. (1997). Did the inheritance of a pathogenicity island modify the virulence of Helicobacter pylori? *Trends in Microbiology*, 5, pp. 205–208.

Covacci, A., Telford, J. L., Del Giudice, G., Parsonnet, J. and Rappuoli, R. (1999). Helicobacter pylori virulence and genetic geography. *Science*, 284, pp. 1328–1333.

Cover, T. L. and Blaser, M. J. (2009). Helicobacter pylori in health and disease. *Gastroenterology*, 136, pp. 1863–1873.

Dian, C., Schauer, K., Kapp, U., McSweeney, S. M., Labigne, A. and Terradot, L. (2006). Structural basis of the nickel response in Helicobacter pylori: crystal structures of

HpNikR in Apo and nickel-bound states. *Journal of Molecular Biology*, 361, pp. 715–730.

Eichman, B. F., O'Rourke, E. J., Radicella, J. P. and Ellenberger, T. (2003). Crystal structures of 3-methyladenine DNA glycosylase MagIII and the recognition of alkylated bases. *EMBO Journal*, 22, pp. 4898–4909.

Esposito, L., Seydel, A., Aiello, R., Sorrentino, G., Cendron, L., Zanotti, G. and Zagari, A. (2008). The crystal structure of the superoxide dismutase from Helicobacter pylori reveals a structured C-terminal extension. *Biochimica et Biophysica Acta - Proteins and Proteomics*, 1784, pp. 1601–1606.

Evans, D. G., Karjalainen, T. K., Evans, D. J., Jr, Graham, D. Y. and Lee, C. H. (1993). Cloning, nucleotide sequence, and expression of a gene encoding an adhesin subunit protein of Helicobacter pylori. *Journal of Bacteriology*, 175, pp. 674–683.

Fonvielle, M., Coincon, M., Daher, R., Desbenoit, N., Kosieradzka, K., Barilone, N., Gicquel, B., Sygusch, J., Jackson, M. and Therisod, M. (2008). Synthesis and biochemical evaluation of selective inhibitors of class II fructose bisphosphate aldolases: towards new synthetic antibiotics. *Chemistry*, 14, pp. 8521–8529.

Foster, J. W. and Moat, A. G. (1980). Nicotinamide adenine dinucleotide biosynthesis and pyridine nucleotide cycle metabolism in microbial systems. *Microbiology Review*, 44, pp. 83–105.

Freigang, J., Diederichs, K., Schafer, K. P., Welte, W. and Paul, R. (2002). Crystal structure of oxidized flavodoxin, an essential protein in Helicobacter pylori. *Protein Science*, 11, pp. 253–261.

Gangwer, K. A., Mushrush, D. J., Stauff, D. L., Spiller, B., McClain, M. S., Cover, T. L. and Lacy, D. B. (2007). Crystal structure of the Helicobacter pylori vacuolating toxin p55 domain. *Proceedings of the National Academy of the Sciences USA*, 104, pp. 16293–16298.

Gustafsson, T. N., Sandalova, T., Lu, J., Holmgren, A. and Schneider, G. (2007). High-resolution structures of oxidized and reduced thioredoxin reductase from Helicobacter pylori. *Acta Crystallographica D Biological Crystallography*, 63, pp. 833–843.

Ha, N. C., Oh, S. T., Sung, J. Y., Cha, K. A., Lee, M. H. and Oh, B. H. (2001). Supramolecular assembly and acid resistance of Helicobacter pylori urease. *Nature Structural Biology*, 8, pp. 505–509.

Hare, S., Fischer, W., Williams, R., Terradot, L., Bayliss, R., Haas, R. and Waksman, G. (2007). Identification, structure and mode of action of a new regulator of the Helicobacter pylori HP0525 ATPase. *EMBO Journal*, 26, pp. 4926–4934.

Hong, E., Lee, H. M., Ko, H., Kim, D. U., Jeon, B. Y., Jung, J., Shin, J., Lee, S. A., Kim, Y., Jeon, Y. H., Cheong, C., Cho, H. S. and Lee, W. (2007). Structure of an atypical orphan response regulator protein supports a new phosphorylation-independent regulatory mechanism. *Journal of Biological Chemistry*, 282, pp. 20667–20675.

Hu, T., Wu, D., Chen, J., Ding, J., Jiang, H. and Shen, X. (2008). The catalytic intermediate stabilized by a "down" active site loop for diaminopimelate decarboxylase from Helicobacter pylori. Enzymatic characterization with crystal structure analysis. *Journal of Biological Chemistry*, 283, pp. 21284–21293.

Hung, C. L., Liu, J. H., Chiu, W. C., Huang, S. W., Hwang, J. K. and Wang, W. C. (2007). Crystal structure of Helicobacter pylori formamidase AmiF reveals a cysteine-glutamate-lysine catalytic triad. *Journal of Biological Chemistry*, 282, pp. 12220–12229.

Ishiyama, N., Creuzenet, C., Miller, W. L., Demendi, M., Anderson, E. M., Harauz, G., Lam, J. S. and Berghuis, A. M. (2006). Structural studies of FlaA1 from Helicobacter pylori reveal the mechanism for inverting 4,6-dehydratase activity. *Journal of Biological Chemistry*, 281, pp. 24489–24495.

Kang, G. B., Kim, Y. S., Im, Y. J., Rho, S. H., Lee, J. H. and Eom, S. H. (2005). Crystal structure of NH3-dependent NAD+ synthetase from Helicobacter pylori. *Proteins*, 58, pp. 985–988.

Kim, D. Y. and Kim, K. K. (2008). The structural basis for the activation and peptide recognition of bacterial ClpP. *Journal of Molecular Biology*, 379, pp. 760–771.

Kim, M. K., Im, Y. J., Lee, J. H. and Eom, S. H. (2006) Crystal structure of quinolinic acid phosphoribosyltransferase from Helicobacter pylori. *Proteins*, 63, pp. 252–255.

Kim, S. H., Kang, G. B., Song, H. E., Park, S. J., Bea, M. H. and Eom, S. H. (2008). Structural studies on Helicobacter pyloriATP-dependent protease, FtsH. *Journal of Synchrotron Radiation*, 15, pp. 208–210.

Koharyova, M. and Kolarova, M. (2008) Oxidative stress and thioredoxin system. *General Physiology and Biophysics*, 27, pp. 71–84.

Kuo, C. J., Guo, R. T., Lu, I. L., Liu, H. G., Wu, S. Y., Ko, T. P., Wang, A. H. and Liang, P. H. (2008). Structure-based inhibitors exhibit differential activities against Helicobacter pylori and Escherichia coli undecaprenyl pyrophosphate synthases. *Journal of Biomedicine and Biotechnology*, 841312.

Lee, B. I., Kwak, J. E. and Suh, S. W. (2003). Crystal structure of the type II 3-dehydroquinase from Helicobacter pylori. *Proteins*, 51, pp. 616–617.

Lee, B. I. and Suh, S. W. (2004) Crystal structure of the schiff base intermediate prior to decarboxylation in the catalytic cycle of aspartate alpha-decarboxylase. *Journal of Molecular Biology*, 340, pp. 1–7.

Lee, B. I. and Suh, S.W. (2003) Crystal structure of UDP-N-acetylglucosamine acyltransferase from Helicobacter pylori. *Proteins* 53, pp. 772–774.

Lee, H. H., Moon, J. and Suh, S. W. (2007). Crystal structure of the Helicobacter pylori enoyl-acyl carrier protein reductase in complex with hydroxydiphenyl ether compounds, triclosan and diclosan. *Proteins* 69, pp. 691–694.

Lewis, H. A., Furlong, E. B., Laubert, B., Eroshkina, G. A., Batiyenko, Y., Adams, J. M., Bergseid, M. G., Marsh, C. D., Peat, T. S., Sanderson, W. E., Sauder, J. M. and Buchanan, S. G. (2001). A structural genomics approach to the study of quorum sensing: crystal structures of three LuxS orthologs. *Structure*, 9, pp. 527–537.

Liu, J. S., Cheng, W. C., Wang, H. J., Chen, Y. C. and Wang, W. C. (2008). Structure-based inhibitor discovery of Helicobacter pylori dehydroquinate synthase. *Biochemical and Biophysical Research Communication*, 373, pp. 1–7.

Loewen, P. C., Carpena, X., Rovira, C., Ivancich, A., Perez-Luque, R., Haas, R., Odenbreit, S., Nicholls, P. and Fita, I. (2004). Structure of Helicobacter pylori catalase, with and without formic acid bound, at 1.6 A resolution. *Biochemistry*, 43, pp. 3089–3103.

Lu, P. K., Tsai, J. Y., Chien, H. Y., Huang, H., Chu, C. H. and Sun, Y. J. (2007). Crystal structure of Helicobacter pylori spermidine synthase: a Rossmann-like fold with a distinct active site. *Proteins*, 67, pp. 743–754.

Lundqvist, T., Fisher, S. L., Kern, G., Folmer, R. H., Xue, Y., Newton, D. T., Keating, T. A., Alm, R. A. and de Jonge, B. L. (2007). Exploitation of structural and regulatory diversity in glutamate racemases. *Nature*, 447, pp. 817–822.

Luthy, L., Grutter, M.G. and Mittl, P.R. (2004) The crystal structure of Helicobacter cysteine-rich protein C at 2.0 A resolution: similar peptide-binding sites in TPR and SEL1-like repeat proteins. *Journal of Molecular Biology*, 340, pp. 829–841.

Luthy, L., Grutter, M. G. and Mittl, P. R. (2002). The crystal structure of Helicobacter pylori cysteine-rich protein B reveals a novel fold for a penicillin-binding protein. *Journal of Biological Chemistry*, 277, pp. 10187–10193.

Mahdavi, J., Sonden, B., Hurtig, M., Olfat, F. O., Forsberg, L., Roche, N., Angstrom, J., Larsson, T., Teneberg, S., Karlsson, K. A., Altraja, S., Wadstrom, T., Kersulyte, D., Berg, D. E., Dubois, A., Petersson, C., Magnusson, K. E., Norberg, T., Lindh, F., Lundskog, B. B., Arnqvist, A., Hammarstrom, L. and Boren, T. (2002). Helicobacter pylori SabA adhesin in persistent infection and chronic inflammation. *Science*, 297, pp. 573–578.

Martinez-Julvez, M., Cremades, N., Bueno, M., Perez-Dorado, I., Maya, C., Cuesta-Lopez, S., Prada, D., Falo, F., Hermoso, J. A. and Sancho, J. (2007). Common conformational changes in flavodoxins induced by FMN and anion binding: the structure of Helicobacter pylori apoflavodoxin. *Proteins*, 69, pp. 581–594.

Morrow, A. L., Williams, K., Sand, A., Boanca, G. and Barycki, J. J. (2007). Characterization of Helicobacter pylori gamma-glutamyltranspeptidase reveals the molecular basis for substrate specificity and a critical role for the tyrosine 433-containing loop in catalysis. *Biochemistry*, 46, pp. 13407–13414.

Moss, S. F. and Sood, S. (2003). Helicobacter pylori. *Current Opinion in Infectious Diseases*, 16, pp. 445–451.

Natrajan, G., Hall, D. R., Thompson, A. C., Gutsche, I. and Terradot, L. (2007). Structural similarity between the DnaA-binding proteins HobA (HP1230) from Helicobacter pylori and DiaA from Escherichia coli. *Molecular Microbiology* 65, pp. 995–1005.

Nikaido, H. (1996). Outer membrane. In *Escherichia coli and Salmonella: Cellular and Molecular Biology* ed. Neidhart, F.C. pp. 29-47. Washington, DC: American Society for Micribiology.

Odenbreit, S., Puls, J., Sedlmaier, B., Gerland, E., Fischer, W. and Haas, R. (2000). Translocation of Helicobacter pylori CagA into gastric epithelial cells by type IV secretion. *Science*, 287, pp. 1497–1500.

Oh, J. D., Kling-Backhed, H., Giannakis, M., Xu, J., Fulton, R. S., Fulton, L. A., Cordum, H. S., Wang, C., Elliott, G., Edwards, J., Mardis, E. R., Engstrand, L. G. and Gordon, J. I. (2006). The complete genome sequence of a chronic atrophic gastritis Helicobacter pylori strain: evolution during disease progression. *Proceedings of the National Academy of the Sciences USA*, 103, pp. 9999–10004.

Papinutto, E., Windle, H. J., Cendron, L., Battistutta, R., Kelleher, D. and Zanotti, G. (2005). Crystal structure of alkyl hydroperoxide-reductase (AhpC) from Helicobacter pylori. *Biochimica et Biophysica Acta - Proteins and Proteomics*, 1753, pp. 240–246.

Parsonnet, J., Friedman, G. D., Vandersteen, D. P., Chang, Y., Vogelman, J. H., Orentreich, N. and Sibley, R. K. (1991). Helicobacter pylori infection and the risk of gastric carcinoma. *New England Journal of Medicine*, 325, pp. 1127–1131.

Parsonnet, J., Hansen, S., Rodriguez, L., Gelb, A. B., Warnke, R. A., Jellum, E., Orentreich, N., Vogelman, J. H. and Friedman, G. D. (1994). Helicobacter pylori infection and gastric lymphoma. *New England Journal of Medicine*, 330, pp. 1267–1271.

Parsonnet, J., Replogle, M., Yang, S. and Hiatt, R. (1997). Seroprevalence of CagA-positive strains among Helicobacter pylori-infected, healthy young adults. *Journal of Infectious Diseases*, 175, pp. 1240–1242.

Pisani, P., Parkin, D. M., Bray, F. and Ferlay, J. (1999). Estimates of the worldwide mortality from 25 cancers in 1990. *International Journal of Cancer*, 83, pp. 18–29.

Robinson, D. A., Stewart, K. A., Price, N. C., Chalk, P.A., Coggins, J. R. and Lapthorn, A. J. (2006). Crystal structures of Helicobacter pylori type II dehydroquinase inhibitor complexes: new directions for inhibitor design. *Journal of Medicinal Chemistry*, 49, pp. 1282–1290.

Roujeinikova, A. (2008). Crystal structure of the cell wall anchor domain of MotB, a stator component of the bacterial flagellar motor: implications for peptidoglycan recognition. *Proceedings of the National Academy of the Sciences USA*, 105, pp. 10348–10353.

Sancho, J. (2006). Flavodoxins: sequence, folding, binding, function and beyond. *Cellular and Molecular Life Science*, 63, pp. 855–864.

Savvides, S. N., Yeo, H. J., Beck, M. R., Blaesing, F., Lurz, R., Lanka, E., Buhrdorf, R., Fischer, W., Haas, R. and Waksman, G. (2003). VirB11 ATPases are dynamic hexameric assemblies: new insights into bacterial type IV secretion. *EMBO Journal*, 22, pp. 1969–1980.

Schoenhofen, I. C., Lunin, V. V., Julien, J. P., Li, Y., Ajamian, E., Matte, A., Cygler, M., Brisson, J. R., Aubry, A., Logan, S. M., Bhatia, S., Wakarchuk, W. W. and Young, N. M. (2006). Structural and functional characterization of PseC, an aminotransferase involved in the biosynthesis of pseudaminic acid, an essential flagellar modification in Helicobacter pylori. *Journal of Biological Chemistry*, 281, pp. 8907–8916.

Sidebotham, R. L. and Baron, J. H. (1990). Hypothesis: Helicobacter pylori, urease, mucus, and gastric ulcer. *Lancet*, 335, pp. 193–195.

Smith, A., Rish, K. R., Lovelace, R., Hackney, J. F. and Helston, R. M. (2009). Role for copper in the cellular and regulatory effects of heme-hemopexin. *Biometals*, 22, pp. 421–437.

Taylor, P. L., Blakely, K. M., de Leon, G. P., Walker, J. R., McArthur, F., Evdokimova, E., Zhang, K., Valvano, M. A., Wright, G. D. and Junop, M. (2008). Structure and function of sedoheptulose-7-phosphate isomerase, a critical enzyme for lipopolysaccharide biosynthesis and a target for antibiotic adjuvants. *Journal of Biological Chemistry*, 283, pp. 2835–2845.

Terradot, L., Bayliss, R., Oomen, C., Leonard, G. A., Baron, C. and Waksman, G. (2005). Structures of two core subunits of the bacterial type IV secretion system, VirB8 from Brucella suis and ComB10 from Helicobacter pylori. *Proceedings of the National Academy of the Sciences USA*, 102, pp. 4596–4601.

Tomb, J. F., White, O., Kerlavage, A. R., Clayton, R. A., Sutton, G. G., Fleischmann, R. D., Ketchum, K.A., Klenk, H. P., Gill, S., Dougherty, B. A., Nelson, K., Quackenbush, J., Zhou, L., Kirkness, E. F., Peterson, S., Loftus, B., Richardson, D., Dodson, R., Khalak, H. G., Glodek, A., McKenney, K., Fitzegerald, L. M., Lee, N., Adams, M. D., Hickey, E. K., Berg, D. E., Gocayne, J. D., Utterback, T. R., Peterson, J. D., Kelley, J. M., Cotton, M.D., Weidman, J.M., Fujii, C., Bowman, C., Watthey, L., Wallin, E., Hayes, W. S., Borodovsky, M., Karp, P. D., Smith, H. O., Fraser, C. M. and Venter, J. C. (1997). The complete genome sequence of the gastric pathogen Helicobacter pylori. *Nature*, 388, pp. 539–547.

Tosi, T., Cioci, G., Jouravleva, K., Dian, C. and Terradot, L. (2009). Structures of the tumor necrosis factor alpha inducing protein Tipalpha: a novel virulence factor from Helicobacter pylori. *FEBS Letters*, 583, pp. 1581–1585.

Tummuru, M. K., Cover, T. L. and Blaser, M. J. (1994). Mutation of the cytotoxin-associated cagA gene does not affect the vacuolating cytotoxin activity of Helicobacter pylori. *Infection and Immunology*, 62, pp. 2609–2613.

Viala, J., Chaput, C., Boneca, I. G., Cardona, A., Girardin, S. E., Moran, A. P., Athman, R., Memet, S., Huerre, M. R., Coyle, A. J., DiStefano, P. S., Sansonetti, P. J., Labigne, A., Bertin, J., Philpott, D. J. and Ferrero, R. L. (2004). Nod1 responds to peptidoglycan delivered by the Helicobacter pylori cag pathogenicity island. *Nature Immunology*, 5, pp. 1166–1174.

Voland, P., Weeks, D. L., Marcus, E. A., Prinz, C., Sachs, G. and Scott, D. (2003). Interactions among the seven Helicobacter pylori proteins encoded by the urease gene cluster. *American Journal of Physiology-Gastrointestinal and Liver Physiology*, 284, pp. G96–G106.

Wang, G., Alamuri, P. and Maier, R. J. (2006). The diverse antioxidant systems of Helicobacter pylori. *Molecular Microbiology*, 61, pp. 847–860.

Williams, K., Cullati, S., Sand, A., Biterova, E. I. and Barycki, J. J. (2009). Crystal structure of acivicin-inhibited gamma-Glutamyltranspeptidase reveals critical roles for its c-terminus in autoprocessing and catalysis (dagger) (double dagger). *Biochemistry*, 48, pp. 2459–2467.

Wilson, K. T. and Crabtree, J. E. (2007). Immunology of Helicobacter pylori: insights into the failure of the immune response and perspectives on vaccine studies. *Gastroenterology*, 133, pp. 288–308.

Wu, C. A., Lokanath, N. K., Kim, D. Y., Park, H. J., Hwang, H. Y., Kim, S. T., Suh, S. W. and Kim, K. K. (2005) Structure of inorganic pyrophosphatase from Helicobacter pylori. *Acta Crystallographica D - Biological Crystallography*, 61, pp. 1459–1464.

Wu, D., Zhang, L., Kong, Y., Du, J., Chen, S., Chen, J., Ding, J., Jiang, H. and Shen, X. (2008). Enzymatic characterization and crystal structure analysis of the D-alanine-D-alanine ligase from Helicobacter pylori. *Proteins*, 72, pp. 1148–1160.

Yeo, H. J., Savvides, S. N., Herr, A. B., Lanka, E. and Waksman, G. (2000). Crystal structure of the hexameric traffic ATPase of the Helicobacter pylori type IV secretion system. *Molecular Cell*, 6, pp. 1461–1472.

Zanotti, G., Papinutto, E., Dundon, W. G., Battistutta, R., Seveso, M., Del Giudice, G., Rappuoli, R. and Montecucco, C. (2002). Structure of the neutrophil-activating protein from Helicobacter pylori. *Journal of Molecular Biology*, 323, pp. 125–130.

Zhang, L., Liu, W., Hu, T., Du, L., Luo, C., Chen, K., Shen, X. and Jiang, H. (2008). Structural basis for catalytic and inhibitory mechanisms of beta-hydroxyacyl-acyl carrier protein dehydratase (FabZ). *Journal of Biological Chemistry*, 283, pp. 5370–5379.

Zhang, L., Liu, W., Xiao, J., Hu, T., Chen, J., Chen, K., Jiang, H. and Shen, X. (2007). Malonyl-CoA: acyl carrier protein transacylase from Helicobacter pylori: Crystal structure and its interaction with acyl carrier protein. *Protein Science*, 16, pp. 1184–1192.

CHAPTER 9

OVERALL PROTEOME ALTERATIONS DURING REVERSE TRANSFORMATION OF GROWING CHO-K1 CELLS

Rosanna Spera and Claudio Nicolini
Nanoworld Institute, CIRSDNNOB-University of Genova and Fondazione EL.B.A., Italy

In this chapter we review the effects of cyclic 3',5'-monophosphate adenosine (cAMP) on the Chinese Hamster Ovary (CHO-K1) mammalian tumoral cellular line. cAMP reverse transforms CHO-K1 cells *in vitro*, which lose their characteristics associated with malignancy and adopt characteristics of normal fibroblasts. In particular we focus our attention on the overall proteome alteration before and after the reverse transformation analyzing all identifiable four proteins fractions (cytosol, membrane, cytoskeleton and nuclear) by high performance liquid chromatography, gel electrophoresis and mass spectrometry. A new unique pattern of protein synthesis and processing emerged after reverse transformation involving the whole proteome.

9.1 Introduction

With the development of genomics and proteomics, and their application in basic biological research and biotechnologies, there is an increasing need of functional and structural protein characterization. The complex net of processes at the base of the cellular proliferation in the cell cycle and of the cellular differentiation is perhaps, among all the biological processes, the most interesting.

In the last years, biophysics and structural biology studies have created an increasing consent around the hypothesis of a strong connection between cellular and nuclear morphology and the mechanisms at the base of the cellular cycle regulation. It is known that alterations of the cellular shape are reflected on the nuclear architecture through the information transduction operated by the cytoskeleton; nuclear morphology changes, then, determine changes in the structure of the chromatin causing modifications of its transcriptional activity. Much attention has been focused on the attempt to undertake global analysis of protein expression. By transducing extracellular signals from cell surface to the nucleus the incoming signals influence enzymatic activities, DNA binding and transcription of other proteins, as well as the sub-cellular localization of these actions, and such activities cause G_0/G_1 cells to enter the commitment and differentiation stages of development. In general, cell differentiation or proliferation requires de novo protein synthesis and/or post-translational modifications, in particular phosphorylation of proteins such as transcription factors. Consequently, there is a need to identify the proteins that are being modified, as well as to know when and where in the cell modifications occur.

The proteome in essentially any organism is a collection of somewhere between 30 and 80% of the possible gene products. Most of these proteins are expressed at relatively low levels ($10-10^2$ per cell), although some are expressed at much higher levels (10^4-10^6 per cell). Regardless of the absolute level of expression of the polypeptide gene products, most proteins exist in multiple post-translationally modified forms. This situation poses the greatest challenge for proteomic analysis: we must find ways to detect a large number of distinct molecular species, most of which are present at relatively low levels and many of which exist in multiple modified forms (Liebler, 2002).

In this chapter we describe the effects of cyclic 3',5'-monophosphate adenosine (cAMP) on the Chinese Hamster Ovary (CHO) mammalian tumoral cellular line, a good model to study the cell differentiation. It is known that cAMP transforms CHO cells *in vitro*: they lose characteristics associated with malignancy and adopt characteristics of normal fibroblasts ("reverse transformation") (Puck 1977; Parodi *et al.*, 1979; Nicolini and Beltrame, 1982; Puck and Krystosek, 1992, 1993; Puck *et al.*, 1998). In particular we focus our attention on the proteome alteration before and after the reverse transformation.

To better characterize the cellular protein expression pattern before and after the differentiation we performed a sub-cellular fractionation. In this way we separated the whole cellular proteome in four fractions; we then have the opportunity to analyze how each cellular compartment protein pattern changes. In literature there are several works in which this approach to proteome analysis is adopted (Abdolzade-Bavil *et al.*, 2004; Rios-Doria *et al.*, 2004; Saika *et al.*, 2005; Sabio *et al.*, 2005; Lypowy *et al.*, 2005), but always to analyze the behavior of a restrict number of proteins, for example to investigate changes in sub-cellular localization of regulatory proteins impacted by experimental or disease parameters. We adopt sub-cellular fractionation to attempt the difficult analysis of the whole proteome obtained from CHO-K1 cells before and after cAMP exposure.

We have chosen to couple 1D gel electrophoresis, 2D gel electrophoresis and high-resolution liquid chromatography (HPLC), for the separation, detection and quantification of individual proteins present in each of our four fractions, with mass spectrometry and sequence database searching for the identification of the separated proteins. The mass spectrometric techniques that we have adopted are both ESI and MALDI-TOF MS. Investigation of proteome alterations accompanying well-defined physiologic processes may help to further characterize and understand functional implications of affected proteins.

9.2 General Effects of cAMP on CHO-K1 Cells

Puck *et al.* (2002) found that cyclic 3',5'-monophosphate adenosine (cAMP) transforms CHO cells (Figure 9.1), in vitro: they lose characteristics associated with malignancy and adopt characteristics of a normal fibroblast ("*reverse transformation*"). The morphology changes from compact, randomly oriented cells to elongated fibroblast-like cells growing parallel to their long dimension. Cell surface knobs or blebs disappear. There is an increase in the number of microtubules per unit volume of cytoplasm and a change in their distribution from a random arrangement to an orderly alignment parallel to each other and to the long axis of the cell. The cells acquire contact inhibition of growth so that growth becomes strictly monolayered instead of three dimensional.

The ability to grow in suspension is eliminated without altering the capacity to grow on collagen surfaces and other protein syntheses such as fibronectin are induced (Hsie and Puck, 1971; Nicolini and Beltrame, 1982; Hatzopoulos *et al.*, 1998; Puck *et al.*, 2002). On the other hand it was also found that untreated and cAMP treated cells are quite similar in terms of the amount of polymerised tubulin (Chan *et al.*, 1989; Vergani *et al.*, 2001).

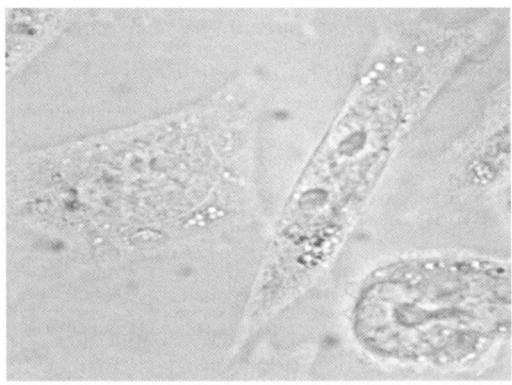

Figure 9.1. CHO-K1 cells.

These changes are consistent with the conversion from a malignant to a normal fibroblastic state. The conversion is under genetic control as demonstrated by the production of specific mutants with altered characteristics with respect to this reaction. The effects of the change are recognizable within an hour and affect cells throughout all or most of the cell cycle (Hsie *et al.*, 1971; Puck 1977; Puck and Krystosek, 1992, 1993; Puck *et al.*, 1998).

Microtubules and chromatin-DNA structure have impact on various biological processes such as the control of gene expression, protein synthesis, cell cycle regulation and cell transformation (Belmont and Nicolini, 1982; Brinkley *et al.*, 1975; Brinkley, 1982; Schliwa 2002). However, the changes in the concentration of the different metabolites and the different protein composition associated with morphological changes observed are largely unknown (Vergani *et al.*, 2001).

Agents like colcemid or cytochalasin B, which disorganize microtubules and microfilaments, respectively, change reverse transformed CHO cells back to the parental CHO form. Therefore, the conversion by cAMP of cells from the transformed to the reverse transformed or normal state is strongly dependent on the cell cytoskeleton (Puck 1977; Chan et al., 1989; Ashall and Puck, 1984; Ashall et al., 1988; Puck and Krystosek, 1992; Puck et al., 1998; Nagasawa et al., 2005).

Puck et al. postulated an intimate association between the cytoskeletal fibers and the nuclear envelope on one end and cell surface components on the other. Malignant cells have lost control of cell reproduction and exhibit a propensity to break away from their original anchorage and set up foci of growth elsewhere in the body. Both of these characteristics are associated with the acquisition of the transformation habitus in which the normal cytoskeletal network has suffered some degree of disorganization (Puck and Krystosek, 1992, 1993).

Transformed cells frequently exhibit instability in chromosome number even in clonal populations. Defects in the cytoskeleton structure could well be reflected in corresponding defects in the spindle, which is made of microtubules and which would result in errors of chromosomal distribution among the cell progeny. Thus, this formulation proposes that cytoskeletal fibers are involved not only with the regulation of cell reproduction, but also in preventing aneuploidy. Both processes are distorted when microtubules are disorganized, as in the case of cancer.

Treatment of CHO cells with cAMP restores the deposition of fibronectin around the cell membrane, a property that has been lost in the transformed cell. Different components of the cytoskeleton appear to exert specific affects on the reverse transformation reaction (Puck et al., 2002).

Work from a number of laboratories has demonstrated that the cytoskeleton exerts a variety of different and complex regulatory metabolic actions in the biochemical economy of the mammalian cell (Puck et al., 2002; Nicolini and Beltrame, 1982; Belmont and Nicolini, 1982).

CHO mutants were selected for unresponsiveness to the reverse transformation reaction of cAMP derivatives. Seven differences in protein phosphorylation in the parental CHO cell were identified as a result of treatment with cAMP. Unresponsive mutants of two kinds were found. One had lost all seven of the phosphorylation changes induced by cAMP in the wild-type cell. However, the other type, equally resistant to reverse transformation by cAMP derivatives, differs from the parental cell in only one (or possibly two) phosphorylation events involving a 55 kDa protein. Phosphorylation of this protein, which is not yet identified, may therefore be directly related to transformation and its reversal in the CHO cell (Gabrielson *et al.*, 1982).

9.2.1 *Involvement of vimentin*

Vimentin is only slightly phosphorylated in transformed CHO cells but is heavily phosphorylated in normal fibroblasts. cAMP addition almost triples the vimentin phosphorylation of CHO cells but does not change that of normal cells. Vimentin phosphorylation is one of the earliest phenomena to occur after addition of cAMP to CHO cells and it precedes the cell stretching and other manifestations of reverse transformation. It seems likely that phosphorylation of cytoskeletal elements initiates a large-scale genetic regulatory action in which a substantial change in the spectrum of genome exposure and sequestration occurs. Early during reverse transformation of CHO cells by cAMP, a new protein, molecular weight 43 kDa and *pI* of 5.3, was rapidly and specifically induced. This protein is a phosphorylated actin homologue whose induction requires new protein synthesis, and which appears to play a critical role in some cancerous processes (Miranti and Puck, 1990).

9.2.2 *cAMP-induced gene expression*

Not all cancers respond to cAMP. The HeLa cell, for example, does not display a typical reverse transformation response. On the other hand, vole cells transformed by avian sarcoma virus carrying the SRC gene display typical reverse transformation when treated with cAMP. The cAMP treatment does not inactivate the SRC gene product. These data imply that cAMP exerts its effect in the same pathway at or after the point in this regulatory chain influenced by the SRC gene product, which mainly phosphorylates tyrosine moieties. The decision to adapt the transformed or the normal state may be determined by the degree to which the SRC gene on the one hand, or cAMP-mediated kinase activities on the other, predominates in the cell. However, the development of all the transformation characteristics noted as a result of introduction of the SRC gene and their coordinate reversal by cAMP support the thesis that in the normal fibroblast all these properties are part of a common cAMP-induced regulatory system involving the cytoskeleton (Puck and Krystosek, 1992, 1993; Puck *et al.*, 2002).

A theory devised to account for these and other facts of differentiation and its distortion in malignancy was developed as follows. There are at least two levels of regulation of gene expression in mammalian cells. The first is conversion of the genes in question from a sequestered state, in which they are unable to interact with molecules in the environment, to an exposed state in which they can be so influenced. The next step is the turning of such genes on and off by appropriate effectors molecules in the nucleus. Puck *et al.* proposed that the cytoskeleton transmits information from the cell membrane to chromosomal loci in the nucleus and is an important element in regulation of the exposure process. Thus, the cytoskeleton is a critical structure both in normal differentiation and in its distortions like that of cancer. Of course, cancer could result in damage to regulatory processes beyond the cytoskeletal stage, and such cancer cells would fail to display the phenomena of reverse transformation when treated with agents like cAMP (Puck and Krystosek, 1992, 1993; Ashall and Puck, 1984).

This picture predicts that the CHO and similar cells that undergo reverse transformation should also display increased genome exposure when treated with reverse transformation agents. Such was, indeed, found to be the case. Experiments have demonstrated that the action of cAMP in increasing genome exposure is gene specific (Ashall and Puck, 1984; Ashall *et al.*, 1988).

The data support the picture that the peripheral nuclear shell of exposed DNA contains differentiation-specific genes that include the specific growth control genes that are functional in normal cells but not in cancer (Puck *et al.*, 2002).

Puck *et al.* found that different regions of the nucleus are specialized for different gene activations. Therefore, a mechanism must exist for transporting specific chromosomal loci to their particular activation sites within the nucleus. They proposed that the nuclear fibers participate in this process, as do the non-coding regions of the nucleus and particularly the repetitive regions; and that the nuclear periphery is the site of action of differentiation genes, while at least some of the housekeeping genes are in the interior, associated with nucleoli.

Different states of differentiation appear to be characterized by different specific sets of exposed genes. By this picture, oncogenes would include genes that regulate cytoskeletal structure and genes that control the level of cAMP within the cell. Less than 10% of the human genes are coding genes. The remainder, which includes a large number of repetitive DNA sequences, contain essential DNA structures, which Puck *et al.* suggested regulate the expression of the coding genes. Their constitution is not randomized, as would be expected if they were functionless, but is rigorously maintained; and, indeed, parts of these non-coding regions can be used to establish individual identity in a fashion even more secure than fingerprinting.

If non-coding repetitive sequences are, indeed, involved in exposure, one might well expect to find specific proteins binding to them in order that they may be drawn to their specific region in the nucleus where appropriate activation can be carried out. Such protein attachment to specific DNA-repetitive sites might well be different in cancer and normal cell states. Analysis by gel retardation assay showed that a protein of 66 kDa binds to a human low-repeat repetitive sequence (LRS) in each of the 10 transformed human cell lines examined. This protein-DNA interaction was not observed in 11 normal human cell cultures (Law *et al.*, 1989).

9.2.3 *Puck model*

Puck *et al.* model proposed that damage to the cytoskeletal and nuclear fiber system can lead to malignant transformation in three ways: by distorting the metabolic patterns necessary for normal biochemical and reproductive regulation; by producing karyotype instability in mitosis, which promotes evolutionary drift toward malignancy through continuous selection of cells that resist signals for reproductive arrest; and by preventing apoptotic death.

This model explains a number of phenomena of malignancy. Transformed cells arising from different tissues may display similar morphological characteristics because they have lost elements of their internal fiber systems that maintain the individual differentiated morphologies and growth behavior. Therefore, they tend toward a similar appearance. The existence of the pre-cancerous state and the frequent development of resistance to therapeutic agents employed in cancer therapy are natural consequences of this picture.

In conclusion cAMP, through control of phosphorylation, regulates the structure of the cell cytoskeleton, which in turn strongly influences many differentiation properties including the assumption of normal or cancerous behavior in mammalian cells. A theory of genome regulation involving migration of specific DNA loci to different regions of the nucleus, involving transfer of information from specific molecules in the cell membrane to the DNA in the nucleus, and involving the cell fiber system as a critical element, was presented by Puck *et al*. However, the changes in the concentration of the different metabolites and the different protein composition associated with morphological changes observed are largely unknown (Hsie *et al.*, 1971).

The mechanism of the transformation in properties here described is as yet obscure. The phenomena make it clear that a fundamental change in properties of the cell membrane has occurred.

9.3 Overall CHO Proteome Investigation

Our studies want to explore more deeply the mechanism of proteome modification in the cAMP-treated differentiated state of CHO cells.

To reverse transform the CHO-K1 cells they were treated with a solution 10^{-3} M cyclic 3',5'-monophosphate adenosine (cAMP, Sigma Chemical Co., St. Louis, MO) that was added to the normal culture medium for six hours before the analysis. It has been shown that single cells of CHO-K1 in the native state grow equally well on plastic surfaces or in suspension. In the presence of reverse transformation conditions, however, excellent growth is still achieved on the plastic surface but no growth whatever occurs in suspension (Hsie and Puck, 1971). We then monitored CHO-K1 reverse transformation monitoring their growing behavior by a phase contrast microscope (Wilovert, Wesco).

For protein extraction we chose to perform a sub-cellular fractionation to separate the whole cellular proteome in four fractions. Due to the complexity of total proteomes, especially of eukaryotic proteomes, and the divergence of protein properties, it is beneficial to prepare standardized partial proteomes of a given organism in order to maximize the coverage of the proteome and to increase the chance to visualize low abundance proteins and make them accessible for subsequent analysis (Abdolzade-Bavil et al., 2004). For the sequential extraction of cell content, we chose to utilize, on its characteristics, a protein extraction kit (Subcellular Proteome Extraction Kit, Calbiochem) that takes advantage of the differential solubility of certain sub-cellular compartments in special reagent mixtures. The four extraction buffers utilized have been optimized to provide reproducible protein isolation from distinct sub-cellular compartments. Upon treatment with the first extraction buffer, cells shrink as a result of the release of the cytosol content but remain intact in their overall structure. After the second extraction step, due to the solubilization of membrane and membrane/organelles, only isolated nuclei and the cytoskeleton remain intact. The treatment of the residual material with the third extraction buffer destroys the structure of the nucleus, solubilizing the nuclear proteins. Finally, the cytoskeleton components are liberated during the fourth and final extraction (Abdolzade-Bavil et al., 2004). In conclusion upon extraction of culture cells, four partial proteomes of the cells are obtained:
- cytosol proteins (fraction 1: CS),
- membrane and membrane/organelles proteins (fraction 2: MM),
- nucleus proteins (fraction 3: NUC),
- cytoskeleton proteins (fraction 4: CK).

The sub-cellular extraction was performed according to the included protocol for freshly prepared adherent cell cultures. We performed extraction from 1×10^6 cells in logarithmic phase at 80% confluence. To monitor the extraction procedure, morphological changes of the cells were examined by a phase contrast microscope (Wilovert, Wesco). We performed extractions from CHO-K1 cells before and after differentiation by cAMP. We performed at least three different extractions for CHO-K1 cells before and after reverse transformation.

A synthesis of the samples obtained and their adopted nomenclature is reported in Table 9.1.

Table 9.1. Samples analyzed.

Cellular line	Fractions			
	Cytosol	Membrane and membrane/organelle	Nuclear	Cytoskeleton
CHO-K1	CS [CHO]	MM [CHO]	NUC [CHO]	CK [CHO]
CHO-K1 differentiated by cAMP	CS [cAMP CHO]	MM [cAMP CHO]	NUC [cAMP CHO]	CK [cAMP CHO]

Sample separation, isolation and preparation for the mass spectrometric techniques generally involves gel electrophoresis. However, proteins cannot easily be eluted from gel without the use of detergents, which are detrimental to the mass spectrometric analysis (Yates 1998); moreover large proteins are quite impossible to be eluted. If a protein is "in-gel" digested with a protease, many of the peptides can be extracted from the gel (Rosenfeld *et al.*, 1992).

In addition, the conversion of a protein into its constituent peptides provides more information than can be obtained from the whole protein itself. For many applications, the peptides recovered from in-gel digestions require concentration and purification before being analysed by mass spectrometry (Ashcroft 2003).

One method of peptide purification commonly employed for this purpose is reverse-phase chromatography, which is available in a variety of formats. Peptides can be purified with ZipTips (Millipore) or by high performance liquid chromatography (HPLC) (Graves and Haystead, 2002).

Reversed phase high performance liquid chromatography (HPLC) is an alternative method to separate proteins that has the advantage to achieve their contamination.

Moreover it has been demonstrated that coupling HPLC to MALDI MS improves the quality of the results. HPLC separations have been captured onto MALDI targets for off-line analysis (Gygi and Aebersold, 2000).

9.3.1 *LC-ESI MS analysis of nucleus fraction*

We performed LC-ESI MS analysis of NUC [CHO] and NUC [cAMP CHO] fractions (Spera and Nicolini, 2007).

The identification of different protein patterns in CHO-K1 cells grown with and without cAMP, is performed with a RP-HPLC-ESI MS apparatus, a Thermo Finnigan (San Jose, CA) Surveyor HPLC connected by a T splitter to a PDA diode-array detector and to Xcalibur LCQ Deca XP Plus mass spectrometer. The mass spectrometer is equipped with an electrospray ion source (ESI).

The chromatographic column is a Vydac (Hesperia, CA) C8 column, with a 5 mm particle diameter (column dimensions 150–2.1 mm). For HPLC-ESI MS analysis the protein solutions, stored at -80 °C, were heated at room temperature and then lyophilized. The lyophilized proteins were immediately dissolved in aqueous 0.2% TFA and centrifuged at 12,000 rpm for 5 min. After centrifugation, the supernatant was analyzed by HPLC-mass spectrometry.

The following solutions were utilized for reversed-phase chromatography: eluent A, 0.056% aqueous TFA, and eluent B, 0.05% TFA in acetonitrile/water 80:20 (v/v). The gradient applied was linear from 0 to 55% in 40 min, at a flow rate of 0.30 ml/min. The T splitter gave a flow rate of about 0.20 ml/min toward the diode array detector and a flow rate of 0.10 ml/min toward the ESI source. The diode array detector was set in the wavelength range 214–276 nm.

Mass spectra were collected every 3 ms in the positive ion mode. MS spray voltage was 4.50 kV and the capillary temperature was 22 °C. The deconvolution of the averaged ESI mass spectra was performed by Mag-Tran1.0 software. A preliminary proteins identification from the mass values of the intact proteins was obtained through a search in Swiss-Prot Data Bank (http://www.expasy.org).

The total ion current (TIC) and the UV absorbance (214 nm) profiles obtained during chromatographic analysis of the two nuclear fractions are reported in Figure 9.2 (Spera and Nicolini, 2007). TIC and UV profiles did not reveal significant signals between 3 and 14 min.

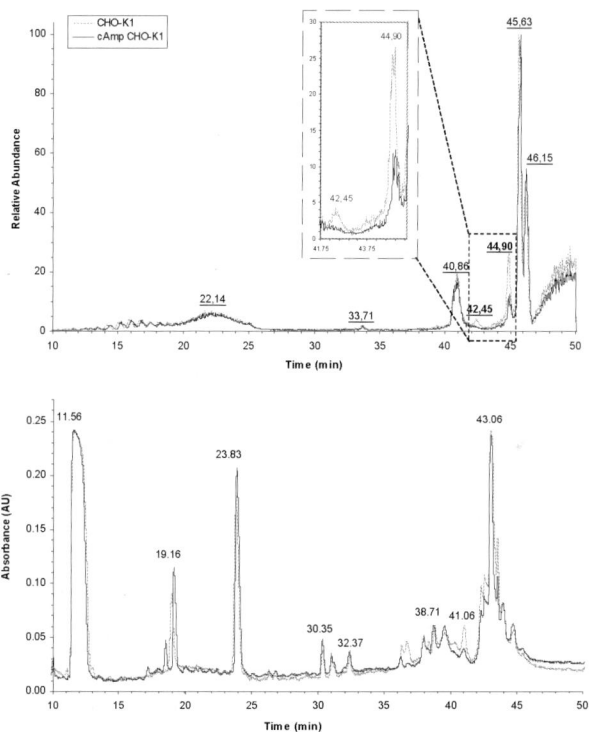

Figure 9.2. HPLC-ESI-MS TIC and UV profiles of nuclear protein fraction of CHO-K1 cells (grey line) and of CHO-K1 treated with cAMP (black line). *Upper plot*: HPLC-ESI-MS TIC profile collected by ion-trap mass spectrometer (elution time 10–50 min). Enlargement of the 41.7–45.5 min elution range, showing the TIC profile, is represented in the box where the greater differences between the two plots are shown: the 42.45 min peak (absent in the treated CHO-K1 plot) and the 44.90 min peak (its relative abundance is double in the untreated CHO-K1 plot). *Bottom plot*: UV (214 nm) profile (elution time 10–50 min) (Reprinted with the permission from Spera and Nicolini, cAMP induced alterations of Chinese hamster ovary cells monitored by mass spectrometry, Journal of Cellular Biochemistry 102, 473-482, © 2007, Wiley-Liss, Inc., a subsidiary of John Wiley & Sons, Inc.).

The total ion current of proteins of interest was used for their relative quantification. Since the total ion current depends not only from concentration but also from the charge of the analyte, sample treatment was standardized both in dilution and HPLC-ESI-MS procedures, to ensure similar bias, namely same ion suppression, pH and organic solvent effects on the analyte charge.

Under these conditions, the total ion current can be roughly considered proportional to protein concentration, and it can be used to evidence correlations existing among different proteins in different samples (Messana *et al.*, 2004).

The highest variations are seen in the TIC plots (see Figure 9.2, TIC plots enlargement in the box):
- in the CHO-K1 plot there is a peak at 42.45 min not observable in the cAMP CHO-K1 plot
- the intensity of the 44.90 min peak in the untreated CHO-K1 TIC plot is double regarding the cAMP treated cells plot peak.

Regarding the first identified variation we analyzed the averaged ESI mass spectrum (300–2000 *m/z* range) recorded in the elution range 42.25–42.68 min, correspondent to 42.45 min TIC peak, shown in the 2nd panel of Figure 9.3 (Spera and Nicolini, 2007).

From the Gaussian deconvolution of spectrum of Figure 9.3 A 2nd panel, performed by MagTran1.0, we identified three protein species (see Figure 9.3B) of molecular weights: 31133, 42024, 52136 Da.

We performed a first protein identification on the basis of these weights through a search in Swiss-Prot data bank (http://www.expasy.org/uniprot). We restricted our search to rodent proteins. In Table 9.2, we report the experimental and theoretical weights, the Swiss-Prot accession number and the function of the identified proteins.

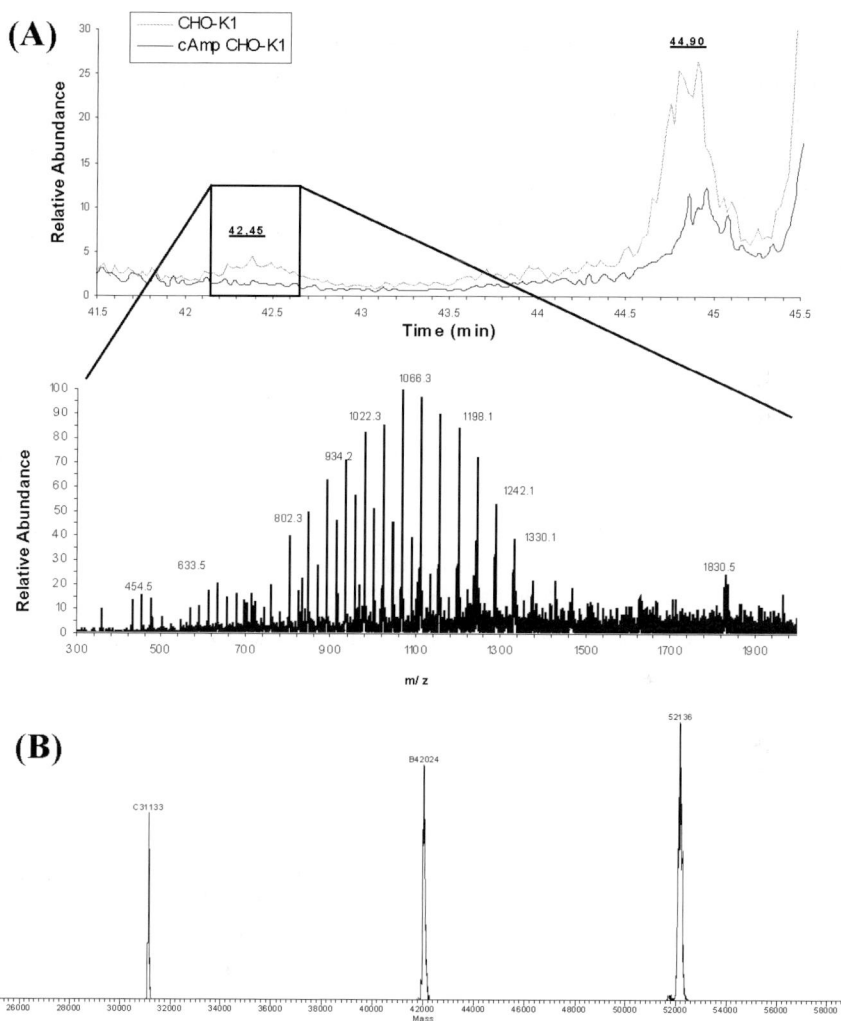

Figure 9.3. (A) 1st panel (from the top): HPLC-ESI-MS TIC profile (elution range 41.5–45.5 min). 2nd panel, averaged ESI mass spectrum (300–2000 m/z range) of CHO-K1 nuclear protein fraction recorded in the elution range 42.25–42.68 min. (B) Deconvoluted spectrum of averaged ESI mass spectrum reported in the second panel (Reprinted with the permission from Spera and Nicolini, cAMP induced alterations of Chinese hamster ovary cells monitored by mass spectrometry, Journal of Cellular Biochemistry 102, 473-482, © 2007, Wiley-Liss, Inc., a subsidiary of John Wiley & Sons, Inc.).

Another difference between the untreated and the cAMP treated CHO-K1 TIC plots is the 44.90 min peak intensity variation: the relative abundance passed from 26.6 for the untreated cells plot to 12.3 for the cAMP treated cells plot (Figure 9.4A, 1st panel). We analyzed the averaged ESI mass spectra recorded in the elution range 44.40–45.20 min (Figure 9.4A, 2nd panel). The two spectra are clearly identical: the difference of intensity in the TIC peaks can be explained by a different amount of the same protein(s) in the two samples.

Table 9.2. Experimental and theoretical weights, Swiss-Prot accession number and function of the proteins identified as present in the NUC [CHO-K1] sample and not in the NUC [cAMP CHO-K1] one and responsible of the peak at 42,45 min in the CHO-K1 plot (Reprinted with the permission from Spera and Nicolini, cAMP induced alterations of Chinese hamster ovary cells monitored by mass spectrometry, Journal of Cellular Biochemistry 102, 473-482, © 2007, Wiley-Liss, Inc., a subsidiary of John Wiley & Sons, Inc.).

Exp. mass (Da)	Th. mass (Da)	Protein	Swiss-Prot Accession number	Function
31133	31133	Myh10 (fragment B) Myosin heavy chain 10, non-muscle	Q9JLT0	Cellular myosin appears to play a role in cytokinesis, cell shape, and specialized functions such as secretion and capping.
42024	42024	Gna11 Guanine nucleotide-binding protein	P21278	Guanine nucleotide-binding proteins (G proteins) are involved as modulators or transducers in various transmembrane signaling systems. Acts as an activator of phospholipase C.
52136	52136	Sc61a_Haran Protein transport protein Sec61 alpha subunit	Q7T278	Appears to play a crucial role in the insertion of secretory and membrane polypeptides into the ER. It is required for assembly of membrane and secretory proteins. Found to be tightly associated with membrane-bound ribosomes, either directly or through adaptor proteins

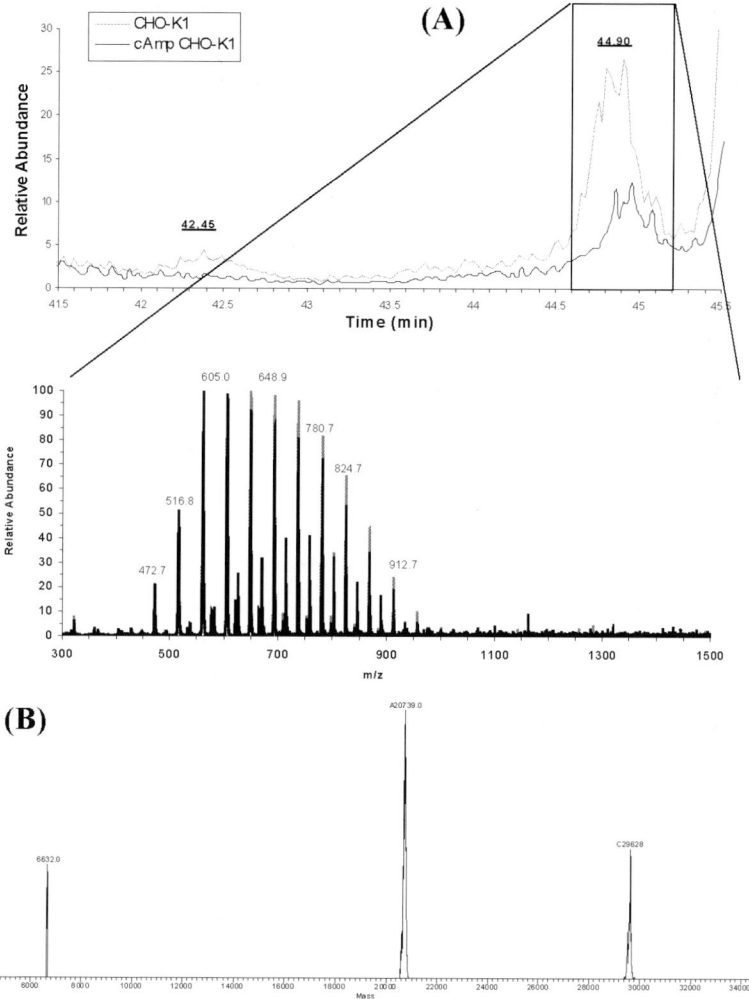

Figure 9.4. (A) 1st panel (from the top), HPLC-ESI-MS TIC profile (elution range 41.5–45.5 min). 2nd panel, enlargement of averaged ESI mass spectra (300–2000 m/z range) of CHO-K1 (grey line) and cAMP CHO-K1 (black line) nuclear protein fraction recorded in the elution range 44.40–45.20 min. The spectra are identical. (B) deconvoluted spectrum of averaged ESI mass spectra reported in the second panel (Reprinted with the permission from Spera and Nicolini, cAMP induced alterations of Chinese hamster ovary cells monitored by mass spectrometry, Journal of Cellular Biochemistry 102, 473-482, © 2007, Wiley-Liss, Inc., a subsidiary of John Wiley & Sons, Inc.).

We performed a Gaussian deconvolution of this spectrum and identified three protein species (Figure 9.4B) of molecular weights: 6632, 20739 and 29628 Da.

Through a search in Swiss-Prot data bank, we identified the 29628 Da peak as CD82_MOUSE protein (Accession number P40237). This protein, associates with CD4 or CD8 glycoprotein delivers co stimulatory signals for the TCR/CD3 pathway (Itoh and Adelstein, 1995) involved in apoptosis regulation. For 6632 Da peak and 20739 peak it was not possible any identification.

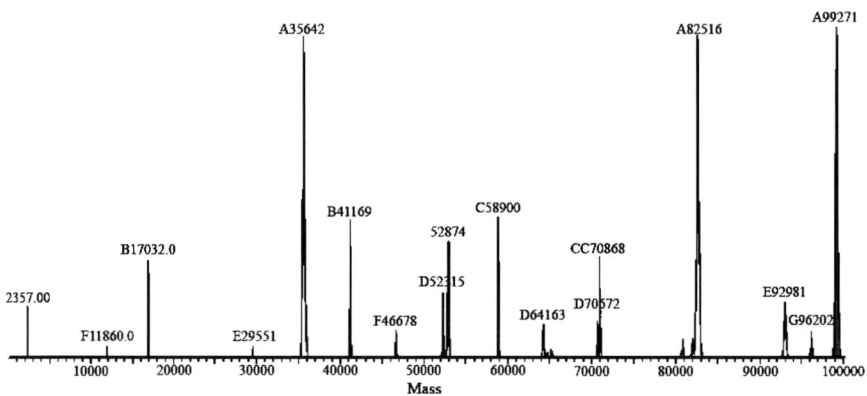

Figure 9.5. Sum of the deconvoluted spectra relative to TIC plot 22 min. peak, 45.79 min and 46.21 min peak. These proteins are present in the same amount in the cells before and after cAMP exposure (as is visible in Figure 9.3) (Reprinted with the permission from Spera and Nicolini, cAMP induced alterations of Chinese hamster ovary cells monitored by mass spectrometry, Journal of Cellular Biochemistry 102, 473-482, © 2007, Wiley-Liss, Inc., a subsidiary of John Wiley & Sons, Inc.).

We analyzed seventeen other proteins of the nuclear envelope (Figure 9.5) present in the same content in the cells before and after cAMP exposure. The relative molecular weights are 2.4 kDa, 11.9 kDa, 17.3 kDa, 29.55 Da, 35.6 kDa, 41.2 kDa, 46.7 kDa, 52.3 kDa, 52.9 kDa, 58.9 kDa, 60.9 kDa, 64.2 kDa, 70.7 kDa, , 82.5 kDa, 93 kDa, 96.2 kDa and 99.3 kDa. It has been possible to identify none of these proteins only from their molecular weight.

9.3.2 HPLC analysis

We employed HPLC of intact proteins to purify and fractionate protein sub-cellular fractions before proteolytic digestion and MALDI-TOF MS. This step is essential to purify the sample solutions that contain a lot of contaminants (salts, detergents, *etc.*) that could complicate the following mass spectrometric analysis.

The HPLC measures were carried out on a Varian Star HPLC system which includes: 9012 Gradient Solvent Delivery System, 9050 UV-VIS Detector, 9300 Refrigerated AutoSampler (fitted with a 20 µl loop), and a Star Chromatography Workstation software (version 5.31; Varian Inc., USA). Proteins were separated on a C18 (250 × 4.6 mm; 5 µm particle size) reverse phase column (Macherey-Nagel, Germany). The UV detector wavelength was set at 280 nm.

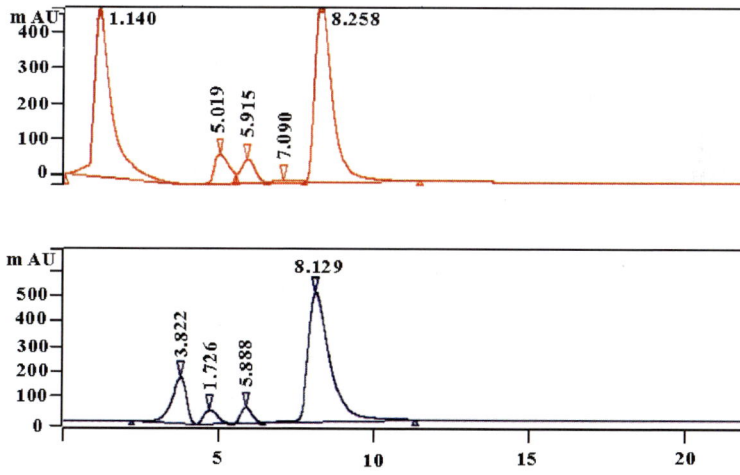

Figure 9.6. Chromatogram of CS [CHO] (lower) and CS [cAMP CHO] (upper) fractions (Spera 2007).

For reversed-phase chromatography we experimented both isocratic elution that gradient elution.

We performed some tests to establish the better parameters (eluents composition, flow rate, elution time) to adopt both for isocratic and

gradient elution. We thus established to adopt the following configuration:
- eluent A: 0.056% aqueous TFA;
- eluent B: 0.05% TFA in 80% acetonitrile and 20% water.

Table 9.3. Elution time, peak area and proteins concentration of CS chromatograms fractions (Figure 9.6). The protein concentration is based on Lysozyme calibration. (*In the calculations of protein concentration we have considered a dilution factor of 10) (Spera 2007).

	Elution time (min)	Peak area (mAU/sec)	Protein concentration* (mg/ml)
Fraction 1			
CHO	1.74	851	0.35
CHO cAMP	3.28	328	0.13
Fraction 2			
CHO	5.19	145	0.06
CHO cAMP	4.73	145	0.06
Fraction 3			
CHO	5.91	105	0.04
CHO cAMP	5.89	128	0.05
Fraction 4			
CHO	8.26	931	0.38
CHO cAMP	8.13	1087	0.44

For isocratic elution we utilized only eluent A, while on gradient elution the gradient applied was linear from 0 to 50% of eluent B in 30 minutes. In both cases the flow rate was of 0.3 ml/min and the total elution time of 30 minutes. Chromatograms were analyzed using Varian Star workstation software.

By HPLC we analyzed both lyophilized and not lyophilized protein samples (stored at –80 °C) to test the better protocol.

The lyophilized protein aliquots were detached in 30 μl of 0.2% TFA aqueous solution and 25 μl of this solution were injected. The not-lyophilized protein aliquots were heated at room temperature and then 25 μl of solutions were injected.

We used lysozyme from chicken egg white (Sigma Chemical Co., St. Louis, MO) powder detached in 0.1% TFA solution at three different concentrations, 0.7–3.3–4.0 mg/ml, as standard to quantify HPLC results (Spera 2007). For each concentration we performed at least three chromatographic analysis.

In the calculations of HPLC fraction concentrations we have considered a dilution factor of 10: 500 μl of sample were lyophilized and then detached in 50 μl of 0.2% TFA for HPLC analysis.

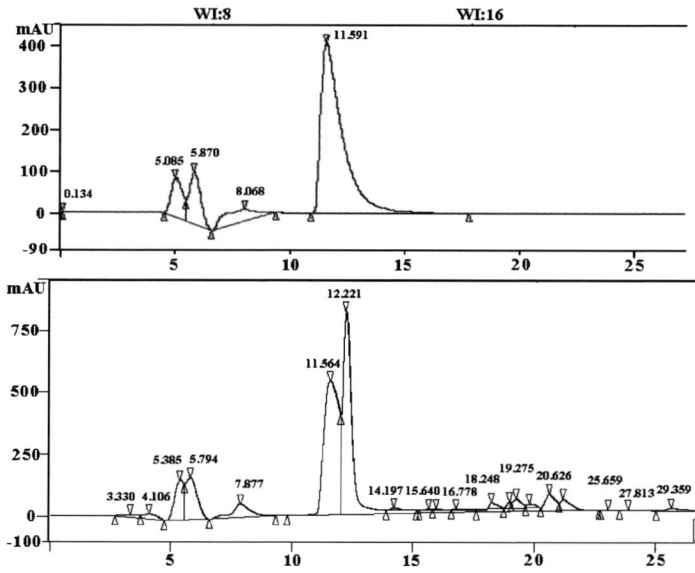

Figure 9.7. Chromatogram of MM [CHO] (upper) and MM [cAMP CHO] (lower) (Spera 2007).

Table 9.4. Elution time, peak area and proteins concentration of MM chromatograms fractions (Figure 9.7). The protein concentration is based on Lysozyme calibration. (*In the calculations of protein concentration we have considered a dilution factor of 10) (Spera 2007).

	Elution time (min)	Peak area (mAU/sec)	Protein concentration* (mg/ml)
Fraction 1			
CHO	5.08	91	0.04
CHO cAMP	5.38	155	0.06
Fraction 2			
CHO	5.87	102	0.04
CHO cAMP	5.80	176	0.07
Fraction 3			
CHO	8.07	49	0.01
CHO cAMP	7.89	118	0.05
Fraction 4			
CHO	-		
CHO cAMP	11.56	1951	0.80
Fraction 5			
CHO	12.13	2587	1.01
CHO cAMP	12.29	1687	0.7

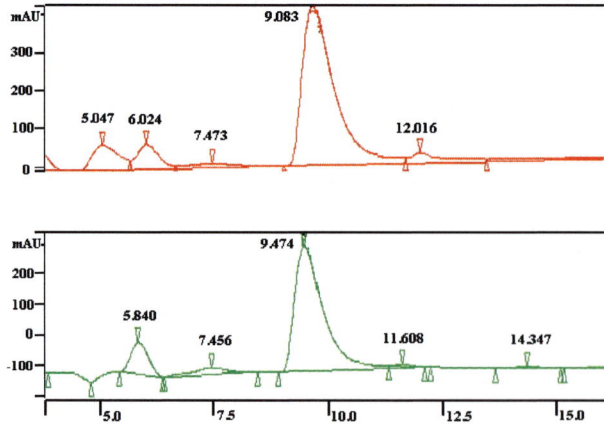

Figure 9.8. Chromatogram of NUC [CHO] (upper) and NUC [cAMP CHO] (lower) fractions (Spera 2007).

Table 9.5. Elution time, peak area and protein concentration of NUC chromatograms fractions (Figure 9.8). The protein concentration is based on Lysozyme calibration. (*In the calculations of protein concentration we have considered a dilution factor of 10) (Spera 2007).

	Elution time (min)	Peak area (mAU/sec)	Protein concentration* (mg/ml)
Fraction 1			
CHO	5.05	128	0.05
CHO cAMP	5.05	0	0
Fraction 2			
CHO	5.84	134	0.05
CHO cAMP	6.02	102	0.04
Fraction 4			
CHO	9.68	902	0.36
CHO cAMP	9.47	622	0.25

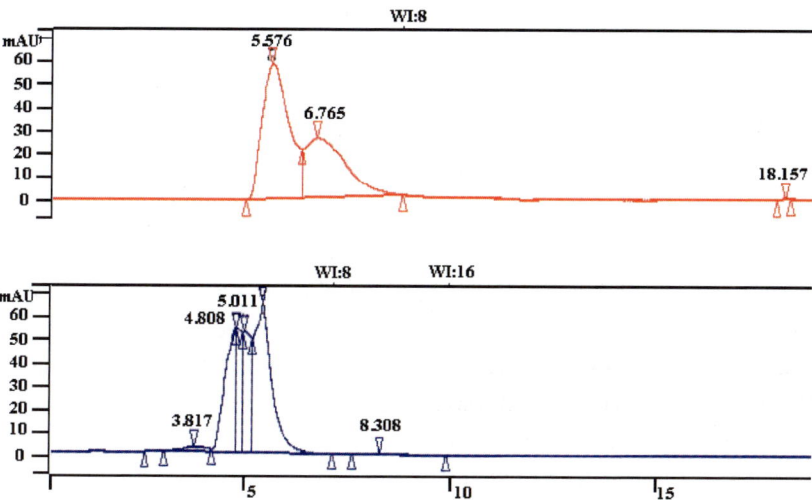

Figure 9.9. Chromatogram of CK [CHO] (upper) and CK [cAMP CHO] (lower) fractions (Spera 2007).

Table 9.6. Elution time, peak area and proteins concentration of CK chromatograms fractions (Figure 9.9). The protein concentration is based on Lysozyme calibration. (*In the calculations of protein concentration we have considered a dilution factor of 10) (Spera 2007).

	Elution time (min)	Peak area (mAU/sec)	Protein concentration (mg/ml)
Fraction 1			
CHO	5.47	72	0.02
CHO cAMP	5.09	68	0.02
Fraction 2			
CHO	6.15	45	0.01
CHO cAMP	5.89	53	0.01

It results a significant variation of the protein content both for nucleus and membrane and membrane/organelle fractions. For membrane and membrane/organelle fractions it is evident a significant increase of the concentration after cAMP exposure while for nucleus fractions it results a decrease of the concentration (Spera 2007).

9.3.3 *Gel electrophoresis*

To analyze the entire profile of the sub-cellular fractions we choose 1D SDS PAGE. Equivalent volumes of the four sub-cellular fractions obtained before and after cAMP differentiation were separated by 1D SDS-PAGE (7.5% gel) (Laemmli, 1970) and stained by Coomassie Blue R-250. We acquired gel images by scanner and processed it by "Image J" (version 1.37; http://rsb.info.nih.gov/ij), a public domain image processing program.

To better separate the proteins of the sub-cellular fractions we utilized the 2D gel electrophoresis that concerns in two consecutive steps: a first dimension separation of proteins that are separated based on their isoelectric point (proteins are applied to polyacrylamide gel strips containing a pH gradient) and a second dimension separation of proteins using SDS-PAGE (proteins are separated based on their molecular weight using denaturing polyacrylamide gel electrophoresis). For 2D gel electrophoresis we utilized the ZOOM® IPGRunner™ System by Invitrogen according to the included protocol. Also for this analysis we utilized equivalent volumes of the four sub-cellular fractions obtained before and after cAMP differentiation. For the detection of proteins on the gel the second dimension gel is stained by SimplyBlue SafeStain (Invitrogen) to visualize the separated proteins as spots on the gel.

We performed a comparative analysis of the four sub-cellular proteomes of CHO-K1 cells before and after cAMP exposure by

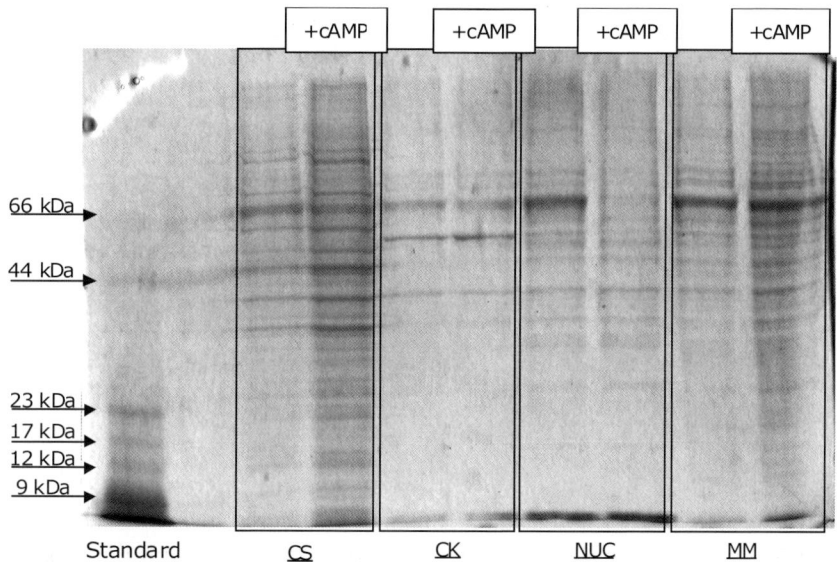

Figure 9.10. 1D SDS PAGE (7.5% acrylamide) of CHO and cAMP CHO protein fractions. The 1st and the 2nd line from the left are for protein standards. The 3rd and 4th are for CS fractions (before [CHO] and then [cAMP CHO] sample), 5th and 6th are for CK fractions ([CHO] and [cAMP CHO]), the 7th and 8th are for NUC fractions ([CHO] and [cAMP CHO]), the 9th and 10th are for MM fractions ([CHO] and [cAMP CHO]) (Spera 2007).

comparing the densitometry profiles obtained from 1D SDS-PAGE analysis. By this way we identified differences in protein contents for each fraction between CHO-K1 and CHO-K1 cAMP samples. The identification of the proteins in these spots of interest was later performed by excision of the spots, in-gel digestion, and MS analysis (Spera 2007).

In Figure 9.11–9.14, gel lines and densitometry profiles for each fraction of CHO-K1 and cAMP CHO-K1 samples are reported. The mass scale was established on the base of standard proteins positions.

On the basis of 1D gel electrophoresis results we performed on the same samples the 2D gel electrophoresis to obtain a better separation of the proteins. In the Figures 9.11C, 9.12C, and 9.13C are reported, together the 1D gel bands, the 2D gel of the correspondent fraction before and after cAMP exposure.

Also in this case there are evident differences in the protein expression after the reverse transformation. The spot analyzed by MS are marked with a circle (Spera et al., 2009).

Densitometry plots analysis shows an inhibition of the expression of at least one high weight protein of cytosol fraction (marked as [a], Figure 9.11) after cAMP exposure. On the contrary for cytosol fraction a small increase of the expression of cAMP sample is evident for at least two low weight proteins (marked as [b] and [c], Figure 9.11).

The apparent molecular mass (kDa) values of the spots identified in Figure 9.11, on the basis of standards, were 67 [a], 23 [b] and 20 [c] (Spera et al., 2009).

It appears, for membrane and membrane/organelle protein fractions, a general increase of protein expression after cAMP exposure. In particular at least three peaks are present only in the cAMP sample densitometry plot (Figure 9.12). The apparent molecular mass (kDa) values of the peaks identified in Figure 9.12, on the basis of standards, were 75 [a], 42 [b], 32 [c] and 22 [d] (Spera et al., 2009).

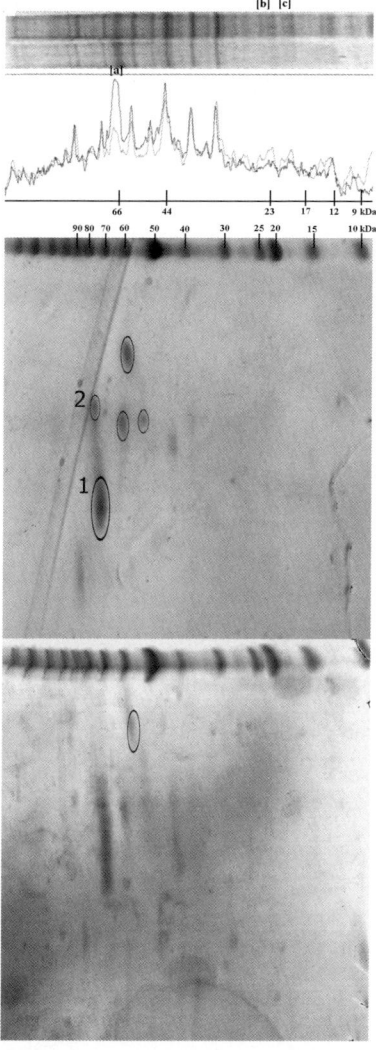

Figure 9.11. Cytosol 1D gel lines, corresponding densitometry profiles and 2D gel image. (A) cAMP CHO (upper) and CHO (lower) gel lines. (B) Corresponding densitometry profiles (in grey cAMP CHO sample profile and in black CHO sample profile): on the x axis the mass scale and on the y axis the spot grey intensity. The letters in square brackets indicate spots whose intensity was altered after treatment with cAMP: the letters above the gel image indicate the spots whose intensity increases after cAMP exposure, the letters under the gel image indicate the spots whose intensity decreases. (C) 2D gel of cAMP CHO fraction (upper) and CHO fraction (lower). The spots marked with a circle have been analyzed by MS (Spera et al., 2009).

Figure 9.12. Membrane and membrane/organelle gel lines and corresponding densitometry profiles. (A) cAMP CHO (upper) and CHO (lower) gel lines. (B) Corresponding densitometry profiles (in grey cAMP CHO sample profile and in black CHO sample profile): on the x axis the mass scale and on the y axis the spot grey intensity. (C) 2D gel of cAMP CHO fraction (upper) and CHO fraction (lower) (Spera et al., 2009).

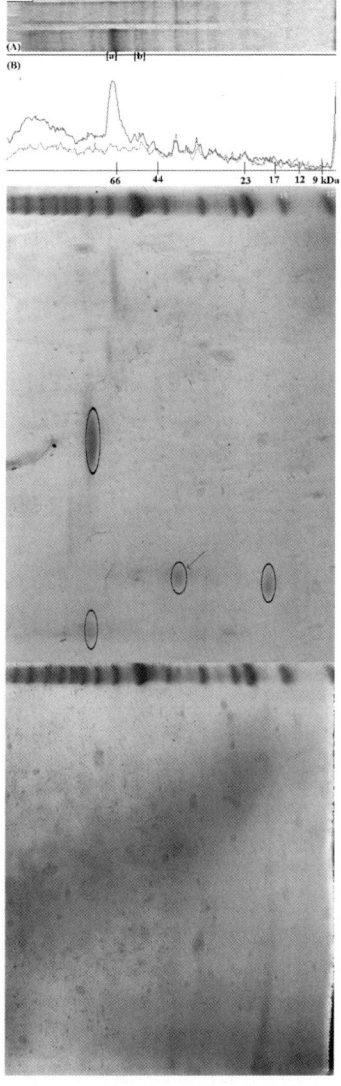

Figure 9.13. Nucleus gel lines and corresponding densitometry profiles. (A) cAMP CHO (upper) and CHO (lower) gel lines. (B) Corresponding densitometry profiles (in grey cAMP CHO sample profile and in black CHO sample profile): on the x axis the mass scale and on the y axis the spot grey intensity. (C) 2D gel of CHO fraction (upper) and cAMP CHO fraction (lower). The spots marked with a circle have been analyzed by MS (Spera et al., 2009).

For the Nucleus protein fractions an inhibition of at least two high weight proteins expression (marked as [a] and [b], Figure 9.13) after cAMP exposure is appreciable. It is evident also a general inhibition of very high weight proteins. It appears a small increase of the expression for at least one heavier protein (marked as [c], Figure 9.13). The apparent molecular mass (kDa) values of the spots identified in Figure 9.13, on the basis of standards, were 67 [a], 55 [b], and 38 [c] (Spera et al., 2009).

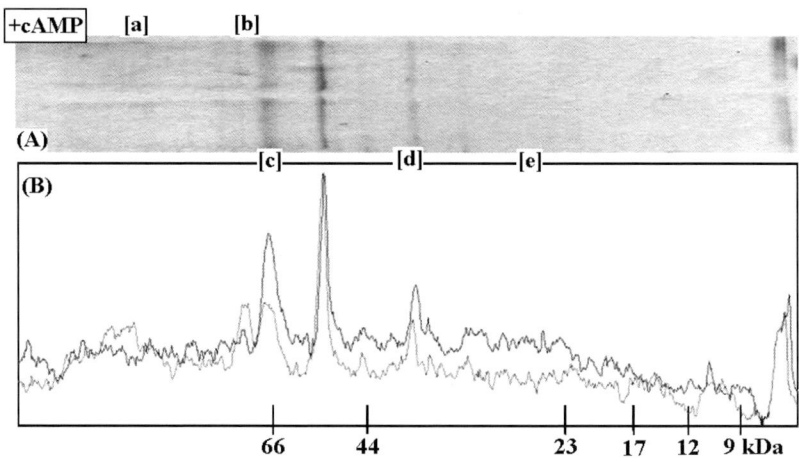

Figure 9.14. Cytoskeleton gel lines and corresponding densitometry profiles. (A) cAMP CHO (upper) and CHO (lower) gel lines. (B) Corresponding densitometry profiles (in grey cAMP CHO sample profile and in black CHO sample profile): on the x axis the mass scale and on the y axis the spot grey intensity. (Spera et al., 2009).

For medium and low weight Cytoskeleton fraction proteins a general expression inhibition appears after cAMP exposure except for at least two high weight proteins (marked as [a] and [b], Figure 9.14) whose peaks intensity appears higher for cAMP fraction. The apparent molecular mass (kDa) values of the spots identified in Figure 9.14, on the basis of standards, were 90 [a], 70 [b], 66 [c], 40 [d] and 25 [e] (Spera et al., 2009). For the Cytoskeleton fraction we did not perform 2D gel electrophoresis due to the few amount of proteins.

9.3.4 MALDI-TOF MS

The proteins extracted with S-PEK sub-cellular extraction kit are dissolved in the extraction buffers that are rich of chemicals not compatible with MALDI-TOF MS analysis. In the case of 1D gel electrophoresis and in gel digestion we did not perform any successive purification. It was, instead, necessary some purification steps for the analysis of intact proteins or when we performed solution digestion.

We tested different purification procedure reported in synthesis in the Table 9.7.

Table 9.7. Purification steps tested to prepare sub-cellular fraction samples for MALDI TOF MS analysis (Spera 2007).

Procedure	Step 1	Step 2	Step 3
I	Dialysis	Lyophilization	HPLC
II	HPLC	Lyophilization	–
III	Dialysis	Lyophilization	–
IV	HPLC	–	–
V	Dialysis	–	–
VI	Lyophilization	–	–
VII	TCA precipitation	–	–

After each one of these procedures a solution of 0.1% TFA was added to the sample to reach the opportune concentration and then MALDI-TOF MS analysis was performed. The procedure that gave the best results was the I. The samples, thereby, were dialyzed against water overnight at 4 °C. The dialysis bags were Spectra/Por cellulose membranes (Medicell, London, UK) with a molecular weight cut off (MWCO) of 8,000 and a flat width of 4 mm. After dialysis the samples were lyophilized (to concentrate) and then detached in 0.2% TFA solution for HPLC analysis (Spera 2007).

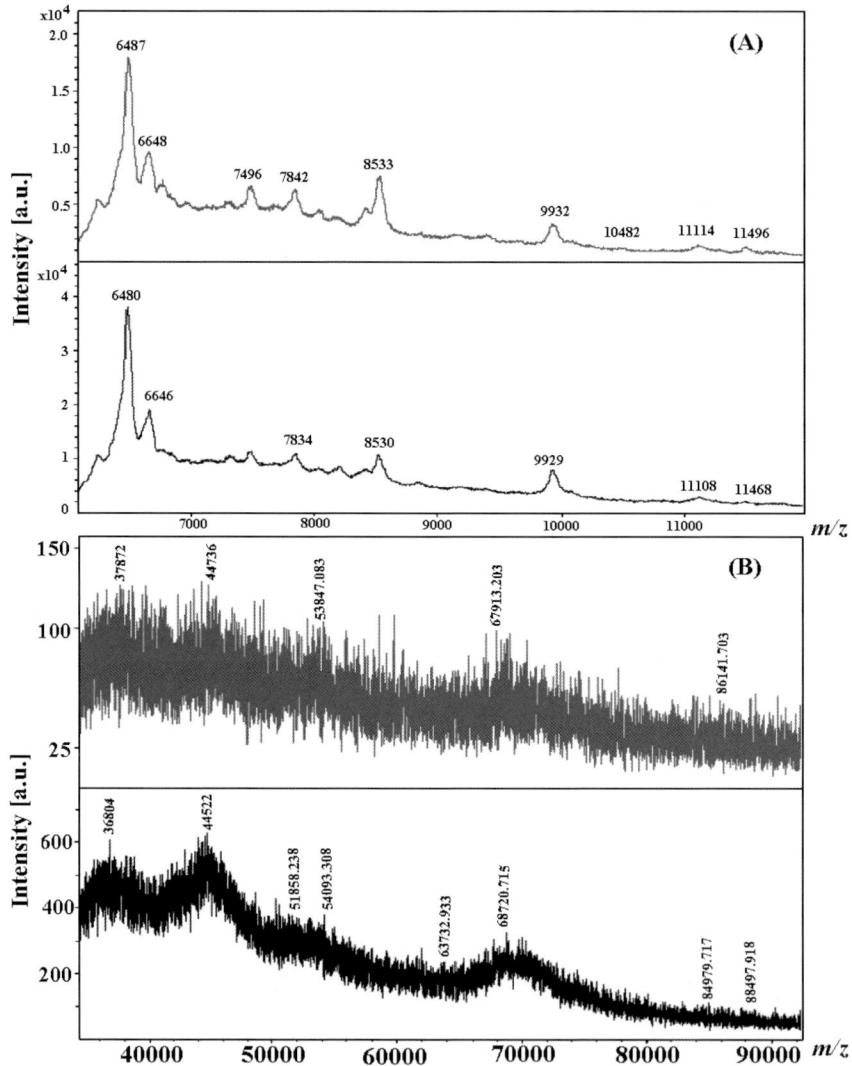

Figure 9.15. Cytosol MALDI TOF mass spectra. cAMP CHO (upper) and CHO (lower) fractions mass spectra. (A) Mass range 6–12 kDa, (B) Mass range 35–90 kDa (Spera et al., 2009).

The MS analysis were performed on an Autoflex MALDI-TOF mass spectrometer (Bruker Daltonics, http://www.bdal.com/) operating in linear and reflector mode. The samples, prepared as above described, were mixed with the appropriate matrix solution and deposited onto the stainless steel target of the MALDI instrument, according to the dried droplet method. To prepare matrix solutions, saturating amounts of either HCCA (α-Cyano-4-hydroxycinnamic acid, for peptides and low molecular weight proteins) or sinapinic acid (for high molecular weight proteins) were added to a solution of 2/3 of 0.1% TFA and 1/3 of acetonitrile.

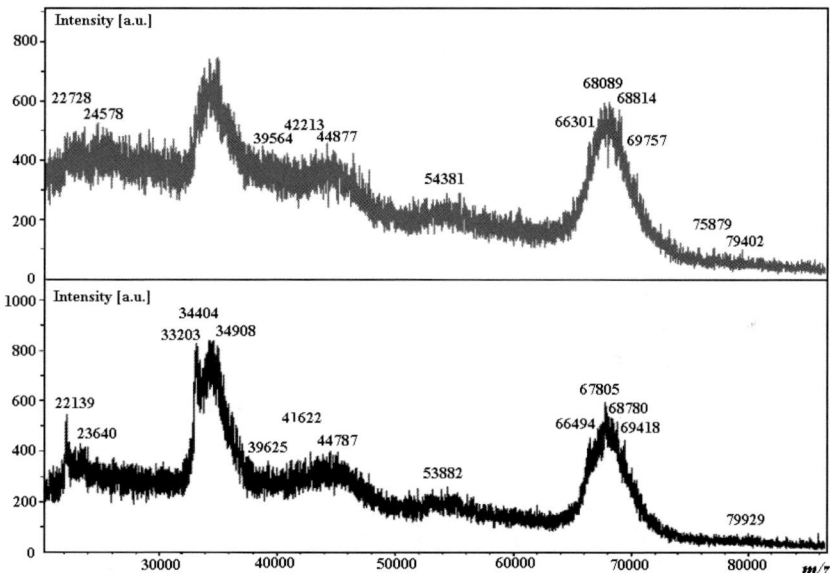

Figure 9.16. Membrane and membrane/organelle MALDI TOF mass spectra. cAMP CHO (upper) and CHO (lower) fractions mass spectra. Mass range 20–85 kDa (Spera et al., 2009).

Calibration was performed using peptide and protein standard mixtures (Bruker Daltonics, Germany), resulting in a mass accuracy < 100 ppm for intact proteins and < 10 ppm for peptides. MALDI TOF mass spectra were acquired with a pulsed nitrogen laser (337 nm) in positive mode. The mass lists obtained were submitted to a data bank search. We used a specific software for protein data interpretation, Biotools (Bruker Daltonics, Germany, www.bdal.de), that allows an automated protein identification via library search and that has a MASCOT Intranet search software (Matrix Sciences Ltd., www.matrixscience.com) fully integrated.

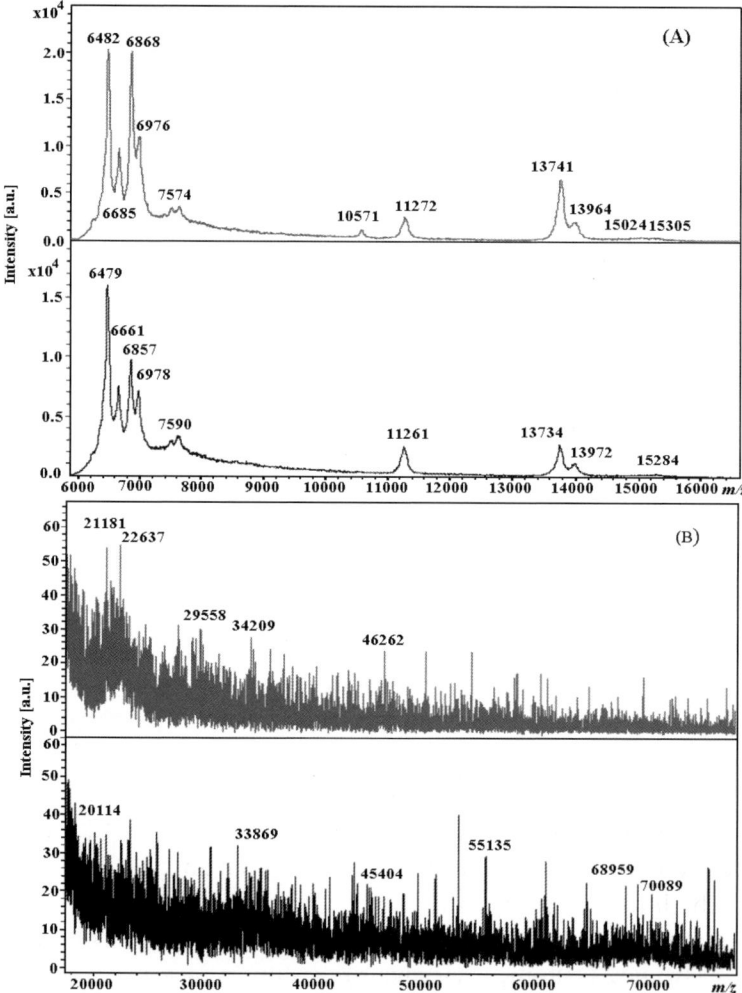

Figure 9.17. Nucleus MALDI TOF mass spectra. cAMP CHO (upper) and CHO (lower) fractions mass spectra. (A) Mass range 6–16 kDa (B) Mass range 20–75 kDa (Spera et al., 2009).

We analyzed by MALDI-TOF MS different samples of cellular fractions. For protein fingerprint we performed both in gel digestion and solution digestion. We fractionated and purified the samples by HPLC, then the obtained fractions have been analyzed before and after digestion in trypsin. For in gel digestion we cut from the gel the more interesting spots and performed digestion with trypsin. In Figures 9.15–9.17 mass spectra of the different fractions of CHO-K1 and cAMP CHO-K1 samples are reported (cytoskeleton fractions mass spectra are not reported) (Spera *et al.*, 2009).

The spectra were acquired in linear mode and the matrix utilized was sinapinic acid. From the observation of the relative spectra in the first two fractions in Figures 9.15 and 9.16, few peaks appear particularly interesting, namely: the two peaks of cytosol [CHO] spectrum at $m/z = 68.7 \pm 0.2$, and $m/z = 44.5 \pm 0.2$; their presence, above all in the cytosol [CHO] spectrum, confirms the SDS PAGE results. In the membrane and membrane/organelle [CHO] spectrum peaks at $m/z = 75680 \pm 60$, $m/z = 33230 \pm 80$ and $m/z = 22140 \pm 50$ that confirms the SDS PAGE results (Spera *et al.*, 2009).

It is well known in literature (Annesley 2003) that the analysis of a mixture of proteins by mass spectrometry is very difficult because the suppression of ion signals could invalidate the results. We then proceeded with the analysis of the proteins, identified by 1D gel electrophoresis, by excision of the spots, in-gel digestion, and MALDI-TOF MS analysis.

Unfortunately when we performed the MS analysis of the 1D gel spots after trypsin digestion we obtained quite always a mass mapping too difficult to be analyzed due to the high number of peptides obtained (presumably due to the number of proteins contained in the spot and again too high for MS analysis).

We obtained a better spectrum (Figure 9.18) by the analysis of the 66 kDa spot present only in the NUC[CHO] fraction. We performed a MASCOT database search for the obtained mass list but unfortunately, due to the large number of peptides obtained, no results with a good score were obtained.

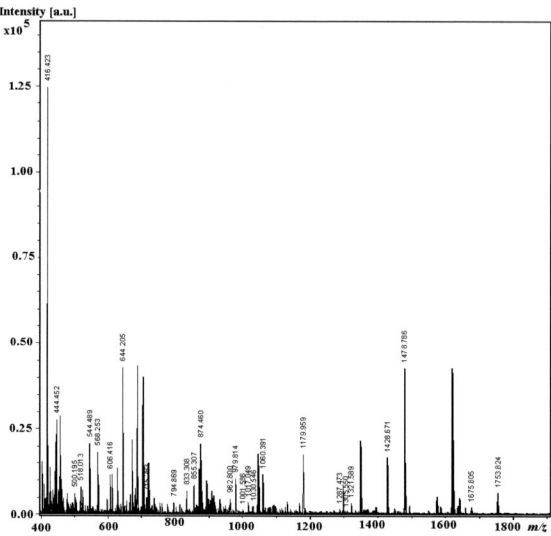

Figure 9.18. MALDI TOF mass spectrum of peptides from in gel digestion of 66 kDa spot of CHO nucleus fraction gel. The spectrum was acquired in reflector mode and the matrix utilized was HCCA. The spectrum was calibrated externally by Peptide standard solution (Bruker Daltonics).

We then proceeded to 2D gel analysis with the aim to obtain a better separation of the proteins. We proceeded to in gel digestion of the spots identified as more interesting by densitometry analysis and comparison of 2D gel images. In this case the best results that we obtained were regarding the analysis of cytosol and nucleus fractions.

In particular the spot of Nucleus fraction identified with the arrow and the circle in the Figure 9.13C has been identified as non-muscle Myosin Heavy Chain-B (Myh 10) with a score of 64, sufficiently high (Figure 9.19) (Spera *et al.*, 2009). Protein score is -10*Log(*P*), where *P* is the probability that the observed match is a random event: scores greater than 64 are significant ($p < 0.05$).

For the cytosol fraction the best results have been obtained for the spots identified as 1 and 2 in the Figure 9.11C. The spot 1 has been identified as Integrin beta-2 with a score of 23 and the spot 2 has been identified as Vesicular Fusion Protein NSF (fragment) with a score of 30 (Figure 9.20).

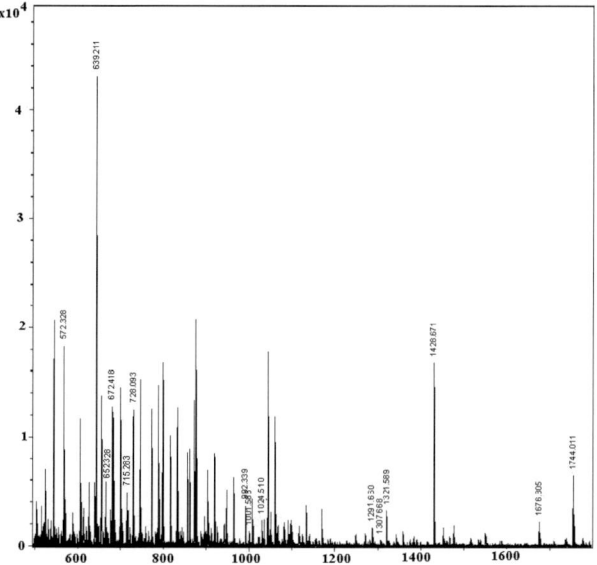

Figure 9.19. Non-muscle myosin heavy chain-B (Fragment): Mascot Search Result. Result of the fingerprint of a 2D gel spot of CHO-K1 Nucleus fraction before cAMP exposure (Spera *et al.*, 2009).

The protein Integrin beta-2 seems to be synthesized only after reverse transformation while the protein Vesicular Fusion Protein NSF is present before and after the differentiation.

Even if these two proteins have been identified the scores obtained are not satisfying for a good fingerprint due to the too high number of fragments not allowing to identify proteins.

Evidently in the analyzed spot were still present more than one protein: the number of proteins in our samples is so high that neither 2D gel electrophoresis is powerful enough to separate them also if it gave us the best results never obtained (Spera *et al.*, 2009).

```
Q8VHL6_SIGHI    Mass: 84918    Score: 23    Expect: 47    Queries matched: 7
Integrin beta-2.- Sigmodon hispidus (Hispid cotton rat).
Observed    Mr(expt)    Mr(calc)    Delta    Start    End    Miss    Peptide
1296.7370   1295.7297   1295.5638   0.1659   286 -   295    1       R.CHLEDNMYKK.S
1619.6510   1618.6437   1618.8103  -0.1666   743 -   755    1       K.LKSQWNNDNPLFK.S
1635.6360   1634.6287   1634.7246  -0.0959   380 -   395    0       K.VTYDSFCSNGASSIGK.S
1711.5410   1710.5337   1709.7825   0.7512   578 -   593    1       R.STDGCLNTRLVECSGR.G
2044.3130   2043.3057   2043.0749   0.2308   206 -   223    1       R.HVLKLTDNSNQFQTVVGK.Q
2070.2610   2069.2537   2068.7860   0.4677   544 -   563    1       R.YDGQVCGGSDRGSCFCGQCK.C
2347.3650   2346.3577   2347.2213  -0.8635   166 -   186    1       R.IGFGSFVDKTVLPFVNTHPEK.L
No match to: 313.6740, 632.1910, 662.6610, 690.4720, 790.4700, 1046.6160, 1347.7560, 1363.7460, 1439.7760, 1567.6190, 1584.0780,
1593.6330, 1694.5610, 1724.5920, 1758.6140, 2092.2730, 2330.9010
```

```
S04235    Mass: 83553    Score: 30    Expect: 9.4    Queries matched: 9
Vesicular fusion protein NSF - Chinese hamster (fragment)
Observed    Mr(expt)    Mr(calc)    Delta    Start    End    Miss    Peptide
644.8320    643.8247    644.3129   -0.4882   170 -   176    0       K.GEPASGK.R
662.6860    661.6787    661.3105    0.3682   575 -   580    0       K.ICSPDK.M
790.4910    789.4837    789.4055    0.0783   226 -   233    0       K.MGIGGLDK.E
888.4510    887.4437    887.5076   -0.0639   700 -   707    0       R.TTIAQQVK.G
902.8900    901.8827    902.4821   -0.5994   476 -   485    0       K.AESLQVTR.G
1046.5890   1045.5817   1045.5702   0.0115   280 -   288    1       R.QIGKMLNAR.E + Oxidation (M)
1312.2070   1311.1997   1310.6830   0.5167   424 -   435    0       R.GHQLLSADVDIK.E
1363.6570   1362.6497   1362.6528  -0.0030   313 -   323    1       K.LFADAEEQRR.L
1939.3500   1938.3427   1937.9954   0.3473    96 -   112    1       K.AKQCIGTMTIEIDFLQK.K
No match to: 314.2180, 354.8370, 453.2540, 618.4740, 632.1230, 659.6720, 690.4980, 917.2850, 1296.6500, 1347.6830, 1593.5530,
1597.5030, 1619.5260, 1673.3740, 1759.5820, 1993.1280, 2350.1530
```

```
NSF_CRIGR    Mass: 82740    Score: 29    Expect: 12    Queries matched: 9
Vesicle-fusing ATPase (EC 3.6.4.6) (Vesicular-fusion protein NSF) (N- ethylmaleimide sensitive fusion protein) (NEM-sensitive fusion p
Observed    Mr(expt)    Mr(calc)    Delta    Start    End    Miss    Peptide
644.8320    643.8247    644.3129   -0.4882   162 -   168    0       K.GEPASGK.R
662.6860    661.6787    661.3105    0.3682   567 -   572    0       K.ICSPDK.M
790.4910    789.4837    789.4055    0.0783   218 -   225    0       K.MGIGGLDK.E
888.4510    887.4437    887.5076   -0.0639   692 -   699    0       R.TTIAQQVK.G
902.8900    901.8827    902.4821   -0.5994   470 -   477    0       K.AESLQVTR.G
1046.5890   1045.5817   1045.5702   0.0115   272 -   280    1       R.QIGKMLNAR.E + Oxidation (M)
1312.2070   1311.1997   1310.6830   0.5167   416 -   427    0       R.GHQLLSADVDIK.E
1363.6570   1362.6497   1362.6528  -0.0030   305 -   315    1       K.LFADAEEQRR.L
1939.3500   1938.3427   1937.9954   0.3473    88 -   104    1       K.AKQCIGTMTIEIDFLQK.K
No match to: 314.2180, 354.8370, 453.2540, 618.4740, 632.1230, 659.6720, 690.4980, 917.2850, 1296.6500, 1347.6830, 1593.5530,
1597.5030, 1619.5260, 1673.3740, 1759.5820, 1993.1280, 2350.1530
```

Figure 9.20. Integrin beta-2 and Vesicular Fusion Protein NSF (fragment): Mascot Search Results. Results of the fingerprint of 2D gel spots (signed as 1 and 2 in the Figure 3C) of CHO-K1 Cytosol fraction after cAMP exposure. Protein score is $-10*Log(P)$, where P is the probability that the observed match is a random event: scores greater than 64 are significant ($p<0.05$) (Spera et al., 2009).

9.4 Conclusions

Employing several biophysical and biochemical techniques this chapter wants to analyze the relationship between protein expression pattern, morphology and gene expression of mammalian cells and how these features change after cellular differentiation. Investigation of proteome alterations accompanying well-defined physiologic processes may help to further characterize and understand functional implications of affected proteins. In particular we analyzed the tumoral Chinese Hamster fibroblast line (CHO), to define how its protein expression profiles change after differentiation by an external chemical agent (cAMP) and which are the cellular morphological and metabolic alterations that accompany this protein pattern change.

It is note that cAMP, through control of phosphorylation, regulates the structure of the cell cytoskeleton, which in turn strongly influences many differentiation properties including the assumption of normal or cancerous behavior in mammalian cells. Work from a number of laboratories has demonstrated that the cytoskeleton exerts a variety of different and complex regulatory metabolic actions in the biochemical economy of the mammalian cell (Puck *et al.*, 2002; Nicolini and Beltrame, 1982; Belmont and Nicolini, 1982).

The CHO-K1 changes after cAMP exposure are consistent with the conversion from a malignant to a normal fibroblastic state (reverse transformation) and this conversion is under genetic control (Puck *et al.*, 1989). There are at least two levels of regulation of gene expression in mammalian cells. The first is conversion of the genes in question from a sequestered state, in which they are unable to interact with molecules in the environment, to an exposed state in which they can be so influenced. The next step is the turning of such genes on and off by appropriate effector molecules in the nucleus. Puck *et al.* proposed that the cytoskeleton transmits information from the cell membrane to chromosomal loci in the nucleus and is an important element in regulation of the exposure process (Puck and Krystosek, 1992, 1993; Ashall and Puck, 1984). The mechanism of the transformation, however, is as yet obscure. The phenomena make it clear that a fundamental change in properties of the cell membrane has occurred but the molecular mechanisms generating these structural changes are largely unknown.

Our studies want to delve more deeply into the mechanism of protein changes in the cAMP-treated differentiated state of CHO cells. Until now the analysis of CHO-K1 cells and their reverse transformation by cAMP has been mainly focused on the role of the cytoskeleton and its transformation. We, on the contrary, want to analyze the whole cellular proteome and how it changes after reverse transformation. We are, in particular, interested in the changes at the levels of nuclear envelope-associated proteins and chromatin-condensing proteins. A description of the changing levels and modifications of nuclear structural proteins constitutes a first step toward developing an understanding of cellular mechanisms for controlling nuclear shape.

To analyze how each cellular compartment changes after differentiation we performed a sub-cellular fractionation. We separated the whole cellular proteome in four fractions (cytosol, membrane and membrane/organelle, nucleus, cytoskeleton). To identify the proteins differentially expressed before and after differentiation we analyzed the four fractions by 1D gel electrophoresis, 2D gel electrophoresis and HPLC. By densitometry analysis of 1D and 2D SDS PAGE we identified marked changes in the protein expression pattern after cAMP exposure (Spera et al., 2009). In particular we identified, for the nuclear fraction, a strong inhibition of a 66 kDa and a 55 kDa proteins and a general inhibition of very high weight proteins. It was also evident, on the contrary, an increase in the expression of at least four proteins of weight: 50, 36, 17 and 14 kDa. It was evident an increasing expression also for a 66 kDa cytoskeleton protein, while it was evident, in the same fraction, a marked expression inhibition for a 70 kDa and a 90 kDa proteins.

We identified, by HPLC-ESI MS three proteins present in the nuclear fraction only before exposure to cAMP and other three proteins whose expression is inhibited after cAMP exposure. To a first identification on the base of their mass we performed a search in Swiss Prot Databank restricting our search to rodent proteins (Spera and Nicolini, 2007). In this way we identify four proteins: Myh10 (fragment B) (33 kDa), Gna11 (42 kDa), Sc61a_Haran (52 kDa) and CD82_Mouse (30 kDa), while for other two it was found no identification (6632 and 20739 Da) (Spera and Nicolini, 2007). Among these proteins, Guanine nucleotide-binding protein (Gna11) and Myosin heavy chain 10, non-muscle (Myh 10), are particularly interesting because they are both involved in cellular regulation. Heterotrimeric guanine nucleotide-binding proteins (G proteins) are integral to the signal transduction pathways that mediate the response of the cell to many hormones, neuromodulators, and a variety of other ligands (Strathmann and Simon, 1990). Within the myosin family of proteins, the conventional myosin II isoform falls into two subgroups: the sarcomeric (skeletal and cardiac muscles) and the nonsarcomeric (smooth muscle and non-muscle). There are evidences that non-muscle myosins play a role in diverse cellular functions, for instance cytokinesis, proliferation, secretion and receptor capping. Myosin molecule consists of a pair of heavy chains and two pairs of light chains. Two isoforms of the non-muscle myosin heavy chain (nmMHC), chain A (nmMHC-A) and chain B (nmMHC-B), have been identified. The nmMHC-B in particular is involved in cell growth regulation and transformation (Yam *et al.*, 2000).

We employed reverse phase HPLC to purify and fractionate protein sub-cellular fractions before MALDI-TOF MS. Moreover, from the HPLC analysis of the CHO-K1 four protein fractions, we found a significant increase of the protein content after cAMP exposure for membrane and membrane/organelle fractions and a decrease for the Nucleus fractions, that is a first indication of the whole cell involvement in the reverse transformation, and of a strong reorganization of the cellular membrane.

We continued our analysis to characterize in detail, by protein fingerprint, these identified proteins. By in gel digestion and MALDI-TOF analysis of the tryptic digest we analyzed different spots of the gel for each fraction. In particular we are interested to the 66 kDa protein present in the nuclear fraction only before the differentiation. Analyzing 10 transformed human cell lines and 11 normal human cell cultures Puck *et al.* identified a 66 kDa protein that they identified as Rp66 that is a nuclear protein that binds low repeat sequence of DNA (LRS). Our hypothesis is that the identified 66 kDa protein is an homologue of this Rp66 that is present only after differentiation when the CHO-K1 cells turns to normal fibroblast. The binding of LRS-1 to Rp66 has, in fact, only been observed in transformed but not normal human cell lines, it is conceivable that this protein is involved in the decision between regulated and unregulated reproduction as in the case in the reverse transformation reaction. Such possibilities are supported by the fact that products of protooncogenes such as *c*-myc rise dramatically when lymphoid cells (Nicolini *et al.*, 2006) or fibroblasts are stimulated to divide, whereas the amount drops when the cells are stimulated to differentiate terminally. Moreover, blocking *c*-myc expression with antisense constructs appears to block cell proliferation and induce differentiation in some systems. It is interesting to note that the *c*-myc gene product is also a DNA-binding protein. Further studies are required to illuminate the role of specific proteins binding to repetitive DNA sequences like LRS-1 in systems like that described here.

Unfortunately by in gel digestion due to the large number of peptides obtained, no results with a good score were obtained. To better separate the proteins of the sub-cellular fraction we then utilized the 2D gel electrophoresis on the same samples. By this technique we identified a spot of the 2D gel of CHO Nucleus fraction as non-muscle myosin heavy chain-B in agreement with our previous results (Spera and Nicolini 2007; Spera et al., 2009). For the cytosol fraction we identified the protein Integrin beta-2, synthesised only after reverse transformation, and Vesicular Fusion Protein NSF (fragment). Unfortunately these two proteins have been identified with scores not satisfying for a good fingerprint. Evidently in the analyzed spot were still present more than one protein: the number of proteins in our samples is so high that neither 2D gel electrophoresis is enough powerful to separate them also if it gave us the best results so far (Spera et al., 2009).

When we will clarify the relationship between protein expression, cellular morphology, metabolism and gene expression we will increase our understanding the basic mechanisms that regulate cell growth and reverse transformation.

Definitively we found that new patterns of protein synthesis and processing are established after reverse transformation and further studies using mass spectrometry and bioinformatics are in progress to enlighten this complex phenomenon (Badino et al., 2009). Immediate experiments would involve the use of NAPPA (Spera and Nicolini, 2008; Ramachandran et al., 2004, 2008) that involves the immobilization of appropriate genes (such c-myc one, projected by bioinformatics employment). In addition, the search for LRS protein binding must be extended to additional normal and transformed cell lines; and relationships between protein binding to repetitive sequences and exposure and sequestration of specific genes must be investigated.

Acknowledgments

This project was supported by grants to Fondazione EL.B.A. by MIUR for "Funzionamento" and by a FIRB International Grant on Proteomics and Cell Cycle (RBIN04RXHS) from MIUR (Ministero dell'Istruzione, Università e Ricerca) to CIRSDNNOB-Nanoworld Institute of the University of Genova.

References

Abdolzade-Bavil, A., Hayes, S., Goretzki, L., Kroger, M., Anders, J. and Hendriks, R. (2004). Convenient and versatile subcellular extraction procedure, that facilitates classical protein expression profiling and functional protein analysis. *Proteomics*, 4, pp. 1397–1405.

Annesley, T. M. (2003). Ion suppression in mass spectrometry. *Clinical Chemistry*, 49, pp. 1041–1044.

Ashall, F. and Puck, T. T. (1984). Cytoskeletal involvement in cAMP-induced sensitization of chromatin to nuclease digestion in transformed Chinese hamster ovary K1 cells. *Proceedings of the National Academy of Sciences USA*, 81, pp. 5145–5149.

Ashall, F., Sullivan, N. and Puck, T. T. (1988). Specificity of the cAMP-induced gene exposure reaction in CHO cells. *Proceedings of the National Academy of Sciences USA*, 85, pp. 3908–3912.

Ashcroft, A. E. (2003). Protein and peptide identification: the role of mass spectrometry in proteomics. *Natural Product Report*, 20, pp. 202–215.

Badino, F., Spera, R., Sivozhelezov, V. and Nicolini, C. (2009). Maldi-Tof detection of NAPPA. *Journal of Mass Spectrometry*, submitted.

Belmont, A. and Nicolini, C. (1982). Cell versus nuclear morphometry of serum stimulated fibroblasts: nuclear changes precede cell changes. *Journal of Cell Science*, 58, pp. 201–209.

Brinkley, B. R. (1982). The cytoskeleton: a perspective. *Methods in Cell Biology*, 24, pp. 1–8.

Brinkley, B. R., Fuller, E. M. and Highfield, D. P. (1975). Cytoplasmic microtubules in normal and transformed cells in culture: analysis by tubulin antibody immunofluorescence. *Proceedings of the National Academy of Sciences USA*, 72, pp. 4981–4985.

Chan, D., Goate, A. and Puck, T. T. (1989). Involvement of vimentin in the reverse transformation reaction. *Proceedings of the National Academy of Sciences USA*, 86, pp. 2747–2751.

Gabrielson, E. G., Scoggin, C. and Puck, T. T. (1982). Phosphorylation changes induced by cAMP derivatives in the CHO cell and selected mutants. *Experimental Cell Research*, 142, pp. 63–68.

Graves, P. R. and Haystead, T. A. J. (2002). Molecular Biologist's Guide to Proteomics. *Microbiology and Molecular Biology Reviews*, 66, pp. 39–63.

Gygi, S. P and Aebersold, R. (2000). Mass spectrometry and proteomics. *Current Opinion in Chemical Biology*, 4, pp. 489–494.

Hatzopoulos, A. K., Folkman, J., Vasile, E., Eiselen, J. K. and Rosenberg, R.D. (1998). Isolation and characterization of endothelial progenitor cells from mouse embryos. *Development*, 125, pp. 1457–1468.

Hsie, A. W. and Puck, T. T. (1971). Morphological transformation of Chinese hamster cells by dibutyryl adenosine cyclic 3':5'-monophospate and testosterone. *Proceedings of the National Academy of Sciences USA*, 68, pp. 358–361.

Hsie, A. W., Jones, C., Puck, T. T. (1971). Further changes in differentiation state accompanying the conversion of chinese hamster cells to fibroblastic form by dibutyryl adenosine cyclic 3':5'-monophosphate and hormones. *Proceedings of the National Academy of Sciences USA*, 68, pp. 1648–1652.

Itoh, K. and Adelstein, R. S. (1995). Neuronal cell expression of inserted isoforms of vertebrate nonmuscle myosin heavy chain ii-B. *Journal of Biological Chemistry*, 270, pp. 14533–14540.

Laemmli, U. (1970). Cleavage of structural proteins during the assembly of the head of bacteriophage T4. *Nature*, 227, pp. 680–685.

Law, M. L., Gao, J. Z. and Puck, T. T. (1989). A nuclear protein associated with human cancer cells binds preferentially to a human repetitive DNA sequence. *Proceedings of the National Academy of Sciences USA*, 86, pp. 8472–8476.

Liebler, D. C. (2002). *Introduction to proteomics*. Humana Press, Totowa, NJ.

Lypowy, J., Chen, I. Y, and Abdellatif, M. (2005). An alliance between Ras GTPase-activating protein, filamin C, and Ras GTPase-activating Protein SH3 Domain-binding protein regulates myocyte growth. *Journal of Biological Chemistry*, 280, pp. 25717–25728.

Messana, I., Cabras, T., Inzitari, R., Lupi, A., Zuppi, C., Olmi, C., Fadda, M. B., Cordaro, M., Giardina, B. and Castagnola, M. (2004). Characterization of the human salivary basic proline-rich protein complex by a proteomic approach. *Journal of Proteome Research*, 3, pp. 792–800.

Miranti, C. and Puck, T. T. (1990). Gene regulation in reverse transformation: cyclic AMP-induced actin homolog in CHO cells. *Somatic Cell and Molecular Genetics*, 16, pp. 67–78.

Nagasawa, S. Y., Takuwa, N., Sugimoto, N., Mabuchi, H. and Takuwa, Y. (2005). Inhibition of Rac activation as a mechanism for negative regulation of actin cytoskeletal reorganization and cell motility by cAMP. *Biochemical Journal*, 385, pp. 737–744.

Nicolini, C. and Beltrame, F. (1982). Coupling of chromatin structure to cell geometry during the cell cycle. Transformed versus reverse-transformed CHO. *Cell Biology International Reports*, 6, pp. 63–71.

Nicolini, C., Spera, R., Stura, E., Fiordoro, S. and Giacomelli, L. (2006). Gene expression in the cell cycle of human T-lymphocytes: II. Experimental determination by Dnaser technology. *Journal of Cellular Biochemistry*, 97, pp. 1151–1159.

Parodi, S., Beltrame, F., Lessin, S. and Nicolini, C. (1979). Morphometric analysis of B_2cAMP induced reverse transformation in synchronized CHO cells. *Cell Biophysics*, 1, pp. 271–292.

Puck, T. T. (1977). Cyclic AMP, the microtubule-microfilament system, and cancer. *Proceedings of the National Academy of Sciences USA*, 74, pp. 4491–4495.

Puck, T. T. and Krystosek, A. (1992). The role of the cytoskeleton in genome regulation and cancer. *International Review of Cytology - A Survey of Cell Biology*, 132, pp. 75–108.

Puck, T. T. and Krystosek, A. (1993). Reverse transformation, genome exposure, and cancer. In: *Advances in Cancer Research*. G.F. Vande & G. Klein, Eds.: pp. 125–151. Academic Press. San Diego, CA.

Puck, T. T., Webb, P., Johnson, R. (1998). Genome exposure and regulation in mammalian cells. *Somatic Cell and Molecular Genetics*, 24, pp. 291–302.

Puck, T. T., Webb, P., Johnson, R. (2002). Cyclic AMP and reverse transformation reaction. *Annals of the New York Academy of Sciences*, 968, pp. 122–138.

Ramachandran, N., Hainsworth, E., Bhullar, B., Eisenstein, S., Rosen, B., Lau, A. Y., Walter, J. C., LaBaer, J. (2004). Self-assembling protein microarrays. Science, 305, pp. 86–90.

Ramachandran, N., Raphael, J.V., Hainsworth, E., Demirkan, G., Fuentes, M.G., Rolfs, A., Hu, Y. and LaBaer, J. (2008). Next-generation high-density self-assembling functional protein arrays. *Nature Methods*, 5, pp. 535–538.

Rios-Doria, J., Kuefer, R., Ethier, S. P. and Day, M. L. (2004). Cleavage of ß-Catenin by calpain in prostate and mammary tumor cells. *Cancer Research*, 64, pp. 7237–7240.

Rosenberger, R. F. (1994). Translational errors during recombinant protein synthesis. *Dev Biol Stand*, 83, pp. 21–26.

Rosenfeld, J., Capdevielle, J., Guillemot, J. C. and Ferrara, P. (1992). In-gel digestion of proteins for internal sequence analysis after one- or two-dimensional gel electrophoresis. *Analytical Biochemistry*, 203, pp. 173–179.

Sabio, G., Arthur, J. S., Kuma, Y., Peggie, M., Carr, J., Murray-Tait, V., Centeno, F., Goedert, M., Morrice, N. A., Cuenda, A. (2005). P38gamma regulates the localisation of SAP97 in the cytoskeleton by modulating its interaction with GKAP. *EMBO Journal*, 24, pp. 1134–1145.

Schliwa, M. (2002). The evolving complexity of cytoplasmic structure. *Nature Reviews Molecular Cell Biology*, 3, pp. 291–295.

Spera, R. and Nicolini, C. (2007). c-AMP induced alterations of chinese hamster ovary cells monitored by mass spectrometry. *Journal of Cellular Biochemistry*, 102, pp. 473–482.

Spera, R. (2007). Proteins and cell structures analysis by mass spectrometry and correlated biophysical techniques. *PhD Thesis*.

Spera, R. and Nicolini, C. (2008). *NAPPA microarrays and mass spectrometry: new trends and challenges*. Nano Article - Taylor & Francis/CRC Press.

Spera, R., Badino F. and Nicolini, C. (2009). CHO-K1 proteome alterations during reverse transformation. *Journal of Cellular Biochemistry*, in press.

Strathmann, M. and Simon, M. I. (1990). G protein diversity: A distinct class of alpha subunits is present in vertebrates and invertebrates. *Proceedings of the National Academy of Sciences USA*, 87, pp. 9113–9117.

Vergani, L., Mascetti, G. and Nicolini, C. (2001). Changes of nuclear structure induced by increasing temperatures. *Journal Biomolecular Structure & Dynamics*, 18, pp. 1–9.

Yam, J. W. P., Chan, K. W., Li, N. and Hsiao, W. L. W. (2000). Molecular cloning and functional analysis of the promoter region of rat non-muscle myosin heavy chain-B gene. *Biochemical and Biophysics Research Communications*, 276, pp. 1203–1209.

Yates, J. R., III. (1998). Mass spectrometry and the age of the proteome. *Journal of Mass Spectrometry*, 33, pp. 1–19.

CHAPTER 10

ORGAN TRANSPLANTS AND GENE MICROARRAYS

Richard Danger[1,2], Jean-Paul Soulillou[1,2],
Sophie Brouard[1,2,§] and Claudio Nicolini [3,4,§]

[1]*Institut National de la Santé Et de la Recherche Médicale (INSERM) – Université de Nantes, UMR 643, CHU Nantes, 30 bd Jean Monnet, 44093 Nantes, France*
[2]*Institut de Transplantation Et de Recherche en Transplantation (ITERT), CHU-Hotel Dieu, Nantes, France*
[3]*Nanoworld Institute and Eminent Biophysics Chair, University of Genova, Corso Europa 30, Genova 16132, Italy*
[4]*Fondazione El.B.A., Piazza SS. Apostoli 66, Rome, Italy*

Application of Microarrays Technology to organ transplantation in human with the aid of statistical and non-statistical analysis, including interaction-based methods, is reviewed here.

10.1 Introduction

Organ transplantation is the treatment of choice for end stage diseases, improving both the patient's quality of life and survival, mainly thanks to advances in HLA knowledge and immunosuppressive treatment occurred in 80–90's (Sayegh and Carpenter 2004). In 2006, a total of 27,578 organ transplantations have been done in United States and 4,428 in France (ABM 2009; Wolfe *et al.*, 2009). Among all the transplantation performed (kidney, heart, lung, liver, …), the incidence of rejection episodes is still a problem. Acute rejection episodes (ARE) can reach a percentage of 55% of lung transplant recipients during the first year post-transplantation (Trulock *et al.*, 2007) and is still the causes of 12% of

---

[§] Both authors contributed equally to this work

graft loses in kidneys graft patients (El-Zoghby *et al.*, 2009). Moreover, ARE are also a negative predictor of graft outcome even if they are treated successfully and early (Opelz and Döhler 2008), increasing the need to a better understood of ARE mechanisms. In the long term, chronic rejection still remains a major concern in organ transplantation. Severe chronic allograft nephropathies are observed in 58.4 percent of 10 year-kidney recipients despites immunosuppressive drugs and patient monitoring (Nankivell *et al.*, 2003). Thus, whereas clinical and biological parameters are used to predict and monitor graft survival, there is heterogeneity of risk factors associated with graft outcome, which need a better understanding of biological mechanisms occurring during early and late stage of organ transplantation. The analysis of gene expression in the field of organ transplantation could help investigators to improve knowledge of such biological mechanisms.

Albeit single gene expression analysis have highlighted molecules like FOXP3, granzyme B and perforin (Strehlau *et al.*, 1997; Muthukumar *et al.*, 2005; Ashton-Chess *et al.*, 2009), they cannot perform a global view of biological changes during transplantation. Thanks to the use of microarrays, which allow the analysis of thousand of gene in a single experiment, gene profiling studies have been performed in last years to assess molecular changes in transplanted organ and in recipient samples. The identification of involved genes or pathways could improve our understanding of immune and rejection mechanisms, and therefore improve organ and patient outcomes. In this chapter, we do an overview of main studies which give a light on organ transplantation mechanisms in human thanks to useful gene microarrays.

10.2 A Glass of Genes?

10.2.1 *Microarray principle*

Because microarray technology allows measuring the expression of thousands genes in a single experiment, this technology is a powerful tool to determine transcription gene changes in cells in responses to a stimulus. Briefly, gene microarray technology, based on the principle of complementary sequences of DNA, consists of a solid support — usually glass slide — on which thousands of microscopic spots of DNA sequences have been deposed; they are probes (Figure 10.1).

Figure 10.1. DNA microarray consists of a solid support – usually glass slide. Adequate scanner generates image for each microarray on which target signal of thousands of microscopic spots of DNA is detected and quantified (Photo credit: Nicolas Degauque).

These gene probes could be complementary DNA (cDNA) sequences used for non-commercially microarrays or short (20–60b) single strand oligonucleotides for commercial arrays (*e.g.* Affymetrix, Agilent, Illumina or Nimblegen). In another hand, messenger RNA (mRNA), extracted with classical techniques from tissue or purified cells, is retrotranscribed into cDNA. However, when starting with small amount of total RNA, linear amplification derived from Eberwine method is perform to generate sufficient amplified antisense RNA (aRNA) (Van Gelder *et al.*, 1990). Thanks to this reliable technique, total RNA quantity needed for microarrays processing has greatly decrease from several micrograms until a few nanograms. Targets sequences, which are mRNA or aRNA, are then labeled with a detectable marker (fluorescent dye). Probe-target hybridization is detected and quantified by detection of the target signal, using adequate scanner which will generate image for each microarray. These images are processed by dedicated software in order to extract numerical data, summarizing spot intensity, shape, location and surrounding background intensity. To compensate the technical differences in overall RNA quantity, labeling efficiency, hybridization conditions and scanner properties a normalization step is processed (Yang et al., 2002). Several normalization methods are available, mainly depending of the nature of the array (Fan and Ren 2006). Finally, in the generated dataset, for each spot a value is assigned which is proportional to the concentration of target mRNA for which the probes matched. In the last few years, significant improvements have been done in microarray manufacturing notably in miniaturization of the spot, allowing the deposit of tens of thousands genes corresponding of entire genomes, the deposit of several samples in a same microarray and a decrease of the experiment price allowing an increase number of sample. Based on complementary sequences of DNA and in addition of gene expression profiling, nucleotide microarrays could also been used in a variety applications in the field of molecular biology such as microRNA profiling, alternative splicing detection, single nucleotide polymorphism (SNP) analysis, comparative genomic hybridization (CGH), chromatin immunoprecipitation (ChIP-chip) or DNA methylation analysis.

10.2.2 Sample source

Classical mRNA sources in organ transplantation are organ biopsy, peripheral blood mononuclear cells (PBMC) or peripheral blood lymphocytes (PBL) and fluids close to the transplanted organ: urines for kidney (Li et al., 2001) or bronchoalveolar lavage (BAL) fluid for lung recipients (Gimino et al., 2003). In addition, recently, the use of bronchial epithelial cells obtained by bronchial brushing has been validated for microarray experiment by Skawran et al., in the case of lung transplantation (Skawran et al., 2009). More difficult to use, laser microdissection has been reported by Ahn et al. to successfully measure gene expression in endogenous islets from B6AF1 mice (Ahn et al., 2007).

10.2.3 *Overview of microarray analysis in organ transplantation*

Many different data analyses have been performed (Figure 10.2) notably depending on microarray nature used (*i.e.* commercial oligonucelotide microarrays or in-house-developed cDNA microarrays) and experimental design (see for review: Roberts 2008). Here, we provide an overview of methods used in organ transplantation studies, which is far from exhaustive. Generally, two main goals are pursued: the identification of differentially expressed genes or gene sets which could provide better explanations of biological mechanisms and the discovery of disease-associated genes or gene sets which could predict disease class or outcome. Actually, these two goals are often interrelated because disease involves chemical and cellular mechanisms.

The identification of differentially expressed genes between two or more conditions (*e.g.* patients with acute rejection versus patients with stable graft function) is the most common analysis which provides gene that are associated with a specific phenotype. However, a high level of multiple testing is reach with the wide number of genes measured, so classical statistic methods cannot be used. Indeed, dedicated powerful statistical approaches have been developed to assess expression data allowing the identification of up or down regulated genes in a particular situation in comparison with another. The most famous one is the SAM software (Significance Analysis of Microarray), initially based on a modified *t*-test, which allows the estimation of a False Discovery Rate (FDR) for the entire data set (Tusher *et al.*, 2001).

Unsupervised clustering method, also commonly used, is generally complementary used to classify microarray data (Eisen *et al.*, 1998). Genes and/or samples with similar expression profiles across a dataset are clustered together; the closer the genes are the greater similarity they have. Moreover, one can identify cluster of genes grouped on the basis of similar expression patterns, with the assumption that clusters of genes are not due to hazard but that these genes participate to a common signaling pathway. Clustering can thus be useful in assigning potential functions to unidentified genes. Many different algorithms have been applied to discover patterns of gene expression common to a particular physiological state such as *K*-means algorithms (Tavazoie *et al.* 1999) which aims to partition n observations into k clusters in which each observation belongs to the cluster with the nearest mean, principle component analysis (Hilsenbeck *et al.* 1999), and self-organizing maps (Tamayo *et al.* 1999).

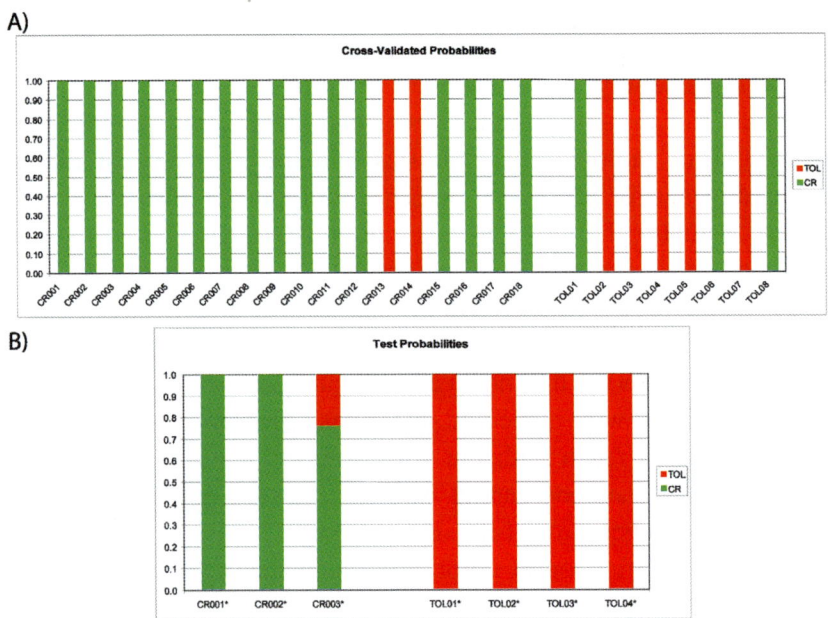

Figure 10.2. Classification probabilities of individual patients by Predictive Analysis of Microarray (PAM) based on the 343 differentially expressed genes between operationally tolerant kidney graft recipients (TOL) and patients with chronic rejection (CR). Each patient sample is shown by a bar, as labelled in the X-axis. The colour codes indicate the probability (0-1, as indicated in Y-axis) that the sample belongs to TOL (red) or CR (green). (A) Cross-validated probabilities on the 8 TOL and 18 CR that were used to set up the 2-class (TOL/CR) classification algorithm. Among the 26 patients, 5 samples (CR013, CR014, TOL01, TOL06 and TOL08) were misclassified. (B) Using the PAM algorithm defined with 8 TOL and 18 CR patients, 7 serially harvested samples (4 TOL and 3 CR) at a time interval of more than 1 year after the first sample were classified. The algorithm correctly classified all samples, with a probability of 100% for TOL and 92.0% for CR. (Reprinted with the permission from Braud C., Baeten D., Giral M., Pallier A., Ashton-Chess J., Braudeau C., Chevalier C., Lebars A., Lèger J., Moreau A., Pechkova E., Nicolini C., Soulillou J.P., Brouard S., Immunosuppressive drug-free operational immune tollerance in human kidney transplant recipient: I. Blood gene expression statistical analysis, Journal of Cellular Biochemistry 103, 1681-1692, ©2008, Wiley-Liss, Inc., a subsidiary of John Wiley & Sons, Inc.).

In addition, unsupervised clustering of sample can also be useful to identify molecular based disease subgroups of a unique syndrome based disease.

This is what performed Sarwal *et al.* in identifying three distinct subgroup of acute rejection in a transplant biopsy study (Sarwal *et al.* 2003). Moreover, hierarchical clustering is also a useful tool for visualizing expression patterns, the typical output being a dendrogram of the genes, from which clusters of closely matched genes can be identified. A heatmap is generated where signal intensity are represented as a color scale, usually from intense red for the highest positive values to intense green for the most negative values.

The identification of disease associated genes or gene sets is frequently performed in order to discover disease-related biomarkers. Ideally, biomarkers should be quantitative, early predictive and noninvasive, and should enable shorter clinical trial survey periods (Brouard *et al.*, 2008). To this end, supervised clustering aids in the discovery of molecular signatures in different disease processes (Alizadeh *et al.*, 2000). As unsupervised clustering, supervised clustering aims in grouping genes in categories that can be associated with function or co-regulation, but it incorporate biological knowledge to label the samples. Therefore, unsupervised clustering can be described as an explanatory process establishing the existence of gene categories without pre-existing assumptions, only based on gene expression similarities, whereas supervised clustering establishes the existence of gene groups associated with defined biological classes. In addition, a class prediction algorithm, PAM (Prediction Analysis for Microarrays*)*, was developed based on an enhancement of the nearest prototype (centroid) classifier. Briefly, the nearest-centroid classification computes a centroid value for each class and compares the distance of the gene expression profile of a new sample; the class whose centroid is the closer to is the predicted class for that new sample (Tibshirani *et al.*, 2002). PAM is useful in identifying minimal subsets of genes that best characterize each class, and has proven to be effective in organ transplantation to identify a pattern of 49 genes able to correctly segregate the tolerant and chronic rejection phenotypes in a cohort of renal transplant patient (Brouard *et al.*, 2007).

Once a list of significantly differentially expressed genes or clustered genes has been obtained, and assuming that similar gene expression profiles means co-regulation or implication in the same biological pathways, the next step to improve biological knowledge is to identify these biological processes involved. With advances in gene annotation particularly in human, biological informations for a particular gene are easily available in many online databases. However, for several genes such a literature mining is quite time-consuming and is clearly impossible for broad gene lists issue from microarray. Hopefully, tools have been developed to query databases to obtain all gene-related information and return lists of ontologies, with statistical measures of their significance; this is the case of GOminer software (Zeeberg et al., 2003), or Expression Analysis Systematic Explorer (EASE) (Hosack et al., 2003) (more details about ontological analysis in Khatri and Drăghici 2005). Among other tools, Gene Set Enrichment Analysis (GSEA) determines whether defined set of genes, mainly issue from DNA microarray publications, show statistically significant concordant differences between two biological states. However, if, and only if, the significantly different genes or clustered genes belong to the same biological process, one can hope that the observed gene expression patterns are relevant to the given process. Indeed, if a regulator gene Z, not changing its expression strongly, of a gene X related to the given process, regulates also the gene Y which is not related, then the expression pattern will show X,Y as related to the process, rather than the correct X,Z. Therefore it seems necessary to explicitly involve interactions, including regulatory ones, into interpreting the expression pattern. An attempt to do this is the LeaderGene approach which has been proposed recently as a non-statistical technique in order to make a hierarchy among genes and to identify genes involved in a certain process (Sivozhelezov et al., 2006). This leader gene approach is based upon the number and the confidence of interactions of every single gene with all the other genes of the set according to databases such as STRING (Search Tool for the Retrieval of Interacting Genes/Proteins) (von Mering et al. 2003). This method was applied to a dataset of tolerance of unrelated kidney graft in human (Sivozhelezov et al., 2008). This approach has proved capacity to identify the key genes responsible for kidney graft rejection and/or tolerance, which may offer new opportunity of monitoring or modulating immune response against allograft (Figure 10.2). Finally, most nucleotide-microarray interpretation techniques are based on expression levels rather than function based, and involve function (mostly, interactions) only implicitly. Interactions based methods like the LeaderGene approach should enhance biological knowledge pointing out interactions of gene networks, which actually is integrated in functional genomics.

The possibility to measure gene expression of the entire human genome at one time had allowed to identified pathways and specific gene expression pattern from organ transplanted or from biological samples of patients. Hence, genome wide expression analysis might enhance the clinical assessment of both the organ outcome and patient outcome, defining subgroups of clinical syndrome and identifying disease associated biomarkers. Main studies that have identified gene expression changes in injured graft, notably following the donor death, the ischemia/reperfusion, acute rejection episode or during chronic rejection are presented. However, reader has to keep in mind that current molecular signature studies have to be validated in independent and large cohorts of patients.

10.3 Organ Transplantation Under Gene Magnifying Glass

10.3.1 *Pre- or peri-transplantation related factors*

The success of graft transplantation is linked to the quality of donor organ, however, facing the growing waiting list extended criteria, such as older donors, are accepted (Barshes *et al.*, 2007). Kidneys from cadaveric donors with reasonable function are considered acceptable for transplantation only on clinical criteria such as histological analysis that are not fully reliable. Thus, clinicians need molecular markers to predict whether a potential graft has a high probability of recovering function in a new host following the stresses of death, storage, implantation, *etc.* Among donor-related factors, donor age remains a major determinant for donor selection and graft survival (Toma *et al.*, 2001). It is why Melk A *et al.* analyzed gene expression of kidney of various ages (Melk *et al.*, 2005). Among older kidneys (> 70 years), class prediction analysis identified a list of 50 genes which segregate old kidneys together with adults (31 to 46 years) or away from the adults, suggesting that gene expression analysis could provide information about donor quality. Then, using implantation biopsies, Hauser *et al.* identified 132 genes that separate living and deceased donors, which main functional roles were cell communication, apoptosis and inflammation (Hauser *et al.*, 2004). The particular enrichment of complement system between living and deceased donor was recently confirmed by Naesens *et al.* (2009). In addition, they demonstrates a significant correlation between longer cold ischemia time and higher expression of 6 complement genes in deceased donor, which suggests that longer cold ischemia time is contributing to the over expression of complement genes. Furthermore, Naesens and colleagues also observed a significant increase in expression of complement genes in post-transplantation biopsies from well-functioning kidneys compared with implantation biopsies, irrespective of donor source, indicating a relation to immediate peri-transplantation phenomena. These findings enhance the interest of treatments which would prevent complement activation before organ recovery, during organ storage and in the post-transplantation period.

Using donor biopsies, some studies identified genes set that classified graft in function of immediate graft outcome, for renal (Hauser *et al.*, 2004; Mas *et al.*, 2008; Mueller *et al.*, 2008; Naesens *et al.*, 2009), liver (Defamie *et al.*, 2008) and lung transplantation (Anraku *et al.*, 2008). Among them, Mueller *et al.* assessed gene expression in 87 implanted donor organs in function of graft outcome (Mueller *et al.*, 2008). Unsupervised clustering method segregate the dataset in 3 groups of patients, one with living donor and two with deceased donor kidneys which can be distinguish by the number of patients with delayed graft function, which was defined by return in dialysis in the first weeks after transplantation. Similar study have been done in lung transplantation by Anraku M. *et al.* who compared 16 lung transplanted patients with good outcome with 10 lung recipients who developed primary graft dysfunction (Anraku *et al.* 2008). They identified four genes validated in an external group of 12 lung recipient to predict poor outcome until

6 months post-transplantation (p=0.0003). Finally, others used donor biopsies to predict graft outcome (Bunnag et al., 2009; Perco et al., 2009). It is the case of Perco and colleagues who identified three biomarkers which can explain 28% of the creatinine variability at 1 year after transplantation (Perco et al., 2009).

Despites additional investigation is required, these studies provided evidences that gene expression profiling performed on donor organ biopsies allow predicting early and medium graft outcome and provide information to the clinicians for the selection of the graft with acceptable quality in a more reliable manner than with histological analysis.

10.3.2 *Microarray analysis in acute allograft rejection*

Currently, the diagnosis of acute rejection is based on the histopathological findings of the biopsy. However, tissue harvests are invasive and subject to sampling error and may not be sufficient to detect early damages that already take place when functional changes are detected by the physician (Thaunat et al., 2007). Thus, gene expression profiling is used to identify potential biomarkers which could improve graft diagnosis, notably to provide early diagnosis of acute rejection. First studies that bring much improvement in acute rejection diagnosis in heart transplantation were focuses on PBMC profiling. Horwitz and colleagues demonstrate that gene expression in PBMC correlates with histological diagnosis of acute cellular rejection (International Society for Heart and Lung Transplantation (ISHLT) grade 3A or higher) in endomyocardial biopsies (Horwitz et al., 2004). They identified 40 transcripts for which expression changed during acute rejection in a case-control study within a cohort of 189 cardiac transplant patients at various time after transplantation (7 days to 5 years post-transplantation). In addition, after rejection treatment, the majority of these discriminatory genes indicates a return toward levels of control but they did not fully reach the baseline level indicating intermediate expression profile that is consistent with the fact that ARE are negative predictor of graft outcome. Thus, these results not only support the utility of gene expression for rejection diagnosis but could be used to validate response to treatment. Deng MC et al. performed similar prospective analysis on 600 heart transplanted recipients in the CARGO (Cardiac Allograft Rejection Gene Expression Observational) study (Deng et al., 2006). Of 97 candidate genes from microarrays, a group of 11 genes was confirm to identified patients with acute rejection (ISHLT grade 3A or higher) at various times after transplantation, using two independent validation sets with quantitative PCR assays. And moreover, they show that patients at more than 1 year post-transplantation with low molecular score have a very low risk of current moderate/severe rejection (negative predictive value > 99%). Therefore, these findings which have led to the development of a commercially available blood-based test, represents a clinically relevant application of microarray in the field of cardiovascular medicine.

Not such results have been developed in other organs, but many studies improved graft acute rejection mechanisms. Currently, for other human organs, nine main experimental studies investigate gene expression changes during acute rejection using microarrays: two after lung transplantation using bronchoalveolar lavage cells (Gimino et al., 2003; Patil et al., 2008), five after kidney transplantation using biopsies (Akalin et al., 2001; Sarwal et al., 2003; Flechner et al., 2004a; Mueller et al., 2007; Reeve et al., 2009) and one also profiling peripheral blood lymphocytes (Flechner et al., 2004b), and two after liver transplantation from biopsy samples (Sreekumar et al., 2002; Inkinen et al., 2005). The results of these individual studies are variable with different genes set being identified as potential biomarkers. Differences between studies are certainly due to heterogeneity of microarray used, various data analysis methods. Moreover, acute rejection is a complex process involving acute T cell-mediated rejection (TCMR), with the recruitment of effectors T lymphocytes and also memory T cells, natural killer cells, monocytes, *etc.* and involving acute antibody-mediated rejection (ABMR) that could play a role in more than 30% of ARE (Racusen et al., 2003). This molecular heterogeneity of acute rejection was demonstrated by Sarwal et al. in biopsy of acute renal rejection (Sarwal et al., 2003). Three distinct subgroups of acute rejection have been distinguish, where one subgroup exhibit T-cell and B-cell signatures, confirm by immunohistochemical analysis. Furthermore, this CD20+B-cell infiltrates was significantly associated with both clinical glucocorticoid resistance (P = 0.01) and graft loss (P < 0.001) highlighting the power and interest of gene microarrays to identify molecular-based disease subgroups.

Interestingly, some comparative analyses have been performed between several independent microarray studies. In one of Saint-Mezard *et al.*, the authors identified a common list of 70 genes for which intragraft expression levels correlated with acute rejection in kidney recipients (Saint-Mezard *et al.*, 2009). Most of these genes were associated with immune process like antigen presenting cells, cytotoxic T lymphocytes or IFN-γ responses, which certainly reflect a predominant acute cellular rejection process. Another comparative study has been done with studies from different organ transplantation field by Morgun and colleagues (Morgun *et al.* 2006). In a first step, they identified set genes that discriminates cardiac biopsies with acute rejection from non-rejection as accurately as does histology. In a second step, they demonstrate that their gene patterns were able to discriminate between acute rejection and non rejection in 51 of 52 biopsies after kidney transplantation (Sarwal *et al.*, 2003), 29 of 31 PBL samples after kidney transplantation (Flechner *et al.* 2004b) and 31 of 34 in samples after lung transplantation (Gimino *et al.* 2003). Exploring the biological common pathways between these different situations, they found a clear common inflammatory signature which was nonspecific of graft rejection events since it is also share by infected transplanted recipients. This study shows that gene profile could reflect similarities of a given pathological process, rather than similarities of the physiology of a particular organ, and demonstrates that such '*in silico*' approach is able to identify relevant genes among independent microarrays studies

In short, gene profiling in acute rejection reflects the complex biological mechanisms and corroborates with pathologic findings in renal allograft biopsies which show mixed lesions of TCMR and ABMR. Currently, broader investigations are needed to obtain, if it is possible, a fully differential molecular diagnosis which could be used in clinic to tailor immunosuppressant regimen in function of the predominant mechanisms of rejection.

10.3.3 *Microarray analysis in chronic allograft rejection*

Chronic allograft rejection is a major concern in intermediate- and long-term organ transplantation resulting of graft dysfunction (Nankivell *et al.*, 2003; Hourmant *et al.*, 2005). Chronic allograft rejection occurs within diverse manifestations considering the transplanted organ like obliterative bronchiolitis in lung graft and interstitial fibrosis and tubular atrophy (IF/TA) in kidney graft.

Many clinical causes of IF/TA have been described, including calcineurin inhibitor toxicity, chronic polyomavirus infection, glomerular or vascular diseases, without evidence of specific etiology (Solez *et al.*, 2007). In addition, chronic alloimmune injury is an important process during IF/TA in the graft that can be chronic T-cell mediated rejection or chronic antibody-mediated rejection. However, chronic rejection is based on clinical and histological observations and gene profiling has been done to highlight associated transcriptome changes that could explain the pathogenesis of both TCMR and ABMR (Donauer *et al.*, 2003; Scherer *et al.*, 2003; Flechner *et al.*, 2004b; Hotchkiss *et al.*, 2006; Mas *et al.*, 2007; Mueller *et al.*, 2007; Maluf *et al.*, 2008; Mengel *et al.*, 2009; Rödder *et al.*, 2009; Skawran *et al.*, 2009).

Among first studies, Scherer and colleagues identified a set of 10 genes that allowed the correct prognosis of chronic rejection 12 months after transplantation (histological diagnosis according to Banff 97 criteria) for 88% of the patient using kidney biopsies at 6 months post-transplantation (Scherer *et al.*, 2003). However, because only 17 samples were used further investigation is required to validate the use of these 10 genes. In contrast, others studies performed on biopsy samples (Mas *et al.*, 2007)(Hotchkiss *et al.*, 2006; Maluf *et al.*, 2008; Rödder *et al.*, 2009) or urine and PBMC (Mas *et al.*, 2007) identified broad lists of genes associated with IF/TA or chronic allograft nephropathy (CAN), the old criteria for IF/TA according to the 2005 Banff classification (Solez *et al.*, 2007). From these large gene sets, biological functions or functional pathways have been described as fibrosis, extracellular matrix deposition, and immune response indicating the valuable use of microarray in such multifactorial diseases. For example, Rödder *et al.*, has demonstrated that deregulation of genes involved in extracellular matrix remodeling in IF/TA, particularly matrix metalloproteinase (MMP)-7, is able to reflect disease progression and to differentiate IF/TA from normal tissue, which was already found by Flechner *et al.* (Flechner *et al.* 2004a; Rödder *et al.*, 2009). Thus, they conclude that the therapeutic use of MMP inhibitors, like the doxycycline which inhibits MMPs including MMP-7, could lead to a decrease of extracellular matrix remodeling in IF/TA. Among differential genes, genes related to both B and T cell types were involved in IF/TA profiles but a predominantly immunologic pattern were not identified (Mas *et al.*, 2007; Maluf *et al.*, 2008). Similarly to acute rejection, these findings are in accordance with histopathological findings indicating the complexity of pathways involved in chronic rejection and the combination of T cell- and antibody-mediated responses during the allograft injury. Furthermore, using a literature gene-set comparison approach, Ashton-Chess *et al.* discovered TRIB1 mRNA was upregulated in biopsies classified as displaying CAN in the studies of Flechner *et al.* and Scherer *et al.* (Ashton-Chess *et al.*, 2008). TRIB1 mRNA was found to be significantly increased in the PBMC of patients with chronic ABMR *versus* patients with stable graft function ($p < 0.01$). ROC curve analysis showed that TRIB1 was a good discriminator of chronic ABMR, with specificity and sensitivity of 88% and 75% respectively. Thus, TRIB1 appear to be a potential minimally invasive biomarker of chronic ABMR. This study emphasis the possibility to define a panel of genes, with TRIB1 ahead, predictive of chronic rejection in renal transplantation as in the case of the CARGO study for heart acute rejection described previously (Deng *et al.*, 2008). Finally, the use of gene-expression profiling data from human renal-protocol biopsies into the clinic could help to improve both treatment and prevention of the progression towards chronic rejection.

10.3.4 *Immunosuppressive drugs in transplantation*

Immunosuppressive drugs and in particular calcineurin inhibitors have wide range of side effect, like cancer induction (Dantal *et al.*, 1995) and nephrotoxicity (Nankivell *et al.*, 2003). Thus identification of gene profile changes they induce can be helpful to understand their mechanisms. But most studies have been done in animal models. In human, Baião *et al.* characterized the gene changes of PBMC treated with cyclosporine A (CsA) as involved in the control of cell-cycle regulation, apoptosis/DNA repair, DNA metabolism/response to DNA damage stimulus, transcription and cell proliferation (Baião *et al.,* 2007). Similarly, Rumberger *et al.* stimulated PBL with CsA, tacrolimus, mycophenolic acid, everolimus and sirolimus, or combinations thereof (Rumberger *et al.* 2009). Gene expression profiles were almost identical in PBL treated with either cyclosporine A or tacrolimus, and with either sirolimus or everolimus. More than 50 genes were synergistically affected by combinations of calcineurin-inhibitors and TOR-inhibitors and drug-specific regulated genes could be identified for both substance groups. Because CsA is able to modify the extracellular matrix composition in connective tissue, Stabellini *et al.* identified 259 genes with significantly gene expression changes (23 up- and 236 down-regulated) in human fibroblast treated with CsA. These genes are related to extracellular matrix turnover, osteoblast differentiation, and lysosomal/endosomal cell compartment, showing some possible relation to tissue alterations that appear in CsA-treated transplant patients, such as IF/TA, gingival overgrowth and Kaposi's sarcoma. Thus, these three studies provided drug-specific clustering. These approaches could improved immunossuppression side-effects comprehension, notably IF/TA and cancers, and may help to reduce these side-effects by an individualization of immunosuppressive drug regimens.

On the other hand, prevention of adverse effects of immunosuppressive drugs by minimization of their use, conversion to another drugs with less side effects or the transplantation's grail: graft "tolerance", are currently attempting strategies. Indeed, tolerant cases have been described in renal and liver transplantation, in which spontaneous, specific to the graft and long-term graft acceptance is observed years after total withdrawal of immunosuppressive drugs (Lerut and Sanchez-Fueyo 2006; Roussey-Kesler *et al.*, 2006). These cases confirm that a clinical operational state of tolerance to a mismatched graft can occur and exist in humans. However, the frequency of this observation in the kidney transplant population is unknown and, currently, we cannot identify patients primed to develop this immune adaptation. Thus, blood from graft recipients with long-term and stable graft function in the absence of immunosuppressant regimen have been analyzed using microarray technology, in the field of renal transplantation (Brouard *et al.*, 2005; Braud *et al.*, 2008; Sivozhelezov *et al.*, 2008) and liver transplantation (Martinez-Llordella *et al.*, 2007, 2008), in order to identified biomarkers of tolerance. In a cohort of renal transplant patient with tolerance, stable graft function and acute and chronic graft injury and of healthy individuals, Brouard *et al.* identified a pattern of 49 genes in PBMC, with the help of PAM algorithm, able to correctly segregate the tolerance and chronic rejection phenotypes with high specificity. Moreover, microarray pathway analysis demonstrated transforming growth factor–β (TGFβ) was found to regulate more than a quarter (27%) of the genes that differentiated operationally tolerant patients from those with chronic rejection. Furthermore, using custom oligonucleotide microarray, 21 of the 49 genes confirm the potential tolerance diagnosis use of this pattern in PBMC from renal recipient with tolerance induction protocol (Sarwal *et al.*, 2009). Serial gene array monitoring showed a change in gene expression in PBMC from non-tolerant pattern to a tolerant pattern in 3 patients in whom immunosuppressive drugs were successfully withdrawal at 6 months whereas no change were observed for 3 patients who failed to meet withdrawal criteria. Thus, tolerant PBMC pattern may diagnosis the stable tolerance status of renal recipients and indicate patient who need to be maintained on immunosuppressant regimen.

Similarly, in 2007, Martinez-Llordella *et al.* performed peripheral blood gene expression profiling on 16 operationally tolerant liver recipients, 16 recipients requiring continuous immunosuppression and 10 healthy subjects (Martinez-Llordella *et al.*, 2007). They identified a pattern of genes which discriminate tolerant patients and non-tolerant patients with high accuracy. This signature included genes encoding for γδ T-cell and NK receptors, and for proteins involved in cell proliferation arrest. Again, a number of these tolerance-specific genes appear to be either regulated by TGF-β or implicated in TGF-β signaling pathways. Same group further extended their studies by designing a clinically applicable molecular test of operational tolerance in liver transplantation (Martinez-Llordella *et al.*, 2008).

Current studies demonstrate that transcriptional profiling may results in the identification of biomarker of tolerance. These biomarkers could help clinicians to identify patients for whom immunosuppression weaning could success and to monitor the stability of this status of tolerance. However, before a clinical use, further investigations validating the predictive power of this signature are needed. Actually, the validation of such molecular tests in clinical trials is limited due to potential risks of rejection that represents the interruption of immunosuppressant regimen, especially for kidney whereas liver has a regenerative capacity. However, the identification of surrogate molecular biomarkers is ongoing notably thanks to large consortiums including the European network Indices of tolerance, the European project RISET (Reprogramming the Immune System for Establishment of Tolerance) (Hernandez Fuentes *et al.*, 2008) and the international clinical research consortium ITN (Immune Tolerance Network) (Newell *et al.*, 2008).

10.4 Conclusion

The use of microarrays has been applied to the organ transplantation field and holds great promise to obtain global view in transcripts changes, especially if associated to new informative non-statistical interaction-based bioinformatic analysis (Figure 10.3 based on connectivity *ab initio* analysis, and Figure 10.4 based on non statistical analysis). Unsupervised clustering and statistical tests identify molecular signatures of the state of activity of organ cells and patient samples and define subgroups of clinical disease. Literature mining and pathway analysis helps in understanding the biological relevance of such signatures. In addition, gene clustering and class prediction tools aid in the discovery of molecular signatures associated to disease processes. Such biomarkers should ideally be minimally invasive, *e.g.* blood or urine, and reliable to diagnose or predict graft outcome. Molecular biomarkers could have a great impact to diagnose or predict graft dysfunction, acute and chronic rejection, and to predict patient for who graft tolerance could be reach. Molecular markers could be also used to help clinicians in an individualization of immunosuppressant regimen in

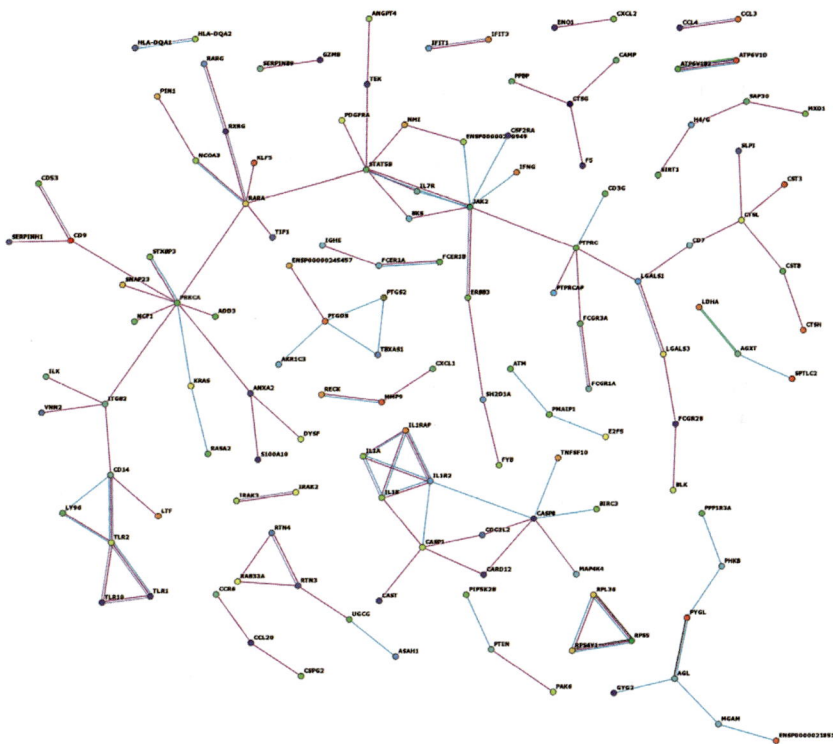

Figure 10.3. Complete interaction map "no textmining" interaction map calculated for the "Class 1" reliable genes and filtered by (CR-TOL) amplitude, as obtained from the new fullchip microarray datasets (see Table 2 for their names and ranking according to their number of interactions) (Reprinted with the permission from Sivozhelezov V., Braud C., Giacomelli L., Pechkova E., Giral M., Soulillou J.P., Brouard S., Nicolini C., Immunosuppressive drug-free operational immune tollerance in human kidney transplant recipient: II Non-statistical gene microarray analysis, Journal of Cellular Biochemistry 103, 1693-1706, ©2008, Wiley-Liss, Inc., a subsidiary of John Wiley & Sons, Inc.).

the purpose to decrease drugs side effects, to decrease acute and chronic rejection incidence and to enhance graft function and quality of life.

However, further investigation is needed to evaluate the generated candidate molecular markers in independent cohorts of patient before any conclusion concerning their diagnostic or prognostic power can be envisaged. In addition, high-throuput technologies in the field of proteomic are in progress, such as antibody microarrays, protein microarrays and more recently NAPPA microarrays. High-throuput proteomic technologies would therefore confirm gene microarray results, and bring new transplantation-related proteomic mechanisms and new surrogate biomarkers. The combination of both data, with the help of methodologies and tools already developed for microarrays, would greatly enhance our understanding of mechanisms that occur after organ transplantation, and promise the identification of reliable biomarkers.

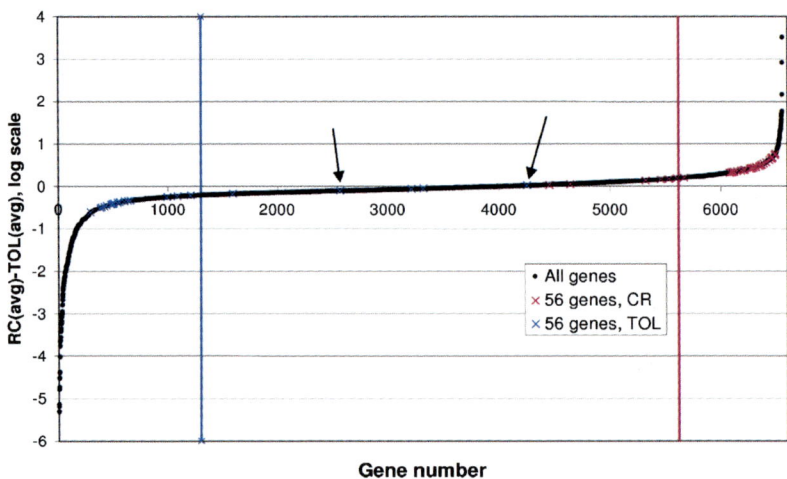

Figure 10.4. Genes from the "fullchip" plotted, using a non-statistical procedure, according to their tolerance or rejection propensity i.e. difference in expression in log scale between RC and TOL genes, with the 56 SAM-identified genes (Brouard et al., 2006) marked. Arrows indicate genes included in the 56 gene dataset but possibly unable to discriminate rejection/tolerance. Lines indicate thresholds used in selecting genes for "pro-tolerance" and "pro-rejection" leader gene calculations (Reprinted with the permission from Sivozhelezov V., Braud C., Giacomelli L., Pechkova E., Giral M., Soulillou J.P., Brouard S., Nicolini C., Immunosuppressive drug-free operational immune tollerance in human kidney transplant recipient: II Non-statistical gene microarray analysis, Journal of Cellular Biochemistry 103, 1693-1706, ©2008, Wiley-Liss, Inc., a subsidiary of John Wiley & Sons, Inc.).

Acknowledgments

This project was supported by grants to Fondazione EL.B.A. by MIUR (Ministero dell'Istruzione, Università e Ricerca) for "Funzionamento" and by a FIRB International Grant on Proteomics and Cell Cycle (RBIN04RXHS) from MIUR to CIRSDNNOB-Nanoworld Institute of the University of Genova. Richard Danger is supported by grants from the CENTAURE foundation (RTRS, France) and by the Fondation de la Recherche Médicale (FRM).

References

ABM (2009). "Agence de la biomédecine: 2008 annual report."

Ahn, Y. B., Xu, G., Marselli, L., Toschi, E., Sharma, A., Bonner-Weir, S. Sgroi, D. C. and Weir, G. C. (2007). Changes in gene expression in beta cells after islet isolation and transplantation using laser-capture microdissection. *Diabetologia*, 50, pp. 334–342.

Akalin, E., Hendrix, R. C., Polavarapu, R. G., Pearson, T. C., Neylan, J. F. Larsen, C. P. and Lakkism F. G. (2001). Gene expression analysis in human renal allograft biopsy samples using high-density oligoarray technology. *Transplantation*, 72, pp. 948–953.

Alizadeh, A. A., Eisen, M. B. Davis, R. E., Ma, C., Lossos, I. S., Rosenwald, A., Boldrick, J. G., Sabet, H., Tran, T., Yu, X., Powell, J. I., Yang, L. M., Marti, G. E., Moore, T., Hudson, J., Lu, L. S., Lewis, D. B., Tibshirani, R., Sherlock, G., Chan, W. C., Greiner, T. C., Weisenburger, D. D., Armitage, J. O., Warnke, R., Levy, R., Wilson, W., Grever, M. R., Byrd, J. C., Botstein, D., Brown, P. O. and Staudt, L. M. (2000). Distinct types of diffuse large B-cell lymphoma identified by gene expression profiling. *Nature*, 403, pp. 503–511.

Anraku, M., Cameron, M. J., Waddell, T. K., Liu, M., Arenovich, T., Sato, M., Cypel, M., Pierre, A. L., de Perrot, M., Kelvin, D. J. and Keshavjee, S. (2008). Impact of human donor lung gene expression profiles on survival after lung transplantation: a case-control study. *American Journal of Transplantation*, 8, pp. 2140–2148.

Ashton-Chess, J., Dugast, E., Colvin, R. B., Giral, M., Foucher, Y., Moreau, A., Renaudin, K., Braud, C., Devys, A., Brouard, S. and Soulillou, J. P. (2009). Regulatory, effector, and cytotoxic T cell profiles in long-term kidney transplant patients. *Journal of the American Society of Nephrology*, 20, pp. 1113–1122.

Ashton-Chess, J., Giral, M., Mengel, M., Renaudin, K., flucher, Y., Gwinner, W., Braud, C., Dugast, E., Quillard, T., Thebault, P., Chiffoleau, E., Braudeau, C., Charreau, B., Soulillou, J. P. and Brouard, S. (2008). Tribbles-1 as a novel biomarker of chronic antibody-mediated rejection. *Journal of the American Society of Nephrology*, 19, pp. 1116–1127.

Baião, A. M. T., Wowk, P. F., Sandrin-Garcia, P., Junta, C. M., Fachin, A. L., Mello, S. S., Sakamoto-Hojo, E. T., Donadi, E. A. and Passos, G. A. S. (2007). cDNA microarray analysis of cyclosporin A (CsA)-treated human peripheral blood mononuclear cells reveal modulation of genes associated with apoptosis, cell-cycle regulation and DNA repair. *Molecular and Cellular Biochemistry*, 304, pp. 235–241.

Barshes, N. R., Horwitz, I. B., Franzini, L., Vierling, J. M. and Goss, J. A. (2007). Waitlist mortality decreases with increased use of extended criteria donor liver grafts at adult liver transplant centers. *American Journal of Transplantation*, 7, pp. 1265–1270.

Braud, C., Baeten, D., Giral, M., Pallier, A., Ashton-Chess, J., Braudeau, C., Chevalier, C., Lebars, A., Lèger, J., Moreau, A., Pechkova, E., Nicolini, C., Soulillou, J. P. and Brouard S., (2008). Immunosuppressive drug-free operational immune tolerance in human kidney transplant recipients: Part I. Blood gene expression statistical analysis. *Journal of Cellular Biochemistry*, 103, pp. 1681–1692.

Brouard, S., Ashton-Chess, J. and Soulillou, J. P. (2008). Surrogate markers for the prediction of long-term outcome in transplantation: Nantes Actualité Transplantation (NAT) 2007 meeting report. *Human Immunology*, 69, pp. 2–8.

Brouard, S., Dupont, A., Giral, M., Louis, S., Lair, D., Braudeau, C., Degauque, N., Moizant, F., Pallier, A., Ruiz, C., Guillet, M., Laplaud, D. and Soulillou, J. P. (2005). Operationally tolerant and minimally immunosuppressed kidney recipients display strongly altered blood T-cell clonal regulation. *American Journal of Transplantation*, 5, pp. 330–340.

Brouard, S., Mansfield, E., Braud, C., Li, L., Giral, M., Hsieh, S. C., Baeten, D., Zhang, M., Ashton-Chess, J., Braudeau, C., Hsieh, F., Dupont, A., Pallier, A., Moreau, A., Louis, S., Ruiz, C., Salvatierra, O., Soulillou, J. P. and Sarwal, M. (2007). Identification of a peripheral blood transcriptional biomarker panel associated with operational renal allograft tolerance. *Proceedings of the National Academy of Sciences USA*, 104, 15448–15453.

Bunnag, S., Einecke, G. Reeve, J., Jhangri, G. S., Muller, T. F., Sis, B., Hidalgo, L. G., Mengel, M., Kayser, D., Kaplan, B. and Halloran, P. F. (2009). Molecular correlates of renal function in kidney transplant biopsies. *Journal of the American Society of Nephrology*, 20, 1149–1160.

Dantal, J., Godfrin, Y. and Soulillou, J. P. (1995). New insight into the pathogenesis of the idiopathic nephrotic syndrome. *Nephrology Dialysis Transplantation*, 10, pp. 1979–1982.

Defamie, V., Cursio, R., Le Brigand, K., Moreilhon, C., Saint-Paul, M. C., Laurens, M., Crenesse, D., Cardinaud, B., Auberger, P., Gugenheim, J., Barbry, P. and Mari, B. (2008). Gene expression profiling of human liver transplants identifies an early transcriptional signature associated with initial poor graft function. *American Journal of Transplantation*, 8, pp. 1221–1236.

Deng, C.-T., El-Awar, N., Ozawa, M., Cai, J. C., Lachmann, N. and Terasaki, P. I. (2008). Human leukocyte antigen class II DQ alpha and beta epitopes identified from sera of kidney allograft recipients. *Transplantation*, 86, pp. 452–459.

Deng, M. C., Eisen, H. J., Mehra, M. R., Billingham, M., Marboe, C. C., Kobashigawa, J., Johnson, F. L., Starling, R. C., Murali, S., Pauly, D. F., Baron, H., Wohlgemuth, J. G., Woodward, R. N., Klinger, T. M., Waither, D., Lai, P. G., Rosenberg, S. and Hunt, S. (2006). Noninvasive discrimination of rejection in cardiac allograft recipients using gene expression profiling. *American Journal of Transplantation*, 6, pp. 150–160.

Donauer, J., Rumberger, B., Klein, M., Faller, D., Wilpert, J., Sparna, T., Schieren, G., Rohrbach, R., Dern, P., Timmer, J., Pisarski, P., Kirste, G. and Walz, G. (2003). Expression profiling on chronically rejected transplant kidneys. *Transplantation*, 76, pp. 539–547.

Eisen, M. B., Spellman, P. T., Brown, P. O. and Botstein, D. (1998). Cluster analysis and display of genome-wide expression patterns. *Proceedings of the National Academy of Sciences of the United States of America*, 95, pp. 14863–14868.

El-Zoghby, Z. M., Stegall, M. D., Lager, D. J., Kremers, W. K., Amer, H., Gloor, J. M. and Cosio, F. G. (2009). Identifying specific causes of kidney allograft loss. *American Journal of Transplantation*, 9, pp. 527–535.

Fan, J. and Ren, Y. (2006). Statistical analysis of DNA microarray data in cancer research. *Clinical Cancer Research*, 12, pp. 4469-4473.

Flechner, S. M., Kurian, S. M., Head, S. R., Sharp, S. M., Whisenant, T. C., Zhang, J., Chismar, J. D., Horvath, S., Mondala, T., Gilmartin, T., Cook, D. J., Kay, S. A., Walker, J. R. and Salomon, D. R. (2004b). Kidney transplant rejection and tissue injury by gene profiling of biopsies and peripheral blood lymphocytes. *American Journal of Transplantation*, 4, pp. 1475–1489.

Flechner, S. M., Kurian, S. M., Solez, K., Cook, D. J., Burke, J. T., Rollin, H., Hammond, J. A., Whisenant, T., Lanigan, C. M., Head, S. R. and Salomon, D. R. (2004a). De novo kidney transplantation without use of calcineurin inhibitors preserves renal structure and function at two years. *American Journal of Transplantation*, 4, pp. 1776–1785.

Gimino, V. J., Lande, J. D., Berryman, T. R., King, R. A. and Hertz, M. I. (2003). Gene expression profiling of bronchoalveolar lavage cells in acute lung rejection. *American Journal of Respiratory and Critical Care Medicine*, 168, pp. 1237–1242.

Hauser, P., Schwarz, C., Mitterbauer, C., Regele, H. M., Muhlbacher, F., Mayer, G., Perco, P., Mayer, B., Meyer, T. W. and Oberbauer, R. (2004). Genome-wide gene-expression patterns of donor kidney biopsies distinguish primary allograft function. *Laboratory Investigation*, 84, pp. 353–361.

Hernandez Fuentes, M., Sawitzki, B. Perucha, E., Sagoo, P., Miqueu, P., Chapman, S., Stephens, D., Craciun, L., Tomiuk, S. and Brouard, S. (2008). Identification of immune tolerance in renal transplants. *8th American Transplant Congress*, 8, pp. 292–292.

Hilsenbeck, S. G., Friedrichs, W. E., Schiff, R., O'Connell, P., Hansen, R. K., Osborne, C. K. and Fuqua, S. A. W. (1999). Statistical analysis of array expression data as applied to the problem of tamoxifen resistance. *Journal of the National Cancer Institute*, 91, pp. 453–459.

Horwitz, P. A., Tsai, E. J., Epstein, J. A., Putt, M. E., Gilmore, J., Lepore, J. J., Parmacek, M. S., Kao, A. C., Desai, S., Goldberg, L. R., Jessup, M. L. and Cappola, T. P. (2004). Detection of cardiac allograft rejection and response to immunosuppressive therapy with peripheral blood gene expression. *Circulation*, 110, pp. 584–585.

Hosack, D. A., Dennis, G., Sherman, B. T., Lane, H. C. and Lempicki, R. A. (2003). Identifying biological themes within lists of genes with EASE. *Genome Biology*, 4, pp. R70.

Hotchkiss, H., Chu, T. T., Hancock, W. W., Schroppel, B., Kretzler, M., Schmid, H., Liu, Y. X., Dikrnan, S. and Akalin, E. (2006). Differential expression of profibrotic and growth factors in chronic allograft nephropathy. *Transplantation*, 81, pp. 342–349.

Hourmant, M., Cesbron-Gautier, A., Terasaki, P. I., Mizutani, K., Moreau, A., Meurette, A., Dantal, J., Giral, M., Blancho, G., Cantarovich, D., Karam, G., Folleam G., Soulillou, J. P. and Bignon, J. D. (2005). Frequency and clinical implications of development of donor-specific and non-donor-specific HLA antibodies after kidney transplantation. *Journal of the American Society of Nephrology*, 16, pp. 2804–2812.

Inkinen, K., Lahesmaa, R., Brandt, A., Katajamaa, M., Halme, L., Hockerstedt, K. and Lautenschlager, I. (2005). DNA microarray-based gene expression profiles of cytomegalovirus infection and acute rejection in liver transplants. *Transplantation Proceedings*, 37, pp. 1227–1229.

Khatri, P. and Drăghici, S. (2005). Ontological analysis of gene expression data: current tools, limitations, and open problems. *Bioinformatics*, 21, pp. 3587–3595.

Lerut, J. and Sanchez-Fueyo, A. (2006). An appraisal of tolerance in liver transplantation. *American Journal of Transplantation*, 6, pp. 1774–1780.

Li, B., Hartono, C., ding, R., Sharma, V., Ramaswamy, R., Qian, B., Serur, D., Mouradian, J., Schwartz, J. E. and Suthanthiran, M. (2001). Noninvasive diagnosis of renal-allograft rejection by measurement of messenger RNA for perforin and granzyme B in urine. *The New England Journal of Medicine*, 344, pp. 947–954.

Maluf, D. G., Mas, V. R., Archer, K. J., Yanek, K., Gibney, E. M., King, A. L., Cotterell, A., Fisher, R. A. and Posner, M. P. (2008). Molecular pathways involved in loss of kidney graft function with tubular atrophy and interstitial fibrosis. *Molecular Medicine*, 14, pp. 276–285.

Martinez-Llordella, M., Lozano, J. J., Puig-Pey, I., Orlando, G., Tisone, G., Lerut, J., Benitez, C., Pons, J. A., Parrilla, P., Ramirez, P., Bruguera, M., Rimola, A. and Sanchez-Fueyo, A. (2008). Using transcriptional profiling to develop a diagnostic test of operational tolerance in liver transplant recipients. *Journal of Clinical Investigation*, 118, pp. 2845–2857.

Martinez-Llordella, M., Puig-Pey, I., Orlando, G., Ramoni, M., Tisone, G., Rimola, A., Latinne, D., Margarit, C., Bilbao, I., Brouard, S., Hernandez-Fuentes, M., Soulillou, J. P. and Sanchez-Fueyo, A. (2007). Multiparameter immune profiling of operational tolerance in liver transplantation. *American Journal of Transplantation*, 7, pp. 309–319.

Mas, V., Maluf, D., Archer, K., Yanek, K., Mas, L., King, A., Gibney, E., Massey, D., Cotterell, A., Fisher, R. and Posner, M. (2007). Establishing the molecular pathways involved in chronic allograft nephropathy for testing new noninvasive diagnostic markers. *Transplantation*, 83, pp. 448–457.

Mas, V. R., Archer, K. J., Yanek, K., Dumur, C. L., Capparuccini, M. L., Mangino, M. J., King, A., Gibney, E. M., Fisher, R., Posner, M. and Malur, D. (2008). Gene expression patterns in deceased donor kidneys developing delayed graft function after kidney transplantation. *Transplantation*, 85, pp. 626–635.

Melk, A., Mansfield, E. S., Hsieh, S. C., Hernandez-Boussard, T., Grimm, P., Rayner, D. C., Halloran, P. F., Sarwal, M. M. (2005). Transcriptional analysis of the molecular basis of human kidney aging using cDNA microarray profiling. *Kidney International*, 68, pp. 2667–2679.

Mengel, M., Reeve, J., Bunnag, S., Einecke, G., Jhangri, G. S., Sis, B., Famulski, K., Guembes-Hidalgo, L. and Halloran, P. F. (2009). Scoring total inflammation is superior to the current banff inflammation score in predicting outcome and the degree of molecular disturbance in renal allografts. *American Journal of Transplantation*, 9, pp. 1859–1867.

Morgun, A., Shulzhenko, N., Perez-Diez, A., Diniz, V. Z., Sanson, G. F., Almeida, D. R., Matzinger, P. and Gerbase-DeLima, M. (2006). Molecular profiling improves diagnoses of rejection and infection in transplanted organs. *Circulation Research*, 98, pp. E74–E83.

Mueller, T. F., Einecke, G., Reeve, J., Sis, B., Mengel, M., Jhangri, G. S., Bunnag, S., Cruz, J., Wishart, D., Meng, C., Broderick, G., Kaplan, B. and Halloran, P. F. (2007). Microarray analysis of rejection in human kidney transplants using pathogenesis-based transcript sets. *American Journal of Transplantation*, 7, pp. 2712–2722.

Mueller, T. F., Reeve, J., Jhangri, G. S., Mengel, M., Jacaj, Z., Cairo, L., et al. (2008). The transcriptome of the implant biopsy identifies donor kidneys at increased risk of delayed graft function. *American Journal of Transplantation*, 8, pp. 78–85.

Muthukumar, T., Dadhania, D., Ding, R. C., Snopkowski, C., Naqvi, R., Lee, J. B., et al. (2005). Messenger RNA for FOXP3 in the urine of renal-allograft recipients. *The New England Journal of Medicine*, 353, pp. 2342–2351.

Naesens, M., Li, L., Yang, L., Sansanwal, P., Sigdel, T. K., Hsieh, S.C., Kambhan, N., Lerut, E., Salvatierra, O., Butte, A. J. and Sarwal, M. M. (2009). Expression of complement components differs between kidney allografts from living and deceased donors. *Journal of the American Society of Nephrology*, 20, pp. 1839–1851.

Nankivell, B. J., Borrows, R. J., Fung, C. L. S., O'Connell, P. J., Allen R. D. M. and Chapman, J. R. (2003). The natural history of chronic allograft nephropathy. *The New England Journal of Medicine*, 349, pp. 2326–2333.

Newell, K. A., Asare, A., Gisler, T., Suthanthiran, M. S., Burlingham, W., Marks, W., Turka, L. and Seyfert-Margolis, V. (2008). A unique B cell signature associated with operational tolerance. *American Transplant Congress*.

Opelz, G. and Döhler, B. (2008). Influence of time of rejection on long-term graft survival in renal transplantation. *Transplantation*, 85, pp. 661–666.

Patil, J., Lande, J. D., Li, N., Berryman, T. R., King, R. A. and Hertz, M. I. (2008). Bronchoalveolar lavage cell gene expression in acute lung rejection: development of a diagnostic classifier. *Transplantation*, 85, pp. 224–231.

Perco, P., Kainz, A., Wilflingseder, J., Soleiman, A., Mayer, B. and Oberbauer, R. (2009). Histogenomics: association of gene expression patterns with histological parameters in kidney biopsies. *Transplantation*, 87, pp. 290–295.

Racusen, L. C., Colvin, R. B., Solez, K., Mihatsch, M. J., Halloran, P. F., Campbell, P. M., et al. (2003). Antibody-mediated rejection criteria - an addition to the Banff 97 classification of renal allograft rejection. *American Journal of Transplantation*, 3, pp. 708–714.

Reeve, J., Einecke, G., Mengel, M., Sis, B., Kayser, N., Kaplan, B. and Halloran, P. F. (2009). Diagnosing rejection in renal transplants: a comparison of molecular- and histopathology-based approaches. *American Journal of Transplantation*, 9, pp. 1802–1810.

Roberts, P. C. (2008). Gene expression microarray data analysis demystified. *Biotechnology Annual Review*, 14, pp. 29.

Rödder, S., Scherer, A., Raulf, F., Berthier, C. C., Hertig, A., Couzi, L., Durrbach, A., Rondeau, E. and Marti, H. P. (2009). Renal allografts with IF/TA display distinct expression profiles of metzincins and related genes. *American Journal of Transplantation*, 9, pp. 517–526.

Roussey-Kesler, G., Giral, M., Moreau, A., Subra, J. F., Legendre, C., Noel, C., Pillebout, E., Brouard, S. and Soulillou J. P. (2006). Clinical operational tolerance after kidney transplantation. American Journal of Transplantation, 6, pp. 736–746.

Rumberger, B., Kreutz, C., Nickel, C., Klein, M., Lagoutte, S., Teschner, S., Timmer, J., Gerke, P., Walz, G. and Donauer, J. (2009). Combination of immunosuppressive drugs leaves specific "fingerprint" on gene expression in vitro. *Immunopharmacology and Immunotoxicology*, 31, pp. 283–292.

Saint-Mezard, P., Berthier, C. C., Zhang, H., Hertig, A., Kaiser, S., Schumacher, M., et al., (2009). Analysis of independent microarray datasets of renal biopsies identifies a robust transcript signature of acute allograft rejection. *Transplant International*, 22, pp. 293–302.

Sarwal, M., Chua, M.-S., Kambham, N., Hsieh, S. C., Satterwhite, T., Masek, M. and Salvatierra, O. (2003). Molecular heterogeneity in acute renal allograft rejection identified by DNA microarray profiling. The New England Journal of Medicine, 349, pp. 125–138.

Sarwal, M., L. Li, et al. (2009). Gene arrays may predict stable clinical tolerance and support safe immunosuppression withdrawal. Oral communication of the Amercian Transplant Congress 2009 Session 43: Immune Monitoring: Tolerance and Immune Modulation (306).

Sayegh, M. H. and Carpenter, C. B. (2004). Transplantation 50 years later--progress, challenges, and promises. *The New England Journal of Medicine*, 351, pp. 2761–2766.

Scherer, A., A. Krause, Walker, J. R., Korn, A., Niese, D. and Raulf, F. (2003). Early prognosis of the development of renal chronic allograft rejection by gene expression profiling of human protocol biopsies. *Transplantation*, 75, pp. 1323–1330.

Sivozhelezov, V., Braud, C., Giacomelli, L., Pechkova, E., Giral, M., Soulillou, J. P., Brouard, S. and Nicolini, C. (2008). Immunosuppressive drug-free operational immune tolerance in human kidney transplants recipients. Part II. Non-statistical gene microarray analysis. *Journal of Cellular Biochemistry*, 103, pp. 1693–1706.

Sivozhelezov, V., Giacomelli, L., Tripathi, S. and Nicolini, C. (2006). Gene expression in the cell cycle of human T lymphocytes: I. Predicted gene and protein networks. *Journal of Cellular Biochemistry*, 97, pp. 1137–1150.

Skawran, B., Dierich, M., Steinemann, D., Hohlfeld, J., Haverich, A., Schlegelberger, B., Welte, T. and von Neuhoff, N. (2009). Bronchial epithelial cells as a new source for differential transcriptome analysis after lung transplantation. *European Journal of Cardio-Thoracic Surgery*, in press.

Solez, K., Colvin, R. B., Racusen, L. C., Sis, B., Halloran, P. F., Birk, P. E., et al. (2007). Banff '05 Meeting Report: differential diagnosis of chronic allograft injury and elimination of chronic allograft nephropathy ('CAN'). *American Journal of Transplantation*, 7, pp. 518–526.

Sreekumar, R., Rasmussen, D. L., Wiesner, R. H. and Charlton, M. R. (2002). Differential allograft gene expression in acute cellular rejection and recurrence of hepatitis C after liver transplantation. *Liver Transplantation*, 8, pp. 814–821.

Strehlau, J., Pavlakis, M., Lipman, M., Shapiro, M., Vasconcellos, L., Harmon, W. and Strom, T. B. (1997). Quantitative detection of immune activation transcripts as a diagnostic tool in kidney transplantation. *Proceedings of the National Academy of Sciences USA*, 94, pp. 695–700.

Tamayo, P., Slonim, D., Mesirov, J., Zhu, Q., Kitareewan, S., Dmitrovsky, E., et al. (1999). Interpreting patterns of gene expression with self-organizing maps: methods and application to hematopoietic differentiation. *Proceedings of the National Academy of Sciences USA*, 96, pp. 2907–2912.

Tavazoie, S., Hughes, J. D., Campbell, M. J., Cho, R. J. and Church, G. M. (1999). Systematic determination of genetic network architecture. *Nature Genetics*, 22, pp. 281–285.

Thaunat, O., Legendre, C., Morelon, E., Kreis, H. and Mamzer-Bruneel, M. F. (2007). To biopsy or not to biopsy? Should we screen the histology of stable renal grafts? *Transplantation*, 84, pp. 671–676.

Tibshirani, R., Hastie, T., Narasimhan, B. and Chu, G. (2002). Diagnosis of multiple cancer types by shrunken centroids of gene expression. *Proceedings of the National Academy of Sciences USA*, 99, pp. 6567–6572.

Toma, H., Tanabe, K., Tokumoto, T., Shimizu, T. and Shimmura, H. (2001). Time-dependent risk factors influencing the long-term outcome in living renal allografts: donor age is a crucial risk factor for long-term graft survival more than 5 years after transplantation. *Transplantation*, 72, pp. 940–947.

Trulock, E. P., Christie, J. D., Edwards, L. B., Boucek, M. M., Aurora, P., Taylor, D. D., Dobbels, F., Rahmel, A. O., Keck, B. M. and Hertz, M. I. (2007). Registry of the International Society for Heart and Lung Transplantation: twenty-fourth official adult lung and heart-lung transplantation report-2007. *The Journal of Heart and Lung Transplantation*, 26, pp. 782–795.

Tusher, V. G., Tibshirani, R. and Chu, G. (2001). Significance analysis of microarrays applied to the ionizing radiation response. *Proceedings of the National Academy of Sciences USA*, 98, pp. 5116–5121.

Van Gelder, R. N., von Zastrow, M. E., Yoola, E., Dement, W. C., Barchas, J. D. and Eberwine, J. H. (1990). Amplified RNA synthesized from limited quantities of heterogeneous cDNA. *Proceedings of the National Academy of Sciences USA*, 87, pp. 1663–1667.

von Mering, C., Huynen, M., Jaeggi, D., Schmidt, S., Bork, P. and Snel, B. (2003). STRING: a database of predicted functional associations between proteins. *Nucleic Acids Research*, 31, pp. 258–261.

Wolfe, R. A., Merion, R. M., Roys, E. C. and Port, F. K. (2009). Trends in organ donation and transplantation in the United States, 1998-2007. *American Journal of Transplantation*, 9, pp. 869–878.

Yang, Y. H., Dudoit, S. Luu, P., Lin, D. M., Peng, V., Ngai, J. and Speed, T. P. (2002). Normalization for cDNA microarray data: a robust composite method addressing single and multiple slide systematic variation. *Nucleic Acids Research*, 30, pp. e15–e15.

Zeeberg, B. R., Feng, W., Wang, G., Wang, M. D., Fojo, A. T., Sunshine, et al. (2003). GoMiner: a resource for biological interpretation of genomic and proteomic data. *Genome Biology*, 4, pp. R28–R28.

Abbreviations

aRNA	amplified antisense RNA
ARE	Acute Rejection Episode
ABMR	Antibody-Mediated Rejection
BAL	Bronchoalveolar Lavage
CARGO	Cardiac Allograft Rejection Gene Expression Observational study
cDNA	complementary DNA
CGH	Comparative Genomic Hybridization
ChIP-chip (also ChIP on chip)	Chromatin ImmunoPrecipitation on microarray
CNI	Calcineurin Inhibitor
CsA	Cyclosporin A
EASE	Expression Analysis Systematic Explorer
FDR	False Discovery Rate
GSEA	Gene Set Enrichment Analysis
IF/TA:	Interstitial Fibrosis and Tubular Atrophy
ISHLT	International Society for Heart and Lung Transplantation
MMP	Matrix MetalloProteinase
mRNA	messenger RNA
PAM	Prediction Analysis for Microarrays
PBMC	peripheral blood mononuclear cells
PBL	Peripheral Blood Lymphocytes
qPCR	quantitative Polymerase Chain Reaction
SAM	Signifiance Analysis of Microarray
SNP	Single Nucleotide Polymorphism
STRING	Search Tool for the Retrieval of Interacting Genes/Proteins
TCMR	T cell-mediated rejection
TGFβ	Transforming Growth Factor-β

CHAPTER 11

SIGNALING NETWORKS. SIMULATIONS OF BIOCHEMICAL INTERACTIONS. APPLICATIONS TO MOLECULAR ONCOLOGY

Lorenzo Tortolina[1,2,*], Nicoletta Castagnino[1,2,*], Roberto Montagna[2], Raffaele Pesenti[3], Anahi Balbi[1,4] and Silvio Parodi[1,5,**]

[1]*Department of Oncology, Biology and Genetics, University of Genoa, Italy; National Cancer Institute of Genoa, Italy;* [2]*Department of Informatics and Information Sciences; University of Genoa, Italy;* [3]*Department of Applied Mathematics; University Ca' Foscari of Venice, Italy;*
[4]*Advanced Biotechnology Center of Genoa, Italy;*
[5]*Research Center for Computational Learning, Italy*

Graphs referred to molecular interaction maps of signaling-proteins, involved for instance in cell cycle control, are not randomly distributed. They display the presence of very few highly interconnected hubs, a vast majority of poorly connected nodes, and situations in between. Path length between two randomly chosen nodes tends toward a scale free distribution. In general biological hubs are interconnected through poorly connected low hierarchy nodes. Motifs and modules with specific structures confer specific properties/behaviors to different regions of a network. We can have a physiological state, but also a cancer state, in which mutated onco-proteins are present. Virtual inhibitors of onco-proteins, onco-proteins affected from excess of function, can also be simulated quite conveniently. We discuss the usage of Molecular Interaction Maps (MIMs) constructed according the syntactic rules proposed by K. W. Kohn. These types of graphs are concise and sufficiently detailed at the same time: relatively large maps can be drawn in a single page. Then we introduce an articulated discussion about some of the main features of dynamic modeling. Concentration of a given molecular species can vary both in time and

---

[*] These authors contributed equally to this work.
[**] Scientific coordinator of the team work, silvio.parodi@unige.it

space. We underline that a network is a new entity with new emerging properties, not present in its individual components. Molecular cancer pathology and more in general most cell behaviors can be understood only at a signaling–network level. A typical character of the new higher level entity is transmission and elaboration of signals (an informatic role), not membership of its components in some specific molecular class. In specific cases it can be important the usage of stochastic simulators. In other cases the treatment of (bio)chemical interactions through ordinary differential equations and non-stochastic models can be entirely appropriate, and the dynamic simulations run much faster. We discuss how to deal with missing information and the parameter space exploration of interpolated data. Signaling–network studies in cancer are a precious support for a rationalization of associations of inhibitors of onco-proteins, according to the individual mutations present in an individual cancer.

11.1 Introduction

Network biology is a relatively young but already vast discipline. It is widely accepted that cell behavior (including the behavior of a cancer cell) requires an understanding of the integrated functioning of a network of molecules. The reductionist level of analysis of properties and interactions of an individual molecule is in a sense overcome, network properties go beyond individual molecules. At the same time network analyses became possible precisely as a consequence of the accumulating wealth of essential information at a more reductionist level.

The amazing features and consequent biological properties of even relatively simple biological networks (for instance the chemotactic behavior of some bacteria (Barkai and Leibler, 1997; Alon et al., 1999), have attracted the attention of more mathematically oriented investigators, who have offered a more general framework to network biology (Barabási and Oltvai, 2004). From this point of view, network biology is part of the more general frame of network theory. I would say that the specific properties of individual molecules are less important for network theory and more important for network biology. Network theory (graph theory) looks especially at the topology of the network, but network topology can have important consequences for the biological behavior of a given network. Looking at a molecule in the more abstract form of a node, a node can have more or less links (degree of connectivity k). Quite often in a sufficiently large biological network we can distinguish rare highly connected nodes (hubs) from common poorly connected nodes, and a gradualism of situations in between; the probabilistic distribution of k takes the form: $P(k) \sim k^{-\gamma}$, where the range of γ is usually between 2 and 3 (scale free networks). A biologic network is usually sensitive to alterations of high hierarchy individual hubs and robust toward individual alterations of low hierarchy nodes. In yeast gene/protein networks, hubs are more essential than poorly connected nodes: elimination of hubs leaves behind disaggregated network fragments, elimination of poorly connected nodes is much less disruptive for the entire network anatomy (Jeong *et al.*, 2001).

In random networks the mean path length (l) from node x to node y grows approximately as $l \sim \ln N$ (where N is the number of nodes); in a scale free network as $l \sim \ln \ln N$. Suppose N moves from 10 to 1000. In the first case we move from $l \sim 2.3$ to $l \sim 6.9$ (scale ratio = 3); in the second case we move from $l \sim .834$ to $l \sim 1.93$ (scale ratio 2.32). People speak of a small world in the first case, of an ultra small world in the second case. A specific unexplained feature of biologic networks (at difference for instance with social networks) is that signaling–networks hubs and biologic hubs in general, are not directly connected. Biologic networks seem organized in modules (regions more densely interconnected). Scale free properties and modularity could seem contradictory, but they coexist in biologic networks (hierarchical networks?). Looking at nodes as abstract connected entities rather than specific molecules, some sub-graph features are more frequent than others (they have been designated as motifs).

An important source of information can be obtained from the analysis of MIM network topology.

MIM networks are expected to enjoy some topology features that characterize other types of cellular networks (often much larger but with a coarser granularity). As an example, MIMs should be scale-free and they should present specific motifs, *i.e.*, subgraphs of highly interlinked groups of nodes with specific topological structures. The Kohn's group (Kohn 1999; Kohn *et al.*, 2006; 2008) has reconstructed several local MIMs, especially within the framework of cell–cycle related networks.

These MIMs could be a precious material for topological analyses, also because they are relatively homogeneous in terms of MIM reconstruction criteria and have a relatively fine degree of granularity (degree of detail). To our knowledge this work on MIMs is substantially still to be done.

The reader is referred to Barabási and Oltvai (2004) for an in depth introduction on topological subjects applied to biological networks.

In a scale-free network the node degree distribution follows a power law, at least asymptotically. Let us here recall that the degree of a node is a measure of the node connectedness, as the degree is defined as the number of links that the node shares with the other nodes of the graph. A scale-free network presents a relatively small number of highly connected nodes (called hubs) and a large number of nodes with a small degree value. In addition, the expected degree of a node is a decreasing function of its distance from the hubs. The greater is the distance of a node from a hub the smaller is its degree. This kind of networks is said scale-free as the above property holds in general for all their induced sub-graphs. In other words, we can observe the same properties even if we are dealing with only a portion of a scale–free network.

From a biological point of view it is reasonable to assume that "damage" or ablation of a hub node should be much more important for the entire network behavior, than "damage" or ablation of a poorly connected node. However this could be true only in probabilistic terms. Alteration of a poorly connected node could sometimes display biologically drastic effects within a local network. Again, MIM networks could be useful objects for this type of exploration. The authors have preliminarily noticed that MIM networks seem to present both hub nodes and perhaps approximate power-law degree distribution. All the same, more than a word of caution must be spent, for instance the MIM networks considered so far have a relatively small number of nodes, nodes downstream of pathways and at MIM borders can be heavily underrepresented in terms of edges: this prevents for the moment definitive statistical conclusions.

Motifs are related to the possible modularity of the cellular networks. Cellular functions are likely to be carried out in a highly modular manner (Hartwell *et al.*, 1999). Accordingly, the topology of networks describing the cellular mechanisms is expected to present motifs indicating the presence of functional modules. The capacity of identifying motifs is critical to allow the breakdown of the cellular network into a set of biologically relevant functional modules.

Indeed, as reported in Barabási and Oltvai (2004), future progress in network biology is expected in many directions, ranging from the development of new theoretical methods, to better characterization of network topologies, to insights into the dynamics of motif clusters and biological functions.

It is worth observing that the possibility of breaking down a complex network in modules would also simplify the analysis and/or the simulations of the system from a mathematical, if not from a biological engineering, point of view. As an example, the presence of specific feedbacks among modules can suggest the possibility of some stationary behaviors of a network, while excluding other ones.

The expected modularity of the cellular networks is also at the basis of the current studies on such networks. As it is not currently possible the representation of a network of all the cellular processes, it would not be worth studying a part of the network if its function could not be understood out of the whole context.

For the time being, there is no general automatic methodology that allows to unequivocally identifying motifs within a network, different methods predict different boundaries between modules that are not sharply separated (see again, Barabási and Oltvai, 2004).

All the same there is already a vast literature and consensus on the functional meaning of some motifs that involve a limited number of nodes (Shen-Orr *et al.,* 2002; Milo *et al.*, 2002; Wuchty *et al.*, 2003; Kholodenko *et al.*, 2006; Alon, 2007).

Different types of software are also available to identify motifs in cellular networks, see, *e.g.*:

- Mfinder
  (http://www.weizmann.ac.il/mcb/UriAlon/groupNetworkMotifSW.html);
- FANMOD
  (http://theinf1.informatik.unijena.de/~wernicke/motifs/index.html);
- MAVisto (http://mavisto.ipk-gatersleben.de/).

Human inspection can identify motifs within MIMs, but to the authors' best knowledge, there is yet no software that can automatically identify them in MIM networks.

When we move from this more abstract level to real signaling networks, reaction rates, catalytic turnover numbers, number of molecules and molecular complexes (concentrations) as a function of time, make their appearance: the network from static becomes dynamic. Network topology and network dynamics are both important, albeit with different focuses, sometimes interconnected.

In molecular oncology we deal with networks of signaling proteins and other signaling molecules. Several of these signaling molecules are subject to mutations (at their coding DNA level). We have dominant oncogenes when the alteration representing one step (one hit) in the process of carcinogenesis is related to an excess of function. We have a recessive oncogene when a loss of function is involved. In this case both alleles need to be altered; however we can also be confronted with a very important reduction in the expression level of a given recessive onco-protein. Alterations in a given signaling-network can depend from molecules functioning too much or too little, because of abundance/scarcity, or because of structural alterations (for instance mutations) of the involved molecules or of molecules affecting their abundance/scarcity. Mutations are important in malignant transformation because they can be somatically inherited from a parental to a daughter cell, and they can be accumulated with time, within a cell undergoing a process of progression toward malignancy.

The reason while a given alteration has the important role of a hit (a step), in the multistep process of carcinogenesis, at the molecular level can only be understood in the framework of a signaling network. Of course at the clinical level we have the counterproof of the more or less frequent presence of a given oncogene mutation in a given type of cancer. If we consider extensively studied different high frequency cancers, for instance non small cells lung cancer (NSCLC), breast cancer (BC), colorectal cancer (CRC), we find, within largely common pathways, for instance pathways related to cell-cycle control, important mutation frequency differences for many important oncogenes. We still don't understand the reasons of these differentiation-related differences.

The study of signaling gives hope that we will start to understand, with the accumulation of experimental reductionist-level knowledge. We need an understanding of signaling networks at the level of interactions of molecules. It is therefore imperative to have coherent approaches for a graphical description of the molecular anatomy of a given network region. Network diagrams are exploited by the most common software programs, such as the Matlab toolbox SimBiology (MATLAB R2009a and SimBiology 3.0; The MathWorks, Inc., Natick, MA) or CellDesigner (Kitano, 2003; Funahashi *et al.*, 2008). They allow to model, simulate and analyze biological systems. The graphical interfaces of these softwares allow the user to build a model using a block diagram editor. As an example, CellDesigner lets the user to describe biochemical networks through process diagrams following a graphical notation system proposed by Kitano (Kitano *et al.*, 2003).

These kinds of graphical representations are then stored using the Systems Biology Markup Language (SBML). The System Biology Markup Language (http://sbml.org), *a universal XML-compliant standard* (eXtensible Markup Language), is a computer–readable format for representing models of biochemical reaction networks via software. It is being developed for exchanging modeling data. It is applicable to models of metabolism, cell–signaling, and many others. Once the model of a biological process is described in terms of a computer-readable format, such as SBML, it is ready to be simulated and analyzed by most of the biological software packages.

The interested reader is referred to the following sites for deeper introduction on the subject: Matlab SimBiology http://www.mathworks.com/products/simbiology/ and CellDesigner http://www.celldesigner.org/, System biology graphical notation http://sbgn.org/.

We have decided to illustrate a type of graphic description with which we are more familiar: the Molecular Interaction Maps (MIMs) proposed by Kurt W. Kohn (Kohn 1999; Aladjem *et al.*, 2004; Kohn *et al.*, 2006), see Figures 11.1–11.6.

Hiroaki Kitano and his group have proposed a similar/different type of diagram interaction map (Kitano 2003; Kitano *et al.*, 2005; Oda *et al.*, 2005), more intuitive but much less synthetic, because the same molecular species can be repeated several times and not only once, as in the case of the Kohn's MIMs. We have to keep in mind that we are still at the description level of a network, on top of (dynamic) simulations of the corresponding network of biochemical interactions and catalytic reactions. Description of the molecular anatomy of a specific region of the signaling network is however the introductory primary essential level of investigation, necessary to become aware of what is going on in a given signaling-network region.

We are much more familiar with a MIM-graph description of a limited network region, than with analogous descriptions using other types of graphs, for instance via Pathway Interaction Database (PID) or other types of databases allowing also the implementation of different graph formats (Reactome, BioCarta among others). However, we find often useful the usage of PID database and other databases, for checking and completing our list of interactions (annotation list and glossary), the background on which we build a MIM. A MIM is then for us a heuristic molecular anatomy network from where we move to a completely detailed writing of our ODEs via Matlab toolbox SimBiology (MATLAB R2009a and SimBiology 3.0; The MathWorks, Inc., Natick, MA).

The multiple phases of cell cycle are among the most extensively studied fields (both in terms of components and connectivity), also at the level of mammalian and human cells. Dynamic simulation models are essentially a translation of familiar pathways maps into mathematical form.

11.2 Molecular Interaction Maps (MIMs)

We start to describe now, at a level of reasonable detail, the type of Molecular Interaction Map (MIM) proposed originally in 1999 by **Kurt W. Kohn** (NIH – NCI – Bethesda, USA).

To make MIMs user friendly, lines and nodes have been colored conventionally as the following:
- lines indicating binding of molecular species: **black**;
- symbols indicating modifications: **blue**;
- symbols indicating activation or stimulation: **green**;

- symbols indicating general or transcriptional inhibition and degradation: **red**;
- symbols indicating transcriptional activation: **purple.**

Examples of the way of working of Kohn's MIMs syntactic rules are reported in Figures 11.2–11.6.

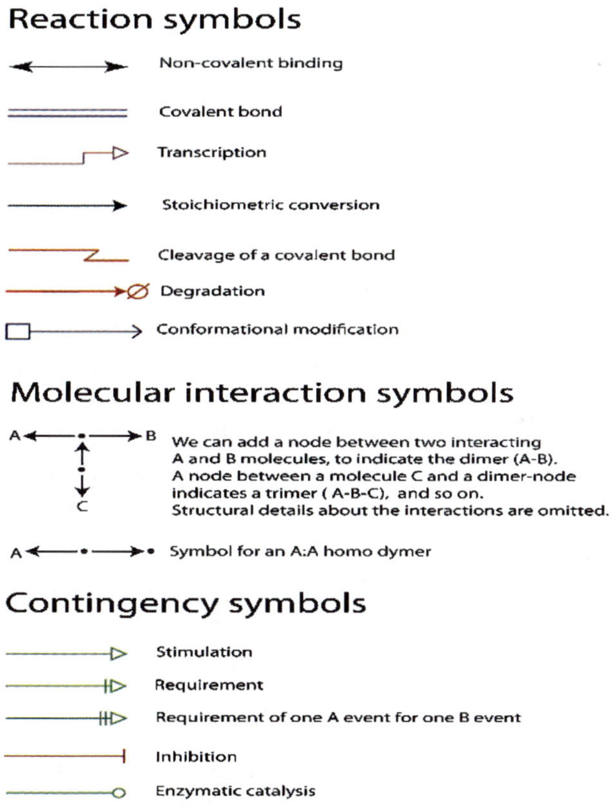

Reaction symbols

- Non-covalent binding
- Covalent bond
- Transcription
- Stoichiometric conversion
- Cleavage of a covalent bond
- Degradation
- Conformational modification

Molecular interaction symbols

A ←•→ B We can add a node between two interacting A and B molecules, to indicate the dimer (A-B). A node between a molecule C and a dimer-node indicates a trimer (A-B-C), and so on. Structural details about the interactions are omitted.

A ←•→• Symbol for an A:A homo dymer

Contingency symbols

- Stimulation
- Requirement
- Requirement of one A event for one B event
- Inhibition
- Enzymatic catalysis

Figure 11.1. Syntactic/logic rules of a Kohn's MIM (Molecular Interaction Map). Reactions operate on molecular species, whereas contingencies operate on reactions or on other contingencies (Slightly modified from "Molecular interaction maps of bioregulatory networks: a general rubric for systems biology". Mol. Biol. Cell 2006, 17, 1-13).

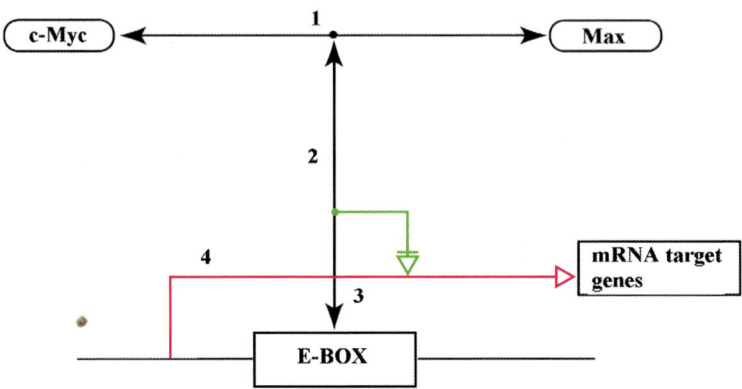

Figure 11.2. c-Myc binds its partner protein Max (interaction **#1**). c-Myc:Max heterodimer can bind to a DNA promoter region called E-Box (interaction **#2**), the trimer c-Myc:Max:E-Box (**#3**) gives origin to an enhanceosome which activates transcription of many genes (**#4**).

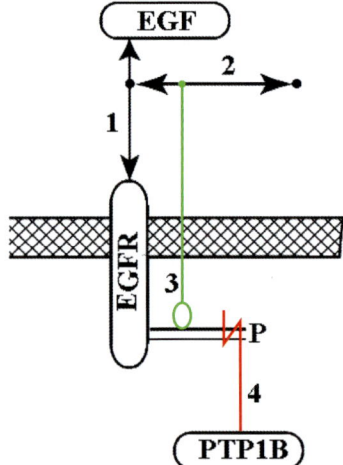

Figure 11.3. EGF binds the extracellular domain of EGFR (**#1**); the EGF:EGFR complex then forms a homodimer (**#2**) which induces reciprocal multiple tyrosine autophosphorylations (**#3**) in the intracellular portion of the receptors, and provides specific docking sites for cytoplasmic proteins. (In this MIM, we pictured only one site of autophosphorylation). De-P of EGFR-P is mediated by PTP1B phosphatase (**#4**).

Dynamic Simulations in Oncology 267

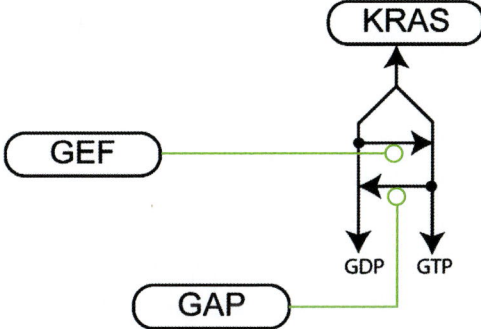

Figure 11.4. GTP and GDP bound state of Ras. GDP and GTP compete with each other for binding to a site on KRAS (the cytoplasmic concentration of GTP is normally 20 times higher than that of GDP). GEF (Guanine Nucleotide exchange Factor, for example SOS) shifts the equilibrium in favor of KRAS-GTP. The GTPase protein GAP converts bound GTP to bound GDP.

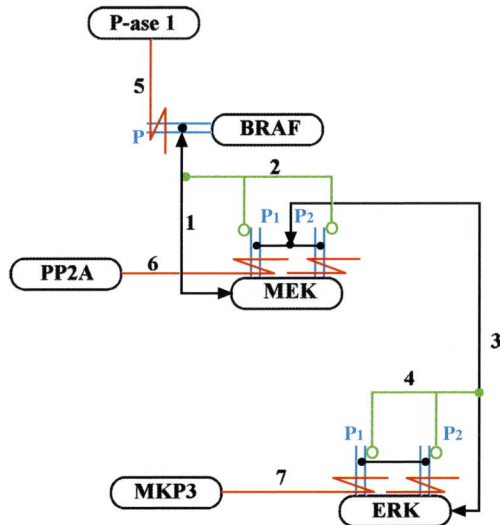

Figure 11.5. Multiple phosphorylations. Phosphorylated BRAF binds **(#1)** and phosphorylates **(#2)** MEK at two sites (P1 and P2); MEK double phosphorylated binds **(#3)** and phosphorylates **(#4)** ERK at two sites. Dephosphorylations by phosphatases **(#5)**, **(#6)**, **(#7)**.

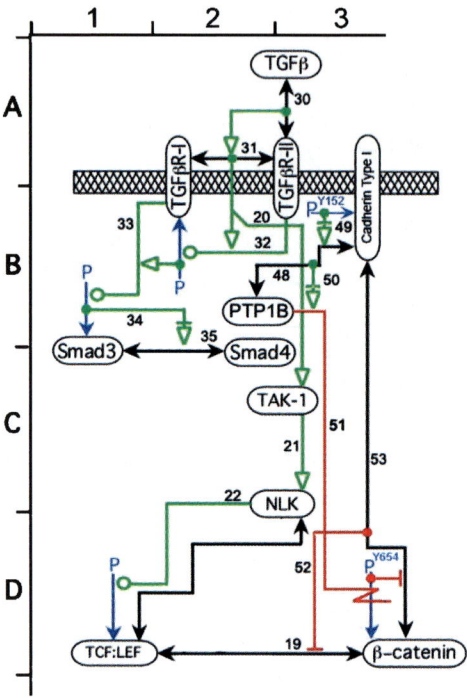

Figure 11.6. A molecular interaction map (MIM) depicting a TGFβ pathway upstream of the transcription factors TCF/LEF: βcatenin and the interaction between cadherin Type I and βcatenin.

Expert interpretation and annotation are an essential phase of the construction of a Molecular Interaction Map (MIM), from which we will move to dynamic simulations.

Therefore a Molecular Interaction Map (MIM) is always accompanied by:

- an **ANNOTATION LIST** (linked to a number conveniently positioned along a reaction or a contingency line; an X,Y coordinate also accompanies each annotation, for a more convenient retrieval).
- a **GLOSSARY:** essential information about each molecular species is given there.

In conclusion, each MIM is always linked to its own Annotation List and its own Glossary.

ANNOTATION LIST related to the previous MIM (Figure 11.6)
(# followed by an Arabic number indicates an interaction or a contingency depicted in the MIM)
#19, coord. 2-3, D: Stabilized β-catenin associates with TCF:LEF transcription factors, which leads to transcription of genes favoring the G0 to G1 transition, like c-Myc and Cyclin D1 among others (Huelsken and Behrens, 2002).
#20-22, coord. 2, A-D: TGF-β activity may involve TAK-1 (#20), which promotes the activity of a second kinase, NLK (#21). NLK (NEMO-like kinase) phosphorylates TCFs (#22). Phosphorylation of TCF is subsequently associated with impaired binding of the β-catenin:TCF complex to DNA (Wong and Pignatelli, 2002; Huelsken and Behrens, 2002).
#30-35, coord. 1-3, A-B: TGFβ binds TGFβR-II on the cell surface (#30), leading to recruitment (#31) and phosphorylation of TGFβR-I (#32). Phosphorylation of TGFβR-I activates its catalytic domain, which can phosphorylate specific sequences on Smad3 (#33). Phosphorylation of Smad3 is required for binding to Smad4 (#34, #35) and for transport of the Smad3:Smad4 complex into the nucleus (not shown) (Massagué and Wotton, 2000).
#48-51, coord. 2-3, B-D: When Cadherin Type I is phosphorylated in Y152 (#49) binds the phosphatase PTP1B (#48). This binding (#50) is mandatory for the activation of the phosphatase. The target of phosphatase PTP1B is the PY654 of β-Catenin (#51). (Lilien and Balsamo, 2005).
#52-53, coord. 3, B-D: Cadherin Type I can bind β-catenin when Y654 is non-phosphorylated (#53). As a consequence of this binding (#54) β-catenin is sequestered from Cadherin type I and cannot translocate to the nucleus and bind TCF/LEF. (Lilien and Balsamo, 2005).

GLOSSARY, in alphabetic order, related to the previous MIM (Figure 11.6).

1. β-catenin — A regulatory protein that integrates cell surface signals with the actin cytoskeleton and transcription factors involved in cell proliferation. — MIM's coord. 3-4D
2. Cadherin Type 1 — Cadherins (**Ca**lcium dependent **ad**hesion molecules) are a class of type-1 <u>transmembrane proteins</u>. They play important roles in <u>cell adhesion</u>. — MIM's coord. 3-A-B
3. NLK — NEMO-like **k**inase is a MAP kinase-related protein and phosphorylates TCF:LEF. — MIM's coord. 2-3C
4. PTP1B — **P**rotein **T**yrosine **P**hospatase **1B** is activated as a hetero-dimer with Cadherin and works as a phosphatase of Y654 of β-catenin. — MIM's coord. 2-3B
5. Smad3, Smad4 — Smad3 and Smad4 are isoforms functionally classified as Receptor activated Smads (R-Smads) that modulate the activity of TGF-β. — MIM's coord. 1-2B-C
6. TAK-1 — **T**GF-β **a**ctivated **k**inase **1** is a member of the MAP3K family. — MIM's coord. 2-3C
7. TCF/LEF — Members of the DNA-binding **T** **c**ell **f**actor/**l**ymphoid **e**nhancer **f**actor (TCF/LEF) family proteins. — MIM's coord. 1D
8. TGFβ — **T**ransforming **g**rowth **f**actor-β is a member of the TGF-β superfamily of cytokines. — MIM's coord. 2A
9. TGFβRI — **TGFβ R**eceptor type **I** is a partner of TGFβRII. — MIM's coord. 2-A
10. TGFβRII — **TGFβ R**eceptor type **II** binds TGFβ on the cell surface, leading to recruitment and phosphorylation of TGFβR-I and activates a downstream cascade. It is often altered in the presence of mismatch repair defects. — MIM's coord. 3-A

11.3 Implementation of a Dynamic Modeling of Cell Signaling Pathways and Networks: What It Is About? What Do We Expect From Dynamic Modeling?

Modeling of signal transduction is about modeling (bio)-chemical interactions: reversible (forward, backward) interactions and interactions associated with catalytic events.

This new field of modern biology, a molecular biology moving from the study of one or very few interacting molecules (signaling-proteins, regulatory RNAs, regulatory DNA sequences, smaller molecules involved in signaling) was made possible from a global progress in identification and characterization of individual molecules.

This "component identification/characterization phase" is quite advanced for some reference species and for some pathways. Examples of extensively studied species are: several bacteria, yeasts, *arabidopsis thaliana* (*a plant*), *caenorhabditis elegans* (*a nematode*), drosophila, mouse, man. At the level of signaling networks, programmed cell growth/division (cell cycle), and programmed cell death, are pathways studied very extensively, often in great detail. The component identification phase is very advanced for these networks.

We are at the breaking point: It becomes more and more difficult for a human mind to deal with the too many components of large pathways. Qualitative expectations become progressively more untractable for the direct intuition of our human mind, semi-quantitative predictions are impossible. On the other end, the quantitative/semi-quantitative component of the information is often an essential part of the knowledge that is required for a realistic understanding of signaling networks.

Pathway modeling has existed for some time, particularly in the field of prokaryotic metabolism. Dynamic models (also named "kinetic"/"reaction" models) work best with pathways in which components and connectivity are relatively well established. Let concentrate only about mammalian signaling-proteins involved in cell–cycle/apoptosis, + alterations of tissue architecture, two of the major functions altered in tumors. We could tentatively guess about signaling–proteins–networks of (let say) 200–400 + 200–400 basic molecular species, for the two aspects of cell behavior mentioned here, respectively. The largest dynamic simulations of present day (Chen *et al.*, 2009) consider only about 5–10% of a hypothetical cell–cycle/apoptosis related signaling-network. It is evident that we are still far from the level of study of a large signaling-network region.

However, we have already moved from the reductionist level of study of an individual protein, to an intermediate level of a protein collectivity.

We now deal with <u>*a new entity*</u>, with <u>*new emerging properties*</u>, not present in the individual components. An amino-acid (aa) has properties not obviously present in the individual constituent atoms. A protein has properties not obviously present in the individual constituent aa. Likewise, a signaling–network (even a relatively small one) has properties not present in the individual constituent components (Barkai and Leibler, 1997). Biochemists analyze the structure/function relationships of biomolecules. However, understanding structure is insufficient for understanding function. Function emerges much more clearly from the network of which that given biomolecule is a component.

Strings, sub-atomic particles, kernels of the atoms, atoms surrounded from their electronic cloud, small organic molecules, macromolecules, *smaller and larger signaling–networks*, interacting cells, tissues, organs, living creatures, ecosystems. I let aside the cosmologic dimensions.

In each case, the human mind reads new properties at the higher level, that was incapable of deducing from the below level. The limited capabilities of a human mind seem capable of reading the immanent Existent, only or mostly as subsequent levels of larger/different scale interactions.

A typical character of the new higher level entity *signaling–network*, is transmission and elaboration of signals (*an informatic role*), not membership of its components in some specific molecular class. Signaling–proteins are a predominant component, but not only proteins are involved in signaling, different classes of RNAs, regulatory DNA sequences, and some small molecules are also involved. All are potentially important components for functions of signal–integration/signal–elaboration, performed within a given network.

11.3.1 *Model construction is error prone*

Careful checking of all details, checking for conservation of species in a list of reactions, exploration of the behavior of simulations applied to small subsets of a given MIM, are recommended.

Mass action kinetics, a direct logic consequence of the IInd thermodynamic law, states that rates of the reactions are proportional to the concentrations of the reacting species and to constants (k) specific for any given reaction.

Reactions are most often dealt with using ordinary and partial differential equations (ODEs and PDEs). We used deterministic ODEs extensively. Unfortunately, the non linear nature of these relatively large sets of differential equations prevents from determining the analytical expressions for the system evolution over time. Nevertheless, it is possible to simulate the system evolution numerically with the help of appropriate software such as the SimBiology toolbox of Matlab (http://www.mathworks.com/products/simbiology). Using our data stored in SBML (http://sbml.org) we sometimes also used CellDesigner (http://celldesigner.org/) integrated with SBML ODE Solver. The two tools gave the same results. These numerical solution approaches have been pursued by different authors, among them (Kholodenko *et al.*, 1999; Wolf *et al.*, 2007) and other authors whose models are available in the BioModels Database (http://www.ebi.ac.uk/biomodels-main/ or http://biomodels.caltech.edu). For the reader not familiar with numerical solutions we explain that the dynamic simulation proceeds ahead advancing through a sufficiently small Δt at each simulation cycle. All the components of the simulated network go through very small adjustments at each cycle. Good approximations can be obtained.

Both SimBiology and CellDesigner make use of standard, also advanced, numerical solvers for stiff and non-stiff ODE systems that allow producing dynamic simulations of the processes that proceed ahead over time. As with all the numerical methodologies, a little caution must be exercised in using such numerical solvers since, under some critical condition, they may get off the correct phase trajectory due to the accumulation of numerical errors. When in doubt, some trivial tricks may be implemented to double check the obtained results. As an example, the same problem can be solved with different numerical solvers or a sensitive analysis of the problem parameters can be performed. Other less trivial checks can all be implemented, but they are out of the scope of this paper and subject of Numerical Analysis studies.

In case we are interested only in stationary solutions (and not in the transient behavior of the system), dynamic modeling and consequent simulations may not be strictly necessary as such solutions can be determined solving algebraic systems instead of a set of differential equations. Then, at least part of the exploration of the parameter space can be performed through a sensitivity analysis on the solutions of the aforementioned algebraic systems.

Here again, a word of caution must be spent, the algebraic system to be solved in order to determine the stationary solutions are in general not linear, then they are solved through numerical methods. Then, again, there is always the risk that not all the stationary solutions are determined, or that the methods individuate a wrong solution due to the accumulation of numerical errors. Finally, an ex-post analysis of the results must be always performed, as we must always double-check that the stationary solutions obtained are reasonable from a biological point of view. As an example, a stationary solution in which all the compound concentrations are equal to 0 usually exists, needless to say that this solution is of scarce interest.

11.3.2 *Stochastic simulation approach*

To our knowledge, a dozen of stochastic simulators have been produced, (COPASI, Gillespie2, Dizzy, among others). A nice and comprehensive review about the state of the art in stochastic modeling is presented in (Wilkinson, 2009). In the framework of a Ph.D. thesis work (R.M.) we built a stochastic simulator, using a modified, relatively user friendly, form of Java language. We used a stochastic approach when we wanted to deal with molecular behaviors in individual cells. In individual cells molecular behavior is not a continuum, but rather a stochastic process. Relatively robust fluctuations have been probably favored during evolution, against sensitive fluctuations.

Compartmentalization is an important player in the dynamic simulation games. We can approach compartmentalization more or less explicitly or implicitly. Extensive sequestering of a transcription factor (TF) in the cytoplasm because of high-affinity binding to a cytoplasmic protein will leave anyway a very small fraction of free TF available for chromatin/DNA binding, even if we don't add details about crossing of pores of the nuclear membrane by our TF.

An important check, when we perform dynamic simulations is comparison of simulated and experimental parameters over different concentrations and time courses. The effects of a virtual exposure to different ligand concentrations for different time lengths will have ideally to display good agreements with parameter concentrations at all different times. Correspondingly, experimental parameter space exploration (times and concentrations) will have to be more extensive than usual, to satisfy questions about the good fitting with a dynamic simulation. When we have a given molecular species which is 10^{-10} M in concentration (not uncommon for signaling-proteins), inside a typical 5 pl ($5 \cdot 10^{-12}$ liters) volume mammalian cell, the number (N) of molecules will be the following: $N \sim 6 \cdot 10^{23} \cdot 10^{-10} \cdot 5 \cdot 10^{-12} = 300$ ($6.022 \cdot 10^{23}$ is the Avogadro's number: N of molecules in 1 mole of molecules).

Consider that, a total of 300 signaling-proteins of a given molecular species inside a mammalian cell could easily have a repartition in five or more different complexes and post-translational modifications. 5 sub-species could have the following distribution: $100 + 75 + 60 + 45 + 20 = 300$. Random fluctuations in species concentration or reaction rates give origin to temporal variations $= (N)^{-1/2}$. $(20)^{-1/2}$ is equivalent to a 22% temporal variation: clearly we are not far from the limit of a temporal variation acceptable for a viable and well coordinated cell, at least in the absence of robust re-equilibrating feedbacks developed during evolution (Sigal et al., 2007).

In implementing dynamic simulations, we have to compromise between two opposite situations:
- a too small region of the signaling network will generate a predictive dynamic simulation, but not an especially interesting one.
- a too large region of the signaling network will generate an unmanageable (practically intractable) situation, probably also affected from overwhelming uncertainties.

The degree of predictive power that can be achieved depends on the quality of the data, and obviously on a formally correct dynamic simulation.

Some lack of knowledge can be surrogated with biologically plausible assumptions.

General default assumptions are better than a multiplicity of "ad hoc" assumptions. In this last instance, apparently good predictive simulations could reflect a retrofitting with already known experimental results, obviously not with new ones. However, should we dispose of a sufficiently large number of experimental data (molecular concentrations and kinetic constants), a systematic exploration of a large parameter space by brute computational force (but astronomical numbers of combinations have still to be avoided), could converge toward a unique or very few satisfactory solutions (satisfactory combinations of parameters) (Chen *et al.*, 2009). The structure of the molecular anatomy of a given network and comprehensive behavioral expectations coming from the experimental world (even if they require a more formal explicitation), probably are already a restraint stronger of what we could think.

11.3.3 *Sufficient/adequate MIM size*

At the present stage of development of this field of molecular systems biology, it's still unclear when a given area of the molecular signaling-network is sufficiently large and inclusive, for interesting dynamic simulations (the hypothesis of semi-autonomous modules).

Obviously to a limited network region you will ask limited and directly pertinent questions.

Similarly, what degree of incompleteness of the reconstruction of the molecular anatomy of a given signaling-network region, is still comprehensive enough for interesting dynamic simulations (predictive/falsifiable)? We can advance the consideration that, when we perform a dynamic simulation, what is missing in a given network is considered implicitly in a permissive state for the functioning of the represented part of the network itself. Suppose a not represented negative feedback could be alternatively on or out, in our simulation this omitted circuit is considered implicitly out (permissive state).

An additional design-decision in physico–chemical/biochemical modeling is the choice of the **degree of detail** (appropriate level of "**shallowness**"), to which we deem appropriate our dynamic simulation. Some people like to speak of "**model granularity**".

If the experimental end behavior of a cascade of events is "on" (for instance the G0 → G1 transition takes indeed place), small parts of the signaling network not represented (not yet known) in a simulated MIM, could be considered in a "permissive state" by default (as discussed above).

Models should be parsimonious, but not too parsimonious: some people proposed the term "**mesoscale models**". Usually little molecular insight is gained when molecular processes are excessively lumped together, to produce too small numbers of molecular entities, complexes, equations.

Hybrid models can be constructed, in which specific biological processes are alternatively modeled in a more detailed or in a more aggregate form. It is not acceptable to fill up missing values with retrofitting procedures tuned toward the most satisfactory biologic plausibility. Each retrofitted value introduces an additional degree of freedom. Given *a discrete number of degrees of freedom, we will be able to fit almost everything.* Retrofitting procedures are disappointing when it comes to predict new experiments. However, if we have enough solid experimental values, satisfactory retrofittings of missing parameters could become exceedingly rare, and perhaps converge toward a unique or very few solutions.

Instead of retrofitting dozens of individual values, an alternative could be to introduce very few default rules.

An unknown molecular concentration, of a protein described as present in a given experimental setting, can be given a median value, median among the known molecular concentrations concerning our module, or more in general for signaling-proteins.

If a good quality experimental paper says, for instance, that protein subspecies A* (after a given post-translational modification*) interacts well with the molecule B■, we can assume that we will have a 95% of [A* - B■]. There will be a free 5% of the less abundant partner. This could be a general default assumption, based on semi-quantitative information. The diffusion limit is $7 \times 10^9 \cdot M^{-1} \cdot sec^{-1}$. However, considering the very crowded cellular environment, the very large size of many multi–protein complexes, the fact that only a fraction of protein–protein encounters will be productive, a median second order association rate constant (k_1) in the order of 10^8–$10^6 \cdot M^{-1} \cdot sec^{-1}$, could be a tentative default value. Similarly, a reasonable first order default dissociation rate constant (k_{-1}) could be in the order of $3 \cdot 10^{-3} \cdot sec^{-1}$.

$K_d = k_{-1}/k_1 = (3 \cdot 10^{-3}/10^6) \cdot M = 3 \cdot 10^{-9} \cdot M$ (a reasonable K_d).

Signaling is a continuum, or is made of, and can be studied as, connected compartments?

Different meta-stable cellular differentiations could represent quasi-separate compartments. Even different phases of the cell cycle could correspond to partially independent modules. The same could be true for apoptotic phenomena, or for mechanisms of DNA-damage and repair. It's not unlikely that limited regions of the global signaling can be studied in a coherent and useful way, even if these regions are not as sharply distinct as protein species or other molecules. We will learn more about signaling modularity in the next few years.

11.3.4 Forward and backward kinetics

$$[AB] = [A] + [B]$$

$$\frac{[AB_t]}{[AB_0]} = e^{-k_{-1}t}$$

$$\ln\left(\frac{[AB_t]}{[AB_0]}\right) = -k_{-1}t$$

$$k_{-1} = -\ln\left(\frac{[AB_t]}{[AB_0]}\right) t^{-1}$$

$$k_{-1} = (\ln[AB_0] - \ln[AB_t]) t^{-1}$$

We made the hypothesis that a reasonable **first order default dissociation rate constant (k_{-1})** could be in the order of $3 \cdot 10^{-3} \cdot \text{sec}^{-1}$.

Then: $3 \cdot 10^{-3} \cdot \text{sec}^{-1} = (\ln [AB_0] - \ln [AB_t]) \cdot t^{-1}$

What will be the half life of $[AB]$?

$3 \cdot 10^{-3} \cdot \text{sec}^{-1} \cdot t_{1/2} = \ln 2 \quad t_{1/2} = (\ln 2) \cdot 333.(3) \cdot \text{sec} \quad t_{1/2} = .693 \cdot 333.(3) \cdot \text{sec} \approx 231 \text{ sec} = 3'51''$

We introduced these simple considerations to give to the reader the filing that dynamic simulations are not only about concentrations of different molecular species, but also about how long it takes to the components of the simulated network to evolve with time.

11.3.5 *Problem of validation of a given dynamic simulation*

In our case, the behavior of a simulation when we have **mutated onco-proteins could represent a validation**. This holds also for the behavior of their **inhibitors** (Castagnino *et al.*, 2009). Several experimental observations can be found in published articles, new ones come out all the times.

However, we are not allowed to readjust parameters a posteriori, only to transform unsatisfactory predictions in satisfactory ones. This procedure is misleading. Ideally, we should have enough known experimental parameters that a systematic exploration of parameter space will be satisfied only by one or very few solutions (Chen *et al.*, 2009).

The pictogram/pathway-diagram of a MIM conveys the general information flow in the network, and also mechanistic details at the qualitative level. It reconstructs a sort of anatomy of the network. However, additional details and quantitative information are required for mathematical modeling. From a pathway diagram we have to move to a complete list of reactions. Some specialized software tools can cooperate to this task.

Let $[E^*]$ be the activated form of an enzyme, and $[S]$ its substrate. If a stationary state for $[E^*S]$ is reached much more rapidly than the turnover number of the enzyme, we can introduce, for a time interval not too long, the simplification of a stationary state assumption ($d[E^*S]/dt = 0$). In this case we can use the Michaelis–Menten approximation to enzyme-substrate kinetics, reducing the number of complexes at work in the model.

Validation of a dynamic simulation model implies a satisfactory agreement between the prediction of the model and experimental results not already retrofitted implicitly in the model.

We presently work toward the possibility of simulating one by one the behavior of a dozen of mutated onco-proteins (associated with gain or loss of function). In the case of onco-proteins affected from gain of function we also simulate virtual inhibitors.

A positive side of the story is that a large variety of experimental results is available for cancer lines bearing most of these mutated onco-proteins. Ideally a given cancer line should bear only the mutation under study, unfortunately a cancer line is a cancer line because of the presence of a multiplicity of mutations. An effort is under way to characterize for their mutational spectrum the 60 standard NCI (Natl. Canc. Inst.) lines they have in their repository.

Hand-writing of two-three hundred equations is at a significant risk of being affected from bugs. We have not explored in our case the usage of rule-based models more amenable to automatic verification. Using for instance Kitano's CellDesigner (Kitano, 2003; Funahashi *et al.*, 2008) or SimBiology's Diagrams (MATLAB R2009a and SimBiology 3.0) it is possible to verify a correct correspondence between graphical representation and associated equations. Obviously concentrations and rates will have to be implemented correctly.

A critical step is determining (obtained from literature) molecular concentrations, reaction rates, initial conditions. Rates are usually known for diluted solutions. We should be always aware that we apply them to a crowded intracellular environment.

Dynamic simulations are a powerful tool for studying cell behavior, gaining a deeper understanding of what is going on, and suggesting new experiments. The simulated data can be compared for instance with experimental data obtained from flow-cytometry, confocal microscopy, and quantitative western blotting.

11.3.6 *Multidimensional sensitivity analysis*

When in a dynamic simulation we change, introducing several discrete intervals, for instance three parameters, we will obtain a "volume" of behaviors. They will often suggest regions of critical behavioral change, of special interest for comparison with experimental data.

We could find more than one stable or unstable region, or oscillatory behaviors. In principle, but not necessarily, different trajectories could bring to different behaviors at the steady state (hysteresis phenomena) (Kholodenko, 2006).

For maps containing 20–40 basic molecular species (the largest maps of these days), applied to a given mammalian tissue, for a cell–cycle related network, only a fraction of the concentrations of the basic species will be known. Similarly, only a fraction of the kinetic rates will be known. The molecular anatomy of the network will impose by itself restraints to the number of possible solutions. Just few dozens of known experimental parameters, in combination with the network structure, will impose stringent limitations to the number of possible solutions. For instance, Chen W.W. (Chen *et al.*, 2009) found that out of one million explored parameter sets only 1/10,000 would be in agreement with the experimental exploration of three basic species, four growth factor treatments and 10 time points. This is an interesting beginning but not a solution to the problem of parameter space exploration. The problem is a very important one and ways to circumvent the obstacles will have to be explored in the next few years. We can think of strategies for improving the situation in terms of number of known parameters (even from not completely homogeneous sources, and through technological improvements), and we can think of strategies for containing to computable sizes the number of parameter combinations. Parameter space exploration is a crucial not yet solved difficulty, and we will leave it here.

Dynamic simulation models could be extremely useful for proposing **rational associations among inhibitors of onco-proteins. This is the frontier of the new cancer therapy.**

A deeper understanding of signaling-networks and rationalization of cancer therapy will go together, especially for associations of selective signaling-molecules inhibitors appropriate for individual tumors.

We are progressively more aware that tumors belonging to the same histological class (for instance CRC, or BC, or NSCLC) can be individually profoundly different at the molecular level. In addition, the probabilistic mutation frequency for different oncogenes differs in the three different epithelial tumors. There is however much more convergence if we examine alterations in terms of pathways.

Simulation of limited signaling-network regions involving 30–50 basic proteins/molecules could take great advantage from an experimental counterpart/technology, capable of giving information about concentrations and interaction K_ds of the molecules involved in the network. To this end some High Throughput (HT)/Medium Throughput (MT) methods involving protein detection and interaction detection could be useful. The development of protein microarrays offers a precious opportunity. Protein microarrays are currently available in two general formats. Antibody arrays contain an array of appropriate antibodies that can measure the abundance of our specific proteins of interest (or other molecules) in experimental samples referred to our network; they could be useful for measuring concentrations in a specific cell type or line (Haab *et al.*, 2001).

To study interaction $K_d s$, target protein arrays, which present arrayed proteins of interest, could be a precious tool. They can be used to examine target protein interactions with other network molecules, but also interactions with inhibitors of onco-proteins affected from excess of function and interactions with inhibitor monoclonal antibodies.

In this perspective, nucleic acid programmable protein arrays (NAPPA) could be a precious tool. In extreme synthesis, the basic principle of NAPPA technology is the following: Spots of c-DNAs of interest are bound to a slide. Through addition of a cell–free expression system, proteins of interest carrying a GST-tag are produced. In the spot area of the corresponding c-DNA, the produced protein is captured from a polyclonal anti-GST antibody. We can query our protein array prepared according to our specific requirements, for interaction with other proteins or molecules (Ramachandran *et al.*, 2004).

11.4 Conclusions

The complexity of signaling-networks (governing behavior of all living creatures) is overwhelming. We will gradually learn the appropriate levels of abstraction/shallowness/simplification, to deal most conveniently with dynamic simulations of signaling-networks.

Pathway dynamic models make it possible to examine in details the **effects of deregulations (mutations)** affecting signaling-molecules (proteins and other molecules).

This is important in general and in **a perspective of pharmacological intervention with selective inhibitors.** When it comes to inhibitors of dominant onco-proteins, affected from excess of function, medicinal chemistry has already produced several drugs of this kind. They are under study both at the preclinical and clinical levels. However, *a rational for their association* **will require dynamic simulations of**

mutations and inhibitors in a signaling-network (Castagnino *et al.*, 2009).

Models and MIMs are also transmittable repositories of knowledge.

The organized background of data they have behind will become more important than simple databases, as vehicles of new knowledge (Kohn *et al.*, 2008).

If we consider past historical experience in fields as different as celestial mechanics, combustion chemistry, semiconductor fabrication, metabolic engineering, we observe the following:

Large sets of observational data, describing complex time-dependent processes, were organized into models. They evolved over time and acquired considerable predictive power.

As a consequence, these models are largely used for scientific and industrial applications.

In a similar way, we can expect that progressively more accurate dynamic models of cell-signaling circuits will have a key role in the **appropriate usage of signaling-molecule-targeted pharmaceutical drugs.**

Accurate dynamic models will direct our attention toward important new molecular targets, and they will become a basic rational for *associations* of *molecule-targeted drugs* in medical treatments.

Acknowledgments

This work was supported from: MIUR PRIN 2007-prot. 2005061408 005; National Project Compagnia di San Paolo 2489 IT/CV, 2006.1997, Genoa Operative Unit (O.U.); Regione Liguria, Cap. 5296 F.S.R. Nr. Prov. 23, Prot. Gen. 616, 2007; Istituto Superiore di Oncologia, O.U.s of an "Istituto Oncologico Veneto" project, and a "Regione Campania project"; project Compagnia di San Paolo 4998 IT/CV, 2007.0087, Genoa University; Project LIMONTE-SP1P 2009 (with Castagnola and Zardi); project CARIGE, Genoa University 2008 (with Patrone, Ballestrero *et al.*).

References

Aladjem, M. I., Pasa, S., Parodi, S., Weinstein, J. N., Pommier, Y. and Kohn, K. W. (2004). Molecular interaction maps a diagrammatic graphical language for bioregulatory networks. *Sci STKE*; (222): pe8.

Alon, U. (2007). *An Introduction to Systems Biology - Design Principles of Biological Circuits*. Chapman & Hall/CRC/Taylor & Francis, Boca Raton, Fl.

Alon, U., Surette, M. G., Barkai, N. and Leibler, S. (1999). Robustness in bacterial chemotaxis. *Nature*, 397, pp. 168–171.

Barabási, A. L. and Oltvai, Z. N. (2004). Network biology: understanding the cell's functional organization. *Nature Review Genetics*, 5, pp. 101–113.

Barkai, N. and Leibler, S. (1997). Robustness in simple biochemical networks. *Nature*, 387, pp. 913–917.

Castagnino, N., Tortolina, L., Montagna, R., Pesenti, R., Balbi, A. and Parodi, S. (2009). Simulations of the EGFR - KRAS - MAPK signaling network in colon cancer. Virtual mutations and virtual treatments with inhibitors have more important effects than a 10 times range of normal parameters and rates fluctuations. *Lecture Notes in Bioinformatics,* Computational Intelligence methods for Bioinformatics and Biostatistics (CIBB) 2009, in press.

Chen, W. W., Schoeberl, B., Jasper, P. J., Niepel, M., Nielsen, U. B., Lauffenburger, D. A. and Sorger, P. K. (2009). Input-output behavior of ErbB signaling pathways as revealed by a mass action model trained against dynamic data. *Molecular Systems Biology*, 5:239 (pp. 1–19).

Funahashi, A., Matsuoka, Y., Jouraku, A., Morohashi, M., Kikuchi, N. and Kitano, H. (2008). CellDesigner 3.5: A versatile modeling tool for biochemical networks. *Proceedings of the IEEE*, 96, pp. 1254–1265.

Haab, B. B., Dunham M. J. and Brown P. O. (2001). Protein microarrays for highly parallel detection and quantitation of specific proteins and antibodies in complex solutions. *Genome Biology,* 2(2).

Hartwell, L. H., Hopfield, J. J., Leibler, S. and Murray, A. W. (1999). From molecular to modular cell biology. *Nature*, 402, pp. C47–C52.

Huelsken, J. and Behrens, J. (2002). The Wnt signalling pathway. *Journal of Cell Science*, 115, pp. 3977–3978.

Jeong, H., Mason, S. P., Barabási, A. L. and Oltvai, Z. N. (2001). Lethality and centrality in protein networks. *Nature*, 411, pp. 41–42.

Kholodenko, B. N. (2006). Cell-signaling dynamics in time and space. *Nature Reviews Molecular Cell Biology*, 7, pp. 165–176.

Kholodenko, B. N., Demin, O. V., Moehren, G. and Hoek, J. B. (1999). Quantification of short term signaling by the epidermal growth factor receptor. *Journal of Biological Chemistry*, 274, pp. 30169–30181.

Kitano, H. (2003). A graphical notation for biochemical networks. *Biosilico*, 1, pp. 169–176.

Kitano, H., Funahashi, A., Matsuoka, Y. and Oda, K. (2005). Using process diagrams for the graphical representation of biological networks. *Nature Biotechnology*, 23, pp. 961–966.

Kohn, K. W. (1999). Molecular interaction map of the mammalian cell cycle control and DNA repair systems. *Molecular Biology of the Cell*, 10, pp. 2703–2734.

Kohn, K. W., Aladjem, M. I., Weinstein, J. N. and Pommier, Y. (2006). Molecular interaction maps of bioregulatory networks: a general rubric for systems biology. *Molecular Biology of the Cell*, 17, pp. 1–13.

Kohn, K. W., Aladjem, M. I., Weinstein, J. N. and Pommier, Y. (2008). Chromatin challenges during DNA replication: a systems representation. *Molecular Biology of the Cell*, 19, pp. 1–7.

Lilien, J. and Balsamo, J. (2005). The regulation of cadherin-mediated adhesion by tyrosine phosphorylation/dephosphorylation of beta-catenin. *Current Opinion in Cell Biology*, 17, pp. 459–465.

Massagué, J. and Wotton, D. (2000). Transcriptional control by the TGF-beta/Smad signaling system. *EMBO Journal*, 19, pp. 1745–1754.

MATLAB R2009a and SimBiology 3.0; The MathWorks, Inc., Natick, MA.

Milo, R., Shen-Orr, S. S., Itzkovitz, S., Kashtan, N. and Alon, U. (2002). Network motifs: simple building blocks of complex networks. *Science*, 298, pp. 824–827.

Oda, K., Matsuoka, Y., Funahashi, A. and Kitano, H. (2005). A comprehensive pathway map of epidermal growth factor receptor signaling. *Molecular Systems Biology*, 1:10 (pp. 1–17).

Ramachandran, N., Hainsworth, E., Bhullar, B., Eisenstein, S., Rosen, B., Lau, A. Y., Walter, J. C. and LaBaer, J. (2004). Self-assembling protein microarrays. *Science*, 305, pp. 86–90.

Shen-Orr, S. S., Milo, R., Mangan, S. and Alon, U. (2002). Network motifs in the transcriptional regulation network of Escherichia coli. *Nature Genetics*, 31, pp. 64–68.

Sigal, A., Danon, T., Cohen, A., Milo, R., Geva-Zatorsky, N., Lustig, G., Liron, Y., Alon, U. and Perzov, N. (2007). Generation of a fluorescently labeled endogenous protein library in living human cells. *Nature Protocols*, 2, pp. 1515–1527.

Wilkinson, D. J. (2009). Stochastic modelling for quantitative description of heterogeneous biological systems. *Nature Review Genetics*, 10, pp. 122–133.

Wolf, J., Dronov, S., Tobin, F. and Goryanin, I. (2007). The impact of the regulatory design on the response of epidermal growth factor receptor-mediated signal transduction towards oncogenic mutations. *FEBS Journal*, 274, pp.729–743.

Wong, N. A. and Pignatelli, M. (2002). Beta-catenin-a linchpin in colorectal carcinogenesis?. *American Journal of Pathology*, 160, pp. 389–401.

Wuchty, S., Oltvai, Z. N. and Barabási, A. L. (2003). Evolutionary conservation of motif constituents within the yeast protein interaction network. *Nature Genetics*, 35, pp. 176–179.

CHAPTER 12

LABEL-FREE DETECTION OF NAPPA: SURFACE PLASMON RESONANCE

Manuel Fuentes, Sanjeeva Svrivastava, Nirosahan Ramachandran,
Eugenie Hainsworth and Josh LaBaer

Harvard Institute of Proteomics,
Harvard Medical School, Cambridge, MA, USA

Our research is focused at coupling to the implementation of Nucleic Acid Programmable Protein Array (NAPPA) technology to real-time label-free SPR-based detection. Here, we show the combination of both technologies that together could lead to the ability to measure binding events in real-time for many protein interactions simultaneously using a label-free technology. This could revolutionize the study of protein interactions networks by enabling quantitative comparisons of binding affinities across many molecular species, as well determining the kinetics rates of binding and release.

12.1 Nucleic Acids Programmable Protein Arrays

In post-genome era having sequence the human genome, one of the most important pursuits is to understand the function of gene-expressed proteins. Despite the immense progress in molecular biology and genetic, only a small fraction of the proteome is understood at the biochemical level.

Proteomics, which refers to the analysis of proteins on a large-scale, has become one of the spotlights in today's biomedical research (MacBeatch and Schreiber, 2000; Zhu *et al.*, 2001).

Defining the function of every protein in the cell is the key for solving a range of health related issues, including better understanding of diseases process, early detection of disease and accelerating drug development. While the human genome is known to contain about 40000 genes, the human proteome is predicted to have a size of 1 million. There is certainly a great demand for developing high-throughput tools for rapid protein analysis. Conventional protein-protein interactions analysis techniques involve protein purification followed by detection step such a sodium dodecyl-sulfate polyacrylamide gel electrophoresis (SDS-PAGE) and Western blots. These procedures are known to be very time consuming, and they also not designed for high-throughput measurements.

In the recent years, the protein microarray technology has become an attractive high-throughput alternative for protein-protein interactions analysis. Hundreds or thousands protein samples are spotted in a single protein microarray and a wide variety of protein interactions and activities are analyzed simultaneously (MacBeatch and Schreiber, 2000; Zhu *et al.*, 2001; Kretschmann and Raether, 1968; Ramachandran *et al.*, 2004).

The protein microarray technology has demonstrated its utility by characterizing biochemical process and providing information which is not obtainable by gene microarrays.

A key remaining challenge for producing protein microarrays has been using high content (many different proteins) with high density and functionality. To address these concerns, Nucleic Acids Programmable Protein Microarrays (NAPPA) developed at Harvard Institute of Proteomics (HIP) revolution protein microarrays by replacing complex process of spotting purified proteins with a simple process of spotting plasmid DNA and simultaneous transcription/translation of all the genes *in situ* at the time of the assay (Ramachandran *et al.*, 2004) (Figure 12.1).

The proteins are translated using a T7-coupled rabbit reticulosyte lysate in vitro transcription/translation system. The expressed proteins are capture locally with an antibody to a C-terminal glutathione S-transferase (GST) tag on each protein (anti-GST). This approach eliminates the need for high-throughput protein isolation and ensures that all the proteins are produced fresh (that is, coincident with or minutes before use) for each experiment (Kretschmann and Raether, 1968; Ramachandran *et al.*, 2004).

Nucleic Acid Programmable Protein Array

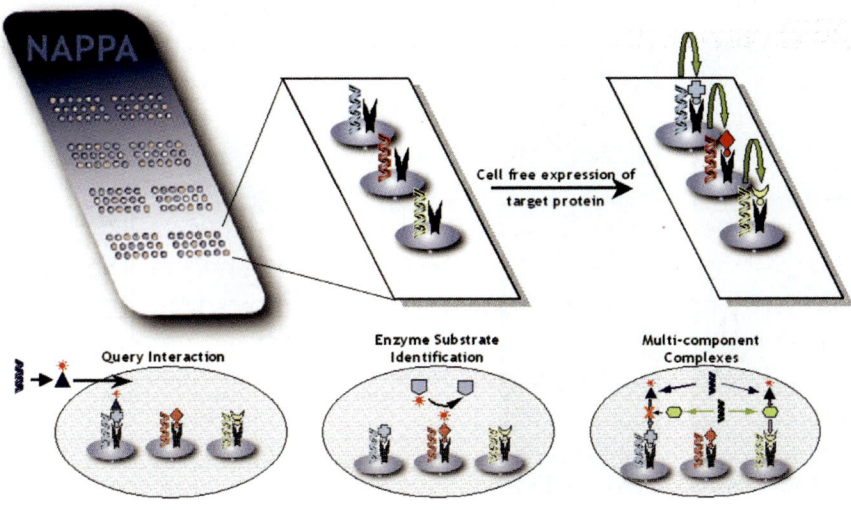

Figure 12.1. Scheme of Nucleic Acids Programmable Protein Array.

Ramachandran *et al.* (2004) demonstrates that a variety of proteins can be displayed by this format, 1000 human genes for which it have isolated plasmid for single colonies, verified the full-length sequence and previously made them readily available through the Plasmid ID Repository. Using an antibody to GST (distinct from the one used to capture the nascent protein) which recognises the C-terminal GST tag genetically encodes in each coding sequence and thus confirms full-length translation. It has been detected protein signal for 96% of the printed genes (3 s.d. above the signal from non-expressing plasmid). Compared to the printed recombinant purified protein (GST), the average protein yield was 9 fmol per feature (4–14 fmol, 10th percentile to 90th percentile) (Figure 12.2).

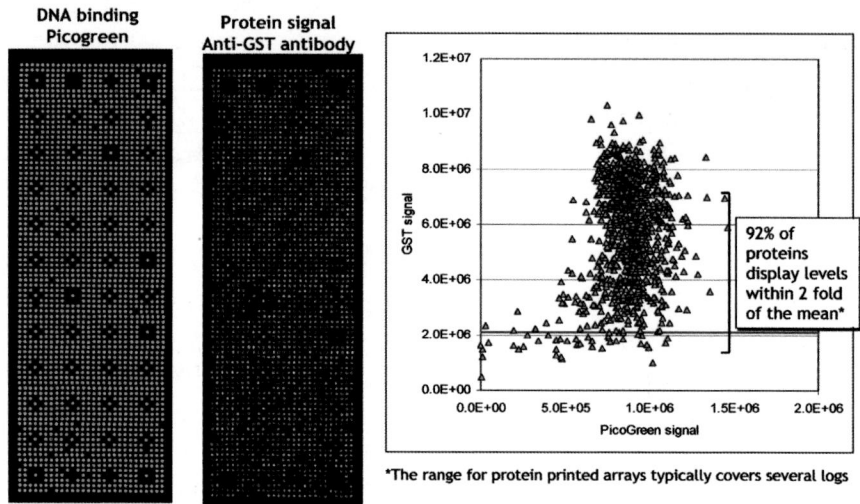

Figure 12.2. High-density NAPPA array. High-density array with cDNA representing 1000 unique genes, each printed in duplicate. The arrays were stained with Picogreen to show DNA binding (left) and proteins were detected with anti-GST after addition of cell free expression lysate (right).

12.2 Surface Plasmon Resonance

Typically, signal transduction is based on fluorescence labelling (MacBeath and Schreiber, 2000; Zhu *et al.*, 2001). However, it is also well-known that labelling tags may modify protein activities, which can lead to cross-sensitivity between sensor elements, unexpected binding or degradation in sensitivity limit.

Surface Plasmon Resonance is a label-free affinity detection technique that relies on detecting the refractive index change caused by immobilization of biomolecules to the metallic sensor surface.

12.2.1 *What is a Surface Plasmon and what is Surface Plasmon resonance?*

When coining the name of the technique, the first-ever question relates to the name of the technique. Surface Plasmon is described as the collective excitation of quantised oscillations of the electrons as the collective excitation of quantised oscillations of the electrons in a metal, and the electromagnetic field near the surface of the metal. Surface Plasmon resonance is the phenomenon which occurs when light is reflected off thin metal films. A fraction of the light energy incident at a sharply defined angle can interact with the delocalised electrons in the metal film (Plasmon) thus reducing the reflected light intensity. The first observation of plasmons date back more than a century when Wood described the discontinuous diffraction pattern after irradiation of a metal film with polychromatic light. However, only in 1968 Otto managers to explain that phenomenon by the excitation of surface Plasmons (Otto 1968). Simultaneously, Kretschmann described the excitation of surface plasmons under the mode of total internal reflection (TIR) and set the basis for the use of this phenomenon for measuring biomolecular interactions (Kretschmann and Raether, 1968). Thus, plasmons can be regarded as a constellation of delocalised electrons in a metal surface that can absorb energy from light (Figure 12.3). Figure 12.3 shows the Kretschmann configuration that is still employed in modern instruments (Kretschmann and Raether, 1968). Monochromatic and plane-polarised light hits a metal surface under an angle where TIR occurs. Under these circumstances, at a particular angle the delocalized electrons absorb part of the energy generating plasmons and an evanescent wave that propagates perpendicular to the plane of incidence.

These two aspects, the reduction of reflected light and the evanescent wave (Figure 12.3), are crucial for the application of the technology to monitoring biomolecular interactions.

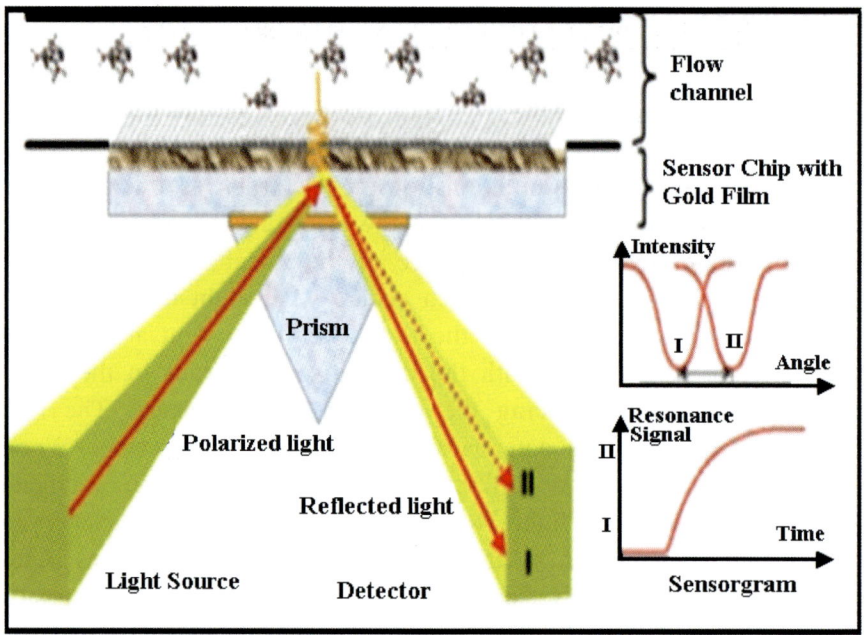

Figure 12.3. Scheme Kretschman configuration of SPR Instrument.

The evanescent wave interacts with the medium into which it penetrates conditioning the extended of which to the nature of this medium. Consequently, the SPR angle θ will directly depend on the nature of the medium.

12.2.2 *The principle of measurement-towards a sensorgram*

In order to apply for SPR phenomenon to biomolecular interaction analysis different instruments employ the setting described in the Figure 12.4.

A clear separation can be devised between the detection module and the environment where the biomolecular interaction occurs (as it is possible to observed in Figure 12.4 how this technology permits real-time monitoring). Depending on the manufacturer the measurements occur either under static or dynamic conditions (Otto, 1968;

A.-Surface Plasmon Resonance

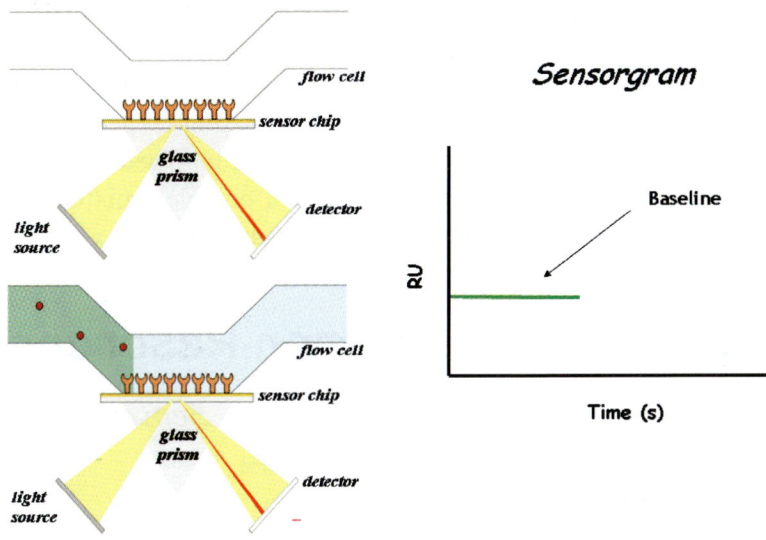

Figure 12.4A. Description of an SPR experiment over several panels. On the left the instrumental Design is depicted with the measurement setting (read-out) and the interaction. On the right the translation of the binding phenomena into sensorgrams from which both qualitative and quantitative data can be derived.

Kretschmann and Raether, 1968), albeit that latter is the most frequently used setting, as it permits the simulating physiological conditions.

In all settings, the metal surface (Au is in the figure but it is possible other metals such as Ag, Cu, Al, ….) is functionalized to permit covalent attachment of a biomolecule such as antibody. Depending on the strategy employed when preparing the metal surface different chemical entities (mostly of them are able to be detected by Infrared) are incorporated that address different functional groups of a protein. As for example: primary amino groups, sulphydril groups, aldehydes groups, carboxyl groups, … may be targeted for specific binding of the biomolecule to the sensor surface.

Upon covalent attachment of the biomolecule to the functionalized surface the refractive index changes which in turn causes a change in the SPR angle. This permanent change provides useful information on the surface coating density that can be used to monitor the state of the surface, calculate the surface orientation of the immobilized molecules, *etc*. Finally, with the prepared surface interaction experiments are performed whereby a specific ligand for the immobilized molecule is injected across (alone or as a component of a complex mixture), very much like a conventional chromatographic experiment.

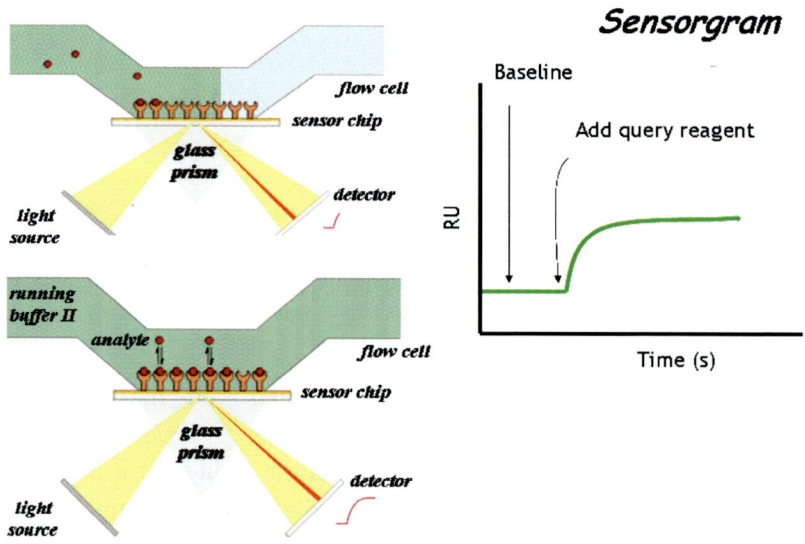

Figure 12.4B. Description of an SPR experiment over several panels. On the left the instrumental Design is depicted with the measurement setting (read-out) and the interaction. On the right the translation of the binding phenomena into sensorgrams from which both qualitative and quantitative data can be derived.

The distinct steps of the interaction experiment are described in Figure 12.4. At the start of a binding experiment the surface the refractive index is basically that of the buffer passing over the surface (situation A). When the sample is injected, the refractive index changes instantly due to the change in buffer composition (θ_1–θ_2). During the injection, binding occurring at the surface will continue changing the refractive index and thus the SPR angle. These angle-changes in time are transformed into a so-called sensorgram with time on the x-axis and a magnitude scale at the y-axis, usually indicated as "resonance units" (RU). When the sample injection stops (situation B), again running buffer is flown across and the resulting RU are due to sample components that still interact at the surface. Finally, through the injection of a regeneration solution (situation C) the basal state can be recovered and the biomolecule immobilized surface reused in a subsequent binding event (situation C).

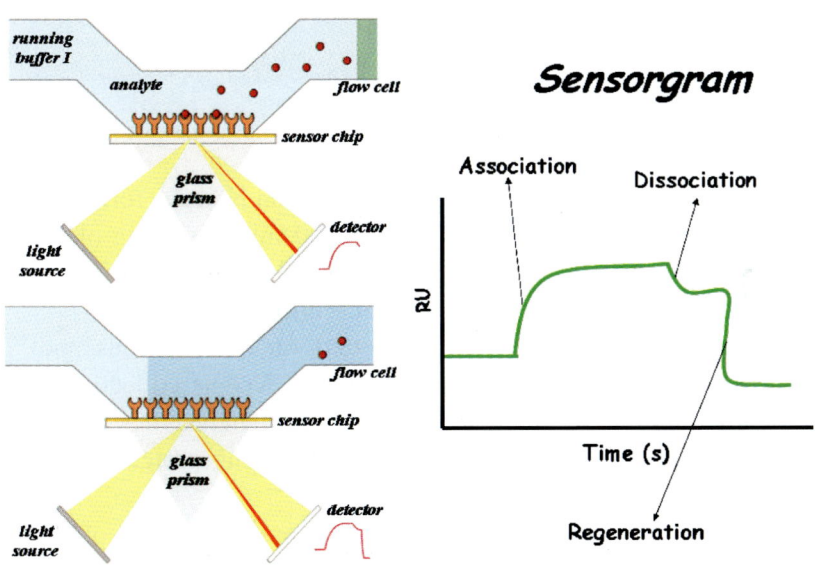

Figure 12.4C. Description of an SPR experiment over several panels. On the left the instrumental design is depicted with the measurement setting (read-out) and the interaction. On the right the translation of the binding phenomena into sensorgrams from which both qualitative and quantitative data can be derived.

12.3 Materials and Methods

As a proof of concept for use with SPR instrument, comprises a support having an electro conductive target surface; this surface is essential where the biomolecules is to be subjected to SPR analysis. The support could be made of an electro conductive material (which term is used herein to include semi conductive materials) it is preferred to coat a rigid support, for example: a glass, with an electro conductive material such as gold. The NAPPA microarray was then produced on a 25 × 76 × 1 mm slide covered with a thin layer of gold (47.5 nm). The gold surface was then functionalized by amino-PEG-thiols generating a homogeneous self-assembled monolayer onto the gold surface.

To analyze NAPPA approach by SPR, it can be mounted on an adapter with a 70 µL adhesive flow cell, which carries it into a SPR instrument. The lodging dimension is exactly the same of the array, guaranteeing the immobility of the glass and flow chamber once putted in the lodging.

12.3.1 *NAPPA array production*

The study is carried out on two different protein on NAPPA format:
1. Human kinase NAPPA spotted on 300 microns spots of kinase genes from HIP FLEXGene human kinase collection; such geometry is same adopted for fluorescence analysis.
2. Transcription factor NAPPA spotted on 300 microns spot of transcription factor genes from HIP FLEXGene collection; such geometry is same adopted for fluorescence analysis.

Cdk2_Human: gene of 897 bp and corresponding protein of 33,930 Da.

Fos_Human: gene of 3402 bp and corresponding protein of 40,695 Da.

12.3.2 *NAPPA expression*

To express proteins we followed the protocol [modified from Ramachandran *et al.* (2008)]:

- Blocking: 1 hr on rocking shaker at room temperature with SuperBlock. We used 30 mL in a pipette box for 3 slides.
- Rinsed with milli-Q water. Dried with nitrogen.
- Applied HybriWell (Grace Biolabs) gasket to the slide.
- Pre-heated the incubator to be used for IVT at 30 °C
- Prepared IVT: 130 μL of IVT lysate mix for slide. Each tube after component addition contained 400 μL of lysate mix and the lysate tubes could not be re-frozen, e.g. 1 tube = 3 slides = 400 μl → 16 μl TNT buffer, 8 μl T7 polymerase, 4 μl of –Met, 4 μl of –Leu or –Cys, 8 μl of RNaseOUT, 160 μl of DEPC water, 200 μl of reticulocyte lysate.
- Added IVT mix. Pipette the mix slowly. Gently massaged the HybriWell to get the IVT mix to spread out and cover all of the area of the array. Applied small round port seals to both ports.
- Placed the slides on a bioassay dish with divider on top of the levelling shelf inside the incubator. Incubated for 1.5 hr at 30 °C for protein expression, followed by 30 min at 15 °C for the query protein to bind to the immobilized protein.
- Removed the HybriWell and immersed each slide in water immediately; washed 3 times, 3 minutes each, in a pipette box. Dried with nitrogen.

12.3.3 *Detection of in situ expressed protein by SPR*

On each spot about 0.2 ng of protein is synthesised (an estimate from unpublished data of HIP), which is enough to analyzed *in situ* expressed protein by SPR instrument.

In general, binding and regeneration protocols have been well studied in SPR experiments. Binding buffers are usually based on PBS or HEPES often with carrier protein and a small amount of non-denaturing detergent. The carrier protein may serve to coat the walls of the fluidic system and flow cell, minimizing the loss of the protein analytes to adsorption. Detergents can also help minimize adsorption and non-specific binding. Regeneration solutions can be based on acid, base, salt, or detergent. An acid such as HCl or HPO_3 in the range of 1–100 mM is common.

In these set of experiments, PBS (137 nM NaCl, 10 mM phosphate, 2.7 mM KCl, pH7.4) + Tween 20 0.01% was the SPR running buffer and 3 mM HCl in deionized water was the regeneration buffer. All solutions were degassed under a 50-kilopascal vacuum for 30 min. The rate of data acquisition was set to 0.2 Hz. A square region of interest 10 pixels high × 16 pixels wide was defined for each spot. Six background regions of interest of 10 × 16 pixels were defined at the top, middle and bottom areas of the array. A simple two point calibration was obtained by injecting 1 × PBS (n = 1.3347 RIU) and 2 × PBS (n = 1.4465 RIU) into the flow cell and recording the signal intensity at each point. The typical calibration factor was 2 camera units/microrefractive index unit (μRIU). All the experiments were performed at 20 °C.

Each sample measurement cycle had four steps: reblocking, sample binding, washing and regeneration. Reblocking consisted of incubating the slides with reblocking solution. Binding occurred with the injection of 200 μL of sample followed by oscillatory flow (50 μL at 2 μL/s) for 5 minutes. The slides were washed for at least 5 minutes with a constant flow of running buffer at 5 μL/s. The array surface was regenerated by injecting 300 μL of regeneration solution, pausing flow for 3 min, and washing with 400 μL at 10 μL/s.

12.4 Results

12.4.1 *Detection of GST purified protein by SPR*

At Harvard Institute of Proteomics have been developed a novel form of protein microarray, called Nucleic Acids Protein Microarray (NAPPA), which replaces the complex process of spotting protein with the simple process of spotting cDNA. The proteins are translated using a T7-coupled rabbit reticulosyte lysate *in vitro* transcription-translation (IVTT) system. The expressed proteins are capture locally with an antibody to a C-terminal glutathione S-transferase (GST) tag on each protein (anti-GST). Using an antibody to GST (distinct from the one used to capture the nascent protein) which recognizes the C-terminal GST tag genetically encodes in each coding sequence and thus confirms the full-length translation. Ramachandran *et al.*, (2008) described using fluorescence dyes that a variety of proteins can be displayed by this format. In the current approach, the first step is the detection using SPR of the binding of an anti-GST antibody to GST purified protein printed on the gold surface.

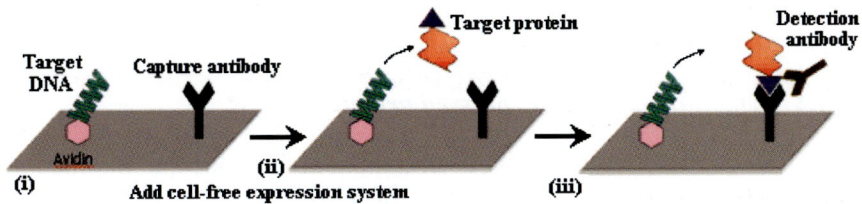

Figure 12.5. Scheme of the detection of *in situ* expressed protein.

In the Figure 12.6 shows the binding of anti-GST antibody onto the immobilized GST. The binding detected is directly dependable to the amount of purified GST protein on the surface. Moreover, the anti-GST antibody binding is highly specific because it is not showing binding on the several controls on the experiment.

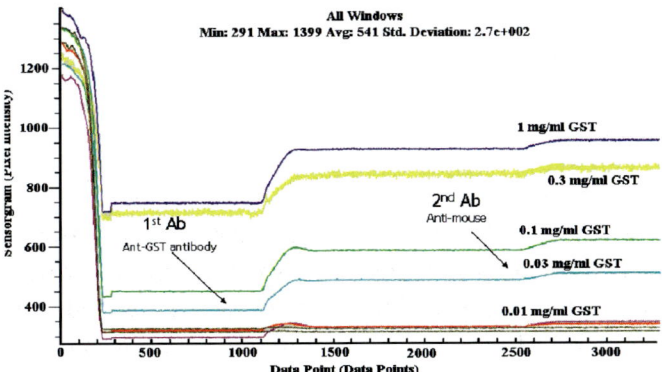

Figure 12.6. Real-time detection by SPR of GST purified proteins at different concentrations from 0.01 mg/mL to 1mg/mL. Binding event of monoclonal anti-GST antibody onto purified protein was determined by SPR device. Binding event of monoclonal anti-mouse antibody onto anti-GST antibody, previously specifically bound to purified GST proteins, was determined by SPR device. The experiments were performed as described in Materials & Methods section.

Besides the detection of bimolecular interaction (purified GST protein *vs* anti-GST antibody), it is also possible to detect multi-protein complexes using the present approach. Figure 12.6 shows the specific binding of secondary antibody (anti-mouse IgG) to anti-GST antibody previously bound to the purified GST. This approach would be the proof of concept for the ability to study multi-protein complex by SPR and NAPPA approach.

12.4.2 *Detection of in situ expressed protein by SPR*

Our main goal will be the characterization of bimolecular interactions by SPR on NAPPA approach. For this purpose, the first step is the detection of the *in situ* expressed protein by SPR using the anti-GST tag antibody.

After incubation with the cell-free protein expression system, the slides have been washed with running buffer and the chamber assembled. The, it was injected some running buffer to obtain a stable baseline. Afterwards, the anti-GST antibody is injected in order to detect binding to the expressed protein using NAPPA approach. Figure 12.7A corresponds to the sensorgram of the detection of the Cdk2 protein by SPR using NAPPA approach.

This kinase, Cdk2, is detected by the binding of the anti-GST antibody to the GST which is on C-terminus. Figure 12.7B shows the sensorgram of the detection of *in situ* expressed Fos protein by SPR. The anti-GST antibody is binding to the protein through the GST tag on the C-terminus.

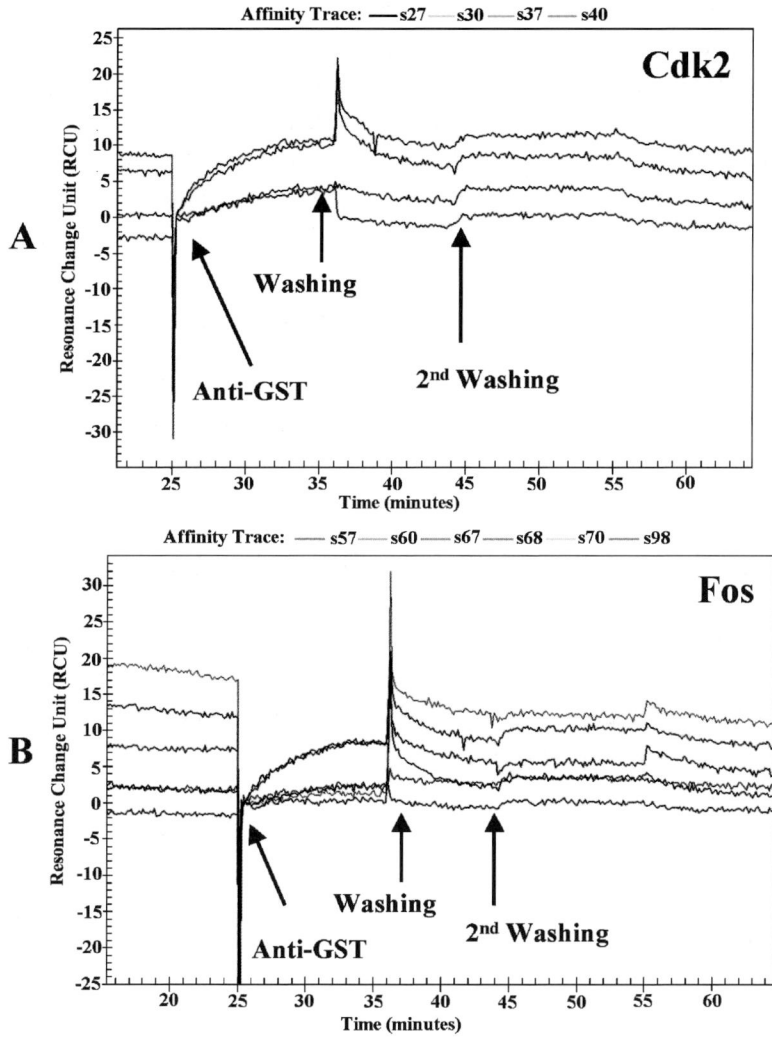

Figure 12.7. Real-time biomolecular interaction studies using NAPPA arrays & Surface Plasmon Resonance Imaging. A. Real-time detection by SPRi of *in situ* expressed proteins Cdk2. Binding event of monoclonal anti-GST antibody onto *in situ* expressed protein was determined by SPR device. B. Real-time detection by SPRi of *in situ* expressed proteins Cdk2. Binding event of monoclonal anti-GST antibody onto *in situ* expressed protein was determined by SPR device. In both, anti-GST antibody was used as query biomolecule. The experiments were performed as described in Materials & Method section.

12.5 Conclusions and Future Directions

In summary, it is possible the label-free detection of the proteins expressed on NAPPA by Surface Plasmon Resonance using an anti-tag antibody. We have presented the results obtained in our research to implement a procedure to analyze biomolecular interactions by SPR using Nucleic Acid Programmable Protein Array (NAPPA).

In conclusion our results on NAPPA approach analysis by SPR are very encouraging being the experiments and the results perfectly repeatable. Having in mind the presented experiments, we are able to detect the *in situ* expressed proteins on NAPPA format but these results are still for studying biomolecular interaction of one single protein.

However, SPR is described as a very low high-throughput technology but it is well-known that NAPPA arrays as a tool for high-throughput experiments. The next step will be to coupling NAPPA arrays to a multiplexed real-time and label-free SPA based detection system (which allows thousands of binding events to be monitored in real-time without any loss in sensitivity).

Acknowledgments

Financial support to Manuel Fuentes from Fondo de Investigación Sanitaria FIS PI081884.

References

Kretschmann, E. and Raether, H. (1968). Radiative decay of non radiative surface plasmon excited by light. *Zeit fur Naturforschung*, 23A, pp. 2135–2136.
MacBeath, G. and Schreiber, S. L. (2000). Printing proteins as microarrays for high-troughput function determination. *Science*, 289, pp. 1760–1763.
Otto, A. (1968). Excitation of nonradiative surface plasma waves in silver by the method of frustrated total reflection. *Zeitschrift fur Physik*, 216, pp. 398–410.
Ramachandran, N., Hainsworth, E., Bhullar, B., Eisenstein, S., Rosen, B., lau, A. Y., Walter, J. C. and LaBaer, J. (2004). Self-assembling protein microarrays. *Science*, 305, pp. 86–90.
Ramachandran, N., Raphael, J. V., Hainsworth, E., Demirkan, G., Fuentes, M. G., Rolfs, A., Hu, Y. and LaBaer, J. (2008). Next-generation high-density self-assembling functional protein arrays. *Nature Methods*, 5, pp. 535–538.

Zhu, H., Bilgin, M., Bangham, R., Hall, D., Casamayor, A., Bertone, P., Lan, N., Jansen, R., Bidlingmaier, S., Houfek, T., Mitchell, T., Miller, P., Dean, R. A., Gerstein, M. and Snyder, M. (2001). Global analysis of protein activities using proteome chips. *Science*, 293, pp. 2101–2105.

INDEX

1D gel electrophoresis, 185, 208, 213, 217, 222
2D gel electrophoresis, 185, 206, 208, 212, 219, 222, 224

ADP, 152, 164, 166, 168
allograft, 230, 237, 239, 241, 242, 249–255
anodic porous alumina (APA), 1, 12, 13, 16–18, 23, 27, 28, 95–104, 106, 107, 121, 122, 124–126, 129–131, 133, 135–143, 146
apoptosis, 200, 238, 244, 249, 271
ATP, 19, 48, 152, 154, 156, 159, 164, 168

ball-crush, 99, 100
bioinformatics, 18, 31, 43, 53, 57, 224
biosynthesis, 155–158, 162, 164, 165, 168, 177, 180, 181
biotin, 8, 83

CagA, 151, 153, 163, 175, 176, 179, 180
CagD, 152, 153, 176
CagS, 152, 153, 176
CagZ, 152, 153, 176
calorimetry, 80

cAMP, 28, 31, 33, 53, 55, 59, 183–185, 187–191, 193–212, 214–217, 219–223, 225, 226
cancer, 21, 25, 26, 29, 151, 172, 187, 189–191, 226, 243, 251, 255, 257, 258, 262, 279, 281, 284
cantilever, 110, 112, 116–118, 129
carcinogenesis, 262, 285
CCD, 10, 32, 34, 35, 102, 107, 137
CCNB1, 40, 41
CCR5, 47, 48
CDC2, 40, 41
CDK4, 1, 11
cell cycle, 1, 21, 23, 24, 26, 28, 31, 33, 40, 41, 53, 55, 56, 58, 183, 186, 226, 254, 257, 264, 271, 278, 285
cell–free expression system, 282
cell transformation, 1, 21, 22, 24, 28, 186
chicken egg white, 135, 203
CHO-K1, 24, 31, 33, 53, 55, 183, 185, 186, 191, 193–199, 207, 208, 217, 219–221, 223, 227
chorismate synthase, 154, 159
chromatin, 33, 58, 184, 186, 221, 225, 226, 232, 274
chromatography, 77, 80, 183, 185, 193, 194, 201

CKD2, 1, 11, 13, 53, 88–90
CRK3, 48
crystal growth, 27, 125, 146
cytoskeleton, 183, 184, 187, 189, 191, 192, 217, 221, 222, 225–227, 270

degradation, 123, 157, 164, 165, 265, 290
dehydroquinate synthase, 158, 159
diffraction, 124, 126, 136–138, 291
dissipation, 58, 83, 84, 86, 90–92, 97, 107, 119
DNA, 2–5, 7, 8, 10, 11, 20, 21, 31–34, 36, 45, 47, 56, 79, 80, 121, 127, 128, 133, 134, 141, 142, 154, 164, 169–172, 175–177, 184, 186, 189–191, 223, 226, 231, 236, 244, 249, 251, 252, 254, 256, 262, 266, 269, 270, 272, 274, 278, 282, 285, 288, 290
DNASER, 1, 10, 12, 18, 24, 26, 28, 31–38, 53, 55, 56, 58, 59, 147

Enoyl-ACP reductase, 158

ferritin, 167, 176
fibroblasts, 183, 184, 188, 223, 225
fluorescence, 1, 4, 12, 17, 18, 24, 28, 34, 55, 64, 95, 102, 106, 107, 110, 290, 296, 299

gene expression, 25, 33, 43, 90, 107, 116, 170, 186, 188, 189, 220, 221, 224, 230, 232–240, 244, 245, 249–255
gene microarrays, 230, 241, 288

gene regulation, 171
glutathione S-transferase (GST), 3, 4, 8, 19, 32, 37, 68, 76, 77, 80, 282, 288–290, 298–301
graft, 230, 233, 234, 236–244, 246, 250, 252, 253, 255
grip test, 99, 100

HeLa, 31, 33, 53, 188
HPLC-ESI-MS, 195–197, 199
human kinase, 21, 34, 37, 51, 53, 55, 64, 68, 296
human kinases, 21, 37, 51, 53
human T lymphocytes, 26, 28, 31, 40, 53, 58, 254

ischemia, 237, 238

JUN, 1, 11, 13, 53, 64, 65, 70, 71, 75, 76, 81, 89–91, 109, 110, 113–116, 119

KBF1, 47, 48
kidney, 25, 28, 33, 229, 232, 234, 236, 238, 240–242, 244, 246–255
kidney graft tolerance, 33

label free, 68, 80, 110
Langmuir-Blodgett (LB), 18, 27, 124, 125, 136, 137, 139, 140
liver, 229, 238, 240, 244, 245, 246, 250, 252, 255
lung, 229, 232, 238, 240, 241, 242, 249, 251, 253, 255, 262
lymphocytes, 33, 40, 50, 54, 55, 226, 232, 240, 241, 251

Index

lymphoma, 31, 33, 43–47, 49–53, 55–58, 149, 167, 180, 249
lysozyme, 27, 125, 126, 135, 137–139, 146, 147, 203

MALDI, 12, 22, 23, 61–63, 67–69, 73, 74, 76, 77, 185, 194, 201, 213–218, 223
MALDI-TOF MS, 77, 185, 201, 213, 217, 223
Malonyl-CoA, 154, 158, 182
mammalian cells, 9, 189, 191, 220, 221, 226
mass spectrometry (MS), 1, 12, 13, 15, 16, 18, 23, 37, 61–63, 65, 66, 68–70, 76, 91, 141, 194, 195, 208, 209, 211, 213, 214, 217, 222
mass spectroscopy, 80
membrane, 20, 78, 123, 135, 144, 150, 152, 153, 156, 163, 168, 179, 183, 187, 189, 191–193, 198, 206, 208, 210, 215, 217, 221–223, 274
membrane/organelles, 192
metabolism, 154, 155, 157, 162, 166, 177, 224, 244, 263, 271
microarrays, 1–3, 7, 8, 10, 12, 19, 21, 22, 25–28, 31–34, 56, 58, 59, 61, 62, 77, 78, 92, 93, 96, 104, 107, 108, 119, 120, 122, 143, 227, 230, 231, 233, 240, 241, 246, 248, 255, 281, 284, 285, 288, 302
microspheres, 112, 113
molecular modeling, 92
mRNA, 56, 133, 141, 231, 232, 243, 256

nanocrystallography, 18, 27, 121, 124, 141–143, 146
nanogravimetry, 12, 25, 90–92
network, 46, 47, 49, 51, 85, 135, 163, 187, 246, 255, 257–262, 264, 271–273, 275, 276, 278–285
Nucleic Acid Programmable Protein Array (NAPPA), 1, 2, 4, 5, 7–9, 11–20, 23–25, 27, 28, 31–34, 36–38, 40, 43, 50, 51, 53–58, 61–70, 73, 74, 76–83, 87, 88, 90, 91, 95, 96, 103, 106, 107, 109–111, 113, 116, 118, 119, 121, 122, 140, 142, 143, 224, 225, 227, 248, 282, 287, 288, 290, 296, 298, 300–302

oncology, 262

P53, 1, 11, 13, 14, 44, 64, 65, 70, 71, 73, 75, 76, 81, 88–91, 109, 110, 113, 115, 116, 119
phosphorylation, 154, 159, 170, 177, 184, 188, 191, 221, 225, 269, 270, 285
PPP2CA, 40, 41
protein crystallization, 124–126, 143, 144, 147
protein expression, 5, 7, 20, 36, 56, 62, 66, 68, 70, 79, 87, 88, 90, 91, 110, 113–115, 121, 134, 171, 184, 185, 208, 220, 222, 224, 225, 297, 300
protein stability, 4, 80
proteome, 3, 8, 25, 29, 150, 183–185, 191, 192, 220–222, 227, 287, 288, 303

PURE system, 126, 128, 129, 134
PURExpress system, 126–128, 133, 134, 140, 142

QCM, 12, 13, 18, 23, 58, 80, 81, 83, 84, 86, 88–90, 92, 107, 119
QCM-D, 83, 84, 86, 89
quality factor, 14, 79, 84, 90

R factor, 138
rejection, 33, 229, 230, 233–235, 237, 239–242, 245, 246, 248–256
reverse transformation, 24, 33, 183–185, 187–189, 191, 193, 208, 219, 221, 223–227

shikimate kinase, 154, 159

signaling–network, 258, 259, 271, 272
stochastic modeling, 274
streptavidin, 8, 71, 83

thaumatin, 135, 136, 139, 140
thioredoxin, 160, 177, 178
trypsin, 67, 68, 217

urea, 150, 157, 166
urease, 150, 166, 167, 171, 177, 180, 181

Vesicular Fusion Protein NSF, 218, 219, 220, 224
Vimentin, 188
VirB11, 152, 180

X-ray, 124, 136–138, 168, 175